Toyohide Watanabe and Lakhmi C. Jain (Eds.)

Innovations in Intelligent Machines – 2

Studies in Computational Intelligence, Volume 376

Further volumes of this series can be found on our homepage: springer.com

Vol. 352. Nik Bessis and Fatos Xhafa (Eds.)
Next Generation Data Technologies for Collective Computational Intelligence, 2011
ISBN 978-3-642-20343-5

Vol. 353. Igor Aizenberg
Complex-Valued Neural Networks with Multi-Valued Neurons, 2011
ISBN 978-3-642-20352-7

Vol. 354. Ljupco Kocarev and Shiguo Lian (Eds.)
Chaos-Based Cryptography, 2011
ISBN 978-3-642-20541-5

Vol. 355. Yan Meng and Yaochu Jin (Eds.)
Bio-Inspired Self-Organizing Robotic Systems, 2011
ISBN 978-3-642-20759-4

Vol. 356. Slawomir Koziel and Xin-She Yang (Eds.)
Computational Optimization, Methods and Algorithms, 2011
ISBN 978-3-642-20858-4

Vol. 357. Nadia Nedjah, Leandro Santos Coelho, Viviana Cocco Mariani, and Luiza de Macedo Mourelle (Eds.)
Innovative Computing Methods and their Applications to Engineering Problems, 2011
ISBN 978-3-642-20957-4

Vol. 358. Norbert Jankowski, Włodzisław Duch, and Krzysztof Grąbczewski (Eds.)
Meta-Learning in Computational Intelligence, 2011
ISBN 978-3-642-20979-6

Vol. 359. Xin-She Yang, and Slawomir Koziel (Eds.)
Computational Optimization and Applications in Engineering and Industry, 2011
ISBN 978-3-642-20985-7

Vol. 360. Mikhail Moshkov and Beata Zielosko
Combinatorial Machine Learning, 2011
ISBN 978-3-642-20994-9

Vol. 361. Vincenzo Pallotta, Alessandro Soro, and Eloisa Vargiu (Eds.)
Advances in Distributed Agent-Based Retrieval Tools, 2011
ISBN 978-3-642-21383-0

Vol. 362. Pascal Bouvry, Horacio González-Vélez, and Joanna Kolodziej (Eds.)
Intelligent Decision Systems in Large-Scale Distributed Environments, 2011
ISBN 978-3-642-21270-3

Vol. 363. Kishan G. Mehrotra, Chilukuri Mohan, Jae C. Oh, Pramod K. Varshney, and Moonis Ali (Eds.)
Developing Concepts in Applied Intelligence, 2011
ISBN 978-3-642-21331-1

Vol. 364. Roger Lee (Ed.)
Computer and Information Science, 2011
ISBN 978-3-642-21377-9

Vol. 365. Roger Lee (Ed.)
Computers, Networks, Systems, and Industrial Engineering 2011, 2011
ISBN 978-3-642-21374-8

Vol. 366. Mario Köppen, Gerald Schaefer, and Ajith Abraham (Eds.)
Intelligent Computational Optimization in Engineering, 2011
ISBN 978-3-642-21704-3

Vol. 367. Gabriel Luque and Enrique Alba
Parallel Genetic Algorithms, 2011
ISBN 978-3-642-22083-8

Vol. 368. Roger Lee (Ed.)
Software Engineering, Artificial Intelligence, Networking and Parallel/Distributed Computing 2011, 2011
ISBN 978-3-642-22287-0

Vol. 369. Dominik Ryżko, Piotr Gawrysiak, Henryk Rybinski, and Marzena Kryszkiewicz (Eds.)
Emerging Intelligent Technologies in Industry, 2011
ISBN 978-3-642-22731-8

Vol. 370. Alexander Mehler, Kai-Uwe Kühnberger, Henning Lobin, Harald Lüngen, Angelika Storrer, and Andreas Witt (Eds.)
Modeling, Learning, and Processing of Text Technological Data Structures, 2011
ISBN 978-3-642-22612-0

Vol. 371. Leonid Perlovsky, Ross Deming, and Roman Ilin (Eds.)
Emotional Cognitive Neural Algorithms with Engineering Applications, 2011
ISBN 978-3-642-22829-2

Vol. 372. António E. Ruano and Annamária R. Várkonyi-Kóczy (Eds.)
New Advances in Intelligent Signal Processing, 2011
ISBN 978-3-642-11738-1

Vol. 373. Oleg Okun, Giorgio Valentini, and Matteo Re (Eds.)
Ensembles in Machine Learning Applications, 2011
ISBN 978-3-642-22909-1

Vol. 374. Dimitri Plemenos and Georgios Miaoulis (Eds.)
Intelligent Computer Graphics 2011, 2011
ISBN 978-3-642-22906-0

Vol. 375. Marenglen Biba and Fatos Xhafa (Eds.)
Learning Structure and Schemas from Documents, 2011
ISBN 978-3-642-22912-1

Vol. 376. Toyohide Watanabe and Lakhmi C. Jain (Eds.)
Innovations in Intelligent Machines – 2, 2012
ISBN 978-3-642-23189-6

Toyohide Watanabe and Lakhmi C. Jain (Eds.)

Innovations in Intelligent Machines – 2

Intelligent Paradigms and Applications

Springer

Editors

Prof. Toyohide Watanabe
Department of Systems and Social Informatics
Graduate School of Information Science
Nagoya University
Japan
E-mail: watanabe@is.nagoya-u.ac.jp

Prof. Lakhmi C. Jain
School of Electrical and
Information Engineering
University of South Australia
Adelaide
Mawson Lakes Campus
South Australia
Australia
E-mail: Lakhmi.Jain@unisa.edu.au

ISBN 978-3-642-23189-6 e-ISBN 978-3-642-23190-2

DOI 10.1007/978-3-642-23190-2

Studies in Computational Intelligence ISSN 1860-949X

Typeset & Cover Design: Scientific Publishing Services Pvt. Ltd., Chennai, India.

Printed on acid-free paper

9 8 7 6 5 4 3 2 1

springer.com

Preface

This research volume is a continuation of our previous volume on intelligent machines. We have laid the foundation of intelligent machines in SCI Series Volume 70 by including the possible and successful applications of computational intelligence paradigms in machines for mimicking the human behaviour.

The present volume includes the recent advances in intelligent paradigms and innovative applications such as document processing, language translation, English academic writing, crawling system for web pages, web-page retrieval technique, aggregate k-Nearest Neighbour for answering queries, context-aware guide, recommendation system for museum, meta-learning environment, case-based reasoning approach for adaptive modelling in exploratory learning, discussion support system for understanding research papers, system for recommending e-Learning courses, community site for supporting multiple motor-skill development, community size estimation of internet forum, lightweight reprogramming for wireless sensor networks, adaptive traffic signal controller and virtual disaster simulation system.

This book is primarily based on the contributions made by the authors to the KES International Conference Series. The original contributions were revised by the authors for inclusion in the book.

This book is directed to engineers, scientists, researchers, professors and the undergraduate/postgraduate students who wish to explore the applications of intelligent paradigms further.

We are grateful to the authors and reviewers for their excellent contributions. We sincerely thank the editorial team of the Springer-Verlag Company for their helpful assistance during the book's preparation.

Toyohide Watanabe, Japan
Lakhmi C. Jain, Australia

Contents

Chapter 1
Advances in Information Processing Paradigms

Jeffrey Tweedale[1] and Lakhmi Jain[2]

[1] Defence Science and Technology Organisation,
PO Box 1500,
Edinburgh SA 5111, Australia
[2] School of Electrical and Information Engineering,
University of South Australia,
Mawson Lakes Campus,
South Australia SA 5095, Australia

Abstract. Information processing plays an important role in virtually all systems. We examine a range of systems, that cover healthcare, engineering, aviation and education. This chapter presents some of the most recent advances in information processing technologies. A brief outline is presented with background about knowledge representation and AI in decision making. A brief outline of each chapters is also included.

Acronyms

AI	Artificial Intelligence
AIP	Advanced Information Processing
ANN	Artificial Neural Network
BDI	Beliefs, Desires, Intentions
CI	Computational Intelligence
DSS	Decision Support System
EA	Evolutionary Algorithm
ES	Evolutionary Strategies
FOPL	First Order Predicate Logic
FNN	Fuzzy Neural Networks
FPGA	Field Programmable Grid or Gate Arrays
FSM	Finite State Machine
FS-NEAT	Feature Selective NeuroEvolution of Augmenting Topologies
FuSM	Fuzzy State Machines
GA	Genetic Algorithm
GOFAI	Good Old-Fashioned Artificial Intelligence
GP	Genetic Programming
HCI	Human Computer Interface
IA	Intelligent Agent
KBS	Knowledge Based System
KIF	Knowledge Interchange Format
k-NN	k-Nearest Neighbour
LGP	Linear Genetic Programming
LMA	Leban Movement Analysis

T. Watanabe and L.C. Jain (Eds.): Innovations in Intell. Machines – 2, SCI 376, pp. 1–17.
springerlink.com

MAS	Multi-Agent System
MEP	Multi Expression Programming
MLP	Multi-Layer Perceptron
NE	Neuro-Evolution
NEAT	Neuro-Evolution of Augmenting Topologies
OODA	Observe Orient Decide and Act
OOPL	Object-Oriented Programming Language
OOPS	Object-Oriented Programming Systems
RBF	Radial Basis Function
RL	Reinforcement Learning
RSK	Rules, Skill and Knowledge
rtNEAT	Real-time Neuro-Evolution of Augmenting Topologies
RTS	Real-Time Strategy
SME	Subject Mater Expert
SODA	Stimulate Observe Decide and Act
SQL	Structured Query Language
SVM	Support Vector Machine

1 Introduction

This book is intended to extend the readers knowledge of information processing and take you on a journey into many of the advanced paradigms currently experienced in this domain. There are as many forms of information as there are methods of prosecuting its sources. To achieve this goal we are required to communicate a collection of acquired facts, goals or circumstances and coalesce into a manageable body of knowledge. We have increasingly become reliant on our ability to prosecute data reliably in order to make decisions about almost everything we do.

Data is the representation of anything that can be meaningfully quantized or represented in digital form as a number, symbol and even text. We process data into information by initially combining a collection of artefacts that are input into a system which is generally stored, filtered and/or classified prior to being translated into a useful form for dissemination. The processes used to achieve this task have evolved over many years and has been applied to many situations using a magnitude of techniques. Accounting and pay role applications take center place in the evolution of information processing. Data mining, expert system and knowledge-based system quickly followed. Today we live in an information age where we collect data faster than it can be processed. This book examines many recent advances in digital information processing with paradigms for acquisition, retrieval, aggregation, search, estimation and presentation.

Technically we could quote the abacus as being the first device used to process information. The calculator, word processors and computing devices had major effects on society. Certainly the Internet became the single most disruptive influence in the modern era. It has provided access to information globally which is doubling exponentially. Our ability to cope with this information continues to provide many challenges. Technology however continues to provide improved access to even more sources of reliable data and faster machines to process information.

1.1 Advanced Information Processing Technology

Very few systems provide complete solutions and for this reason generations of development occur. One goal of re-use is for each new generation to extend rather than replace existing functionality. New technology enables alternative techniques to be developed and it becomes a matter of time before these additions are integrated[1]. This domain grew to significance, although the author of the terminology has since admitted that he would have chosen the term Computational Intelligence (CI) to reflect its true capacity. AI is based predominantly on Object-Oriented Programming Languages (OOPLs). Confusion surfaces when designers use UML descriptions, such as; *aggregation* and *composition* when decomposing problems. Abstraction enables the programmer to aggregate classes[2] which can be composed[3], were inheritance *extends is-a* as a specialized part of object and an *interface* makes that component which *look-like* something else.

As discussed, the design of Object-Oriented Programming Systems (OOPS) uses an iterative process based on a strong system engineering methodology. The design of Advanced Information Processing (AIP) technology uses a structured framework. The fundamental concepts include:

Performance: AIP technologies are generally capable of solving many problems quicker than the time it appears to press a button. When humans are included in the process, the performance and interaction is based on response times provided or accepted by the operator. This form of functionality nolonger relies on the number of instructions the system can process per second. Alternatively, some system based stimuli are time dependant. As time dependent applications need a response within a specified threshold, agent based decision making becomes a viable alternative source of response or clarity.

Reliability: The assistant shall have built-in hardware and software elements that are designed to reduce the risk of a complete system failure. The applied technologies should allow for graceful performance degradation in case of failure.

Modularity: The assistant shall be based on technologies that allow logical decomposition of the system into smaller components (modules) with well-defined interfaces. Modularity facilitates development, enables future upgrades and reduces life-cycle costs by improved maintenance.

Integration: The assistant includes many diverse functions needing different implementation methods and techniques. The technology used should support integration with conventional, as well as advanced, methodologies preserving modularity.

Maturity: The assistant shall be based on mature and proven implementation technologies. This is expressed by the availability of tools, successful prototypes and operational applications.

---

[1] Artificial Intelligence (AI) was born from within the field of mathematics and was manifested using software.

[2] Classes which associate whole-things, that *uses-a* component or data type.

[3] A composition is represented as a *has-a* relationship where the object is part of a larger object.

All information is translated into formation in order to gain knowledge about the subject or goal. The sources and methods of representing the data are critical factors in the methodology employed to acquire and process this into knowledge.

1.2 Knowledge Representation

The Heuristic Computing domain has evolved over the past 60 years [75] with many new fields of study emerging as key obstacles are being solved. Many of these are related to attempts at personifying attributes of human behaviour or the knowledge processes into an Intelligent Agent (IA) system. During this time, AI [8, 26] has made a great deal of progress in many fields, such as knowledge representation, inference, machine learning, vision and robotics. Minsky poses that AI is the science of making machines do things that would require intelligence if done by man [56]. Many researchers regard AI as more than engineering, demanding the study of science about human and animal intelligence be included. Current AI considers cognitive aspects of human behavior, including *reasoning, planning, learning and communication.* AI was initially discussed by Newell and Simon using production systems as an example [83]; however, the field divided into two streams led by John McCarthy and Nil Nillson (considered the *Neats*, who used formal logic as a central tool to achieving AI) on one side and Marvin Minsky and Roger Schanks (considered the *scrufs*, used a psychological approach to AI) on the other. Russel and Norvig entered the argument by describing an *environment* as something that provides input and receives output, using *sensors* as inputs to a program and producing outputs as a result of *acting* on something within that program. The AI community now uses this notion[4] as the basis of definition of an agent [23].

After his football coaching career, Knuth became a mathematician and subsequently a computer scientist. He is acknowledged as being the inventor of the modern computer [46] and has published a significant series of seminal papers based on his wealth of experience in the computing domain. These books document data structures, algorithms and many formalized programming techniques which are still in use today. Wirth formalized the basic requirements of a program. He proposed it embodies data, data structure(s) and re-lated algorithm [90][5]. This approach enables the programmer to represent knowledge in a structured form. Each element of knowledge Rules, Skill and Knowledge (RSK) [66], programmers concentrate on First Order Predicate Logic (FOPL) because it can be used to disprove anything that exists can be false. It contains Axioms based on single argument (Arity) predicates surrounded by one or more universal qualifiers (that can be nested) [20]. Kowalski proved this style of logic (originally conceived by Frege [24]). Herbrand latter used this logic to formulate a model based on a domain or a logical view of the world [76]. Horn minimised this logic by negating the model [58] which led to the development of the first prolog compiler in *Edinburgh* during 1977 [39]. The science of AI stalled as the

[4] Software that creates an environment that reacts to sensing (inputs) and acting (outputs).

[5] This predominantly separates a program; that is *data* and *corporate logic*).

scale of problems being represented started to encompass real-world problems. Graphing and search techniques where being employed with limited success. The complexity of representing knowledge maintained a statistical/mathematical direction. The use of FOPL quickly evolved into *frames*, *semantic-nets* (briefly exploring uncertainty) and again stalling at *neural-nets* (knowledge engineering and machine learning).

Many agree with Rasmussen's definition of knowledge, "as facts, conditions, expertise, ability or understanding associated with the sensors (sight, taste, touch, smell and sound) relating to anything in their environment [67]". This originally confined the analysis and processing of knowledge as a symbolic representation processes being diagnosed [66, 68, 69, 70]. Early systems were forced to store symbolic tables in flat data-bases, however the growth in capability of hierarchical, relational databases has extended the scope of knowledge engineering, especially in expert systems [12, 86].

The concept of knowledge is a collection of facts, principles, and related concepts. Knowledge representation is the key to any communication language and a fundamental issue in AI. The way knowledge is represented and expressed has to be meaningful so that the communicating entities can grasp the concept of the knowledge transmitted among them. This requires a good technique to represent knowledge. In computers symbols (numbers and characters) are used to store and manipulate the knowledge. There are different approaches for storing the knowledge because there are different kinds of knowledge such as facts, rules, relationships, and so on. Some popular approaches for storing knowledge in computers include procedural, relational, and hierarchical representations [5]. Procedural representation method encodes knowledge into program code and sequential instructions. However, encoding the knowledge into the algorithm used to process knowledge makes it difficult to modify knowledge. Therefore, declarative knowledge concept is used to represent facts, rules, and relationships by themselves and separate knowledge from the algorithm used to process it. In relational representation method such as Structured Query Language (SQL), data is stored in a set of fields and columns based on the attributes of items. This method of representing knowledge is flexible but it is not as good as hierarchical representation in stating the relationships and shared attributes of different objects or concepts. Network hierarchical database systems are very strong in representing knowledge and *is-a* relationship between related groups. An *is-a* relationship is when a specific object is linked to a more abstract term such as linking the object apple to the category of fruits. Other forms of knowledge representation used include *Predicate Logic, Frames, Semantic Nets, If-Then rules and Knowledge Inter-change Format.* The type of knowledge representation to be used depends on the AI application and the domain that IA is supposed to function. [5]. In the cases where there are limited numbers of situations that might occur knowledge can be hard-coded into procedural program codes. However, as the number of situations increases IAs would need a broader knowledge base and a more flexible interface. Therefore, knowledge should be separated from the procedural algorithms in order to simplify knowledge modification and processing. For IAs to be capable of solving problems at different levels of abstraction knowledge should be presented in form of frames or semantic nets that can show the *is-a* relationship of objects and concepts. If the IAs are required to find the solution from the existing data, Predicate logic or IF-THEN rules can be used. In the situations where multiple agents interact or perform a task they should use standardised data reading and writing capabilities, such as Knowledge Interchange Format (KIF), in order to share their knowledge.

1.3 Decision Support Systems

The concept of Decision Support System (DSS) emerged in the early 70s and developed over the next decade. A good example of a DSS is a closed system that uses feedback to control its output. According to Russell and Norvig, a thermostat could be regarded as an agent that provides decision support [75]. DSS are computer programs that assist the users in decision making that incorporate data models which support humans [21]. They are more commonly employed to emphasize effectiveness. This gain is generally achieved by degrading the systems efficiency, however using modern computing, this factor is less of an issue. Russell and Norvig also defined an agent as "anything that can be viewed as perceiving its environment through sensors and acting upon that environment through effectors" [75] noting that a DSS generally forms the basis of components within an agent, application or system.

Agent oriented development can be considered as the successor of object oriented development when applied in the AI problem domains. Agents embody a software development paradigm that attempts to merge some of the theories developed in AI research within computer science. The growing *density* of data had an overall effect on the efficiency of these systems. Conversely a series of measures were created to report on the performance of DSS. Factors such as; accuracy, response time and explain-ability were raised as constraints to be considered before specifying *courses of action* [17].

Since the eighties AI applications have concentrated on problem solving, machine vision, speech, natural language processing/translation, common-sense reasoning and robot control [72]. The Windows/Mouse interface currently still dominates as the predominant Human Computer Interface (HCI), although it is acknowledged as being impractical for use with many mainstream AI applications.

Intelligent data retrieval/management relies heavily on the designer and/or programer(s) to provide the sensors, knowledge representation and inference required to provide a meaningful output to stimulate the operator(s). Scholars believes that operators respond symbolically using "Thin slicing" to provide judgement or derive snap decisions [19][6]. Through his work on decision making under pressure situations, Klein [42, 43, 44, 45] extends the concept of processing information gained through human based sensors against the experiential patterns stored in our subconscious mind. To maximize this concept, the combination of both issues (clouded judgement and subconscious expertise), Boyd

[6] This is often impaired where verbal queues are used to describe the symbolic links that are established.

[14, 28] further extends the human though process to enable us to employ a control mechanism to focus on the goal, through Observe Orient Decide and Act (OODA). Being a closed loop system, stimuli can be used in place of observation of OODA in a sensor based OODA system termed Stimulate Observe Decide and Act (SODA), especially in a known contextual environment. Such issues predominantly surface during complex, hostile engagements. Especially in an environment where Beliefs, Desires, Intentions (BDI) can result in mode confusion. Such confusion potentially compromises the desired goal [7, 18, 64] . To reduce this problem Rasmussen postulates that we use an experience ladders [65, 66, 67, 69] based on the RSK associated with the context of the environment. Here the scenarios should be extrapolated by Subject Mater Expert (SME). Vicente studied this approach from the *work domain* perspective, concentrating on the *cognitive domain* to derive the problem scope [88].

1.4 AI in Decision Making

IA is a growing domain that has had made a significant influence in many fields including: disaster recovery, traffic control, space exploration and computer games. During the introduction of AI, researches focused on developing theories or techniques to solve puzzles and implement game strategies. Advances in computer technology have created *super-fast* computers with high quality graphic cards, improved data storage, bandwidth and data transaction speed. These improvements have stimulated the emergence of many new research opportunities within AI. The seamless ubiquity of the digital domain and fast Internet connection has created an environment in which information evolves. In the past, humanity elected to go on-line, today mankind remains connected to an on-line environment and elects to engage with others in a virtual world.

AI is normally associated with human intelligence generated through reasoning or optimisation that is based on experience[7]. Intelligence can be simulated using computers, but each machine must be designed to reason, based on facts and heuristic knowledge. AI is otherwise know as CI[8] and emerged out of code breaking work conducted during 'World War 2'. It is acknowledged that McCarthy first used the term AI during a conference held in 1956 at *Dartmouth* [54][9] however Minsky defines CI as the science or engineering required to make intelligent machines do the tasks that humans are capable of doing [56].

Alan Turing proposed a test to measure computing intelligence and distinguished two different approaches to AI known as Top-Down and Bottom-up [84, 85]. AI began as Top-Down or traditional symbolic AI approach where cognition is a high-level concept, independent of the lower-level details of the implementing mechanism [37]. The Bottom-Up approach aims to emerge cognition

---

[7] This is different to the public's perception of how artificial intelligence is represented in science-fiction movies.

[8] Although some researchers consider CI to be a branch of AI, the textbooks broadly consider CI is a synonym of AI [60, 63, 75].

[9] Later he stated it would be more appropriate to use the term CI [55].

from the operation of many simple elements similar to how human's brain process information. Artificial Neural Network (ANN) is the core of this approach.

Like any domain, AI has evolved in leaps and bounds. For example, the research in ANN was almost ceased after Minsky and Papert showed the limitation of Perceptrons in learning linearly inseparable problems [57]. During the 1980s, researchers [1, 31, 74] realized that those problems can be solved using a new learning method for Multi-Layer Perceptron (MLP) ANN called backpropagation. These developments and many other significant contributions aided the resurgence of ANN research [10, 27, 41, 51, 81]. The renew effort enable researchers achieve their original goals.

AI research began using basic symbolic computation, hence it is referred to as Weak or Good Old-Fashioned Artificial Intelligence (GOFAI) [29]. Bourg and Seeman discussed a broader interpretation for use in games [6]. Since then, a number of techniques have been created to model the cognitive aspects of human behavior. Other developments include: perceiving, reasoning, communicating, planning and learning. Techniques required to solve problems within game evolved. These techniques are related to Search and Optimization, Path Finding, Collision Avoidance, Chasing and Evading, Pattern Movement, Probability, Potential Function-Based Movement, Flocking and Scripted Intelligence. Many address deterministic problems, which are easy to understand, implement and debug. The main pitfall of deterministic methods is that developers have to anticipate all the scenarios and explicitly code all behaviors. This form of implementation becomes predictable after several attempts.

There is a transition period where more Modern AI techniques were progressively introduced. A number of techniques have been integrated or even hybridised as these fields evolved. Some of these techniques include: Rule-Based AI, Finite State Machine (FSM), Fuzzy Logic and even Fuzzy State Machines (FuSM). Rule-Based AI comprises If-Then conditionals that map the actions of the system based on various conditions and criteria. FSM and Fuzzy Logic fall under the general category of Rules-Based AI. The idea in FSM is to specify a group of actions and/or states for agents and execute and make transitions between them. Fuzzy Logic deals with fuzzy concepts that may not have discrete values and allows the representation of conditions in degrees of truth rather than a two-valued binary system [73, 92]. FuSM combines the concept of Fuzzy Logic with FSM to create more realistic and somewhat less predictable behavior.

Some of these techniques have led to the success of Expert Systems, like that used in the chess playing program called *Deep Blue*. This program successfully defeated the world champion in 1997 [9]. Expert Systems are rule-based processing systems that consist of a knowledge base, working memory and inference engine for processing the data with the defined reasoning logic [22, 33]. As the complexity and diversity of problem solving escalated, agents were introduced. Agents have been used to create more sophisticate behavior in anything from the *Ghosts* in the classic arcade game of *Pac-Man* to the creatures and machines in many popular award-winning Real-Time Strategy (RTS) games. During this period, Russell and Norvig redefined AI as the study of creating systems that think or act like humans, or in a rational way, meaning that they do the 'right thing', given what they know of the environment [75]. They preferred to embody rationality into agents that receive inputs from the environment via sensors and provide outputs using effectors respectively. This definition has been adopted by many in the AI community.

The concept of Multi-Agent System (MAS) emerged to tie together the isolated subfields of AI. MAS consists of teams of IA that are able to perceive the environment using their sensory information, process the information with different AI techniques to reason and plan their actions in order to achieve certain goals [35, 91]. IA may be equipped with different capabilities including learning and reasoning. They are able to communicate and interact with each other to share their knowledge and skill to solve problems as a team. MASs have been used to create intelligent system and they have a very promising future.

Advanced AI includes non-deterministic techniques that enable entities to evolve and learn or adapt [6]. Techniques like ANN, Bayesian Networks, Evolutionary Algorithm (EA) and Reinforcement Learning (RL) have become mainstream pre-processors used in hybridised techniques. Bayesian Networks are used to enable reasoning during uncertainty. ANNs provide a relevant computational model used by agents to adapt to changes in the environment. Behaviour is also provided using *Supervised, Unsupervised* and *Reinforcement* learning [75]. Using supervised learning the ANN is presented with a set of input data and corresponding desired target values to train it and find the mapping function between inputs and their correct (desired) outputs. In Unsupervised learning, no specific target outputs are available and the ANN finds patterns in the data without getting any help and feedback from the environment. RL allows the agent to learn by trial-and-error by getting feedback (in a form of reward or punishment) from the environment [81]. Some examples of learning paradigms include: Temporal Difference learning [80] and Q-Learning [89]. EA techniques are within the category of *Evolutionary Computation* and have been used for learning, which include Genetic Algorithm (GA) [30], Genetic Programming (GP) [48], Evolutionary Strategies (ES) [3] and Neuro-Evolution (NE) [61]. GA techniques also offer opportunities for optimise or evolve intelligent game behavior. NE is a machine learning technique that uses EAs to train ANN. Examples of NE techniques include Neuro-Evolution of Augmenting Topologies (NEAT), Feature Selective NeuroEvolution of Augmenting Topologies (FS-NEAT) [79] and Real-time Neuro-Evolution of Augmenting Topologies (rtNEAT) [77, 78].

Many of the advanced techniques use Hybrid AI systems. Here, traditional AI techniques are used to pre-process uncertainty prior to using advanced AI techniques to solve real-world problems. The implementation of advanced AI techniques has provided researchers with many challenges because they are extremely difficult to understand, develop and debug. The lack of advanced AI technique experience by game developer has created a barrier to the expansion of these techniques in commercial games. The aim of this research is to provide appropriate tools in a test-bed to enable researchers investigate all.

The use of AI in decision making is not new. Recent advances in AI techniques provide better accessible to this technology which has resulted in an increased number of applications using DSS based MAS. These applications aid the decision maker in selecting an appropriate action in real-time, especially when under stressful conditions. The net effect is reduced information overload by enabling up-to-date information to be used in providing a dynamic response. Intelligent agents are used to enable the communications required for collaborative decisions and deal with uncertainty. AI researchers possess a comprehensive toolbox to deal with issues such as, architecture and integration [53]. A number of recent topics are listed in Table 1.

Table 1. Examples of Decision Making within AI

Field	Example
Cancer Decision Support	Case-Based reasoning as a decision support system for cancer diagnosis: A case study [16].
Diagnosing Breast Cancer	Using Linear Genetic Programming (LGP), Multi Expression Programming (MEP) and Gene Expression programming [34].
Clinical Healthcare	Using collaborative decision making and knowledge exchange [25].
Medical Decision Making	Choice of antibiotic in open heart surgery [11].
Fault Diagnosis	An agent-based system for distributed fault diagnosis [71].
Power Distribution	Uses Fuzzy Fuzzy Neural Networks (FNN) to predict load forecasting on power distribution networks [13].
Forest Fire Prevention	Based on fuzzy modelling [32].
Manufacturing	Supporting a multi-criterion decision making and multi-agent negotiation in manufacturing systems [82].
Mission Planning & Security	Ubiquitous Command and Control in intelligent decision making technologies [50].
Petroleum Production	Using a bioinformatics Knowledge Based System (KBS) [4, 12].
Production & Manufacturing	FASTCUT is a KBS the assists in optimising high speed machining to cut complex contoured surfaces so accurately that little or no finishing operation is necessary [52].
PCB Inspection	Uses EA to detect if all components have been placed correctly on the board using bioinformatics [15].
Transportation	Transportation Decision Support System in agent-based environments [2].
In Car Navigation	Adaptive route planning based on GA and the Dijkstra search algorithm [38].
Evolvable Hardware	Introduces the use of GA compilation in an aggregated adaptation of hardware in Field Programmable Grid or Gate Arrayss (FPGAs) [59].
Detecting Spam in Email	Created an anti-spam product using a Radial Basis Function (RBF) network [36].
Bankruptcy Detection	Assess an firms imbalanced dataset through the use of a classifier network [47].
Robot Soccer	Using Fuzzy logic vision based system to navigate agents toward a target in real-time system [49].
MAS Research Framework	Web-Based (distributed) MAS architecture to support research with reusable autonomous capabilities in a complex simulated environment [40].

We believe IA is perhaps the mostly widely applied method used for decision making in recent years. This utilization has significantly advanced many applications, particularly Web-based systems [62]. Many forms of machine learning and computational intelligence can now be incorporated into an agent character, which extends the capability of MAS by providing intelligent feedback [87].

2 Chapters Included in the Book

This book includes 19 chapters. Chapter 1 provides an introduction to information processing paradigms. It also presents a brief summary of all chapters included in the book. Chapter 2 is on the extraction of figure-related sentences to understand figures. A weight propagation mechanism is introduced and validated using examples.

Chapter 3 presents an alignment-based translation system for simultaneous Japanese-English spoken dialogue translation. The system is validated and its superiority over the existing reported systems is demonstrated.

Chapter 4 is on the automatic collection of useful phrases for English academic writing. The authors have successfully developed a phrase search system using extracted phrasal expressions and validated their study.

Chapter 5 presents the design and implementation of a focused crawling system for effectively collecting webpages related to specific topics. The authors have demonstrated the merit of their approach using a number of case studies.

Chapter 6 presents a new web-page retrieval technique for finding user preferred web-pages. The scheme infers user preference on the basis of relevant or irrelevant indications for the page. It also reflects the inferred preference into the next retrieval query with a view to improve the retrieved results.

Chapter 7 is on searching aggregate k-Nearest Neighbour (k-NN) on remote spatial databases using representative query points. Author has proposed a system for efficiently answering aggregate k-NN queries. The system is useful for developing a location based service to support a group of mobile users in spatial decision making.

Chapter 8 presents the design and implementation of a context-aware guide application for providing information according to the preference of each user. A Support Vector Machine (SVM) is used for deciding the appropriate information for the user. The authors have used the principal component analysis to generate the input data for the SVM learning. The system is validated in real environment.

Chapter 9 is on human motion retrieval system based on Leban Movement Analysis (LMA) using interactive evolutionary computing useful for movie and video game industries. A number of case studies are presented to validate the usefulness and effectiveness of the system.

Chapter 10 presents an exhibit recommendation system based on semantic networks for museum. The system recommends exhibits according to the interest of the visitors. The system is evaluated using the artwork of Japanese arts such as the pictures related to Buddhism.

Chapter 11 is on presentation based meta-learning environment by facilitating thinking between lines. The authors have justified the novelty of their design by comparing their model with other meta-cognition support schemes.

Chapter 12 presents a case-based reasoning approach for adaptive modelling in exploratory learning. The proposed research enhances the modelling approach with an adaptive mechanism that enriches the knowledge base as new relevant information becomes available. The system is validated and its merit is demonstrated by conducting three experiments.

Chapter 13 presents a discussion support system for understanding re-search papers based on topic visualization. It is claimed by the authors that the proposed system supports collaborative discussion for enhancing the understanding of the research papers. The experiments demonstrate that the visualization of topics is appropriate for grasping the discussion.

Chapter 14 proposes a system for recommending the e-Learning courses matching the learning styles of the learners. The authors have investigated the relationship between learning preferences and e-learning course adaptability by administrating questionnaires to students who were enrolled in e-learning courses.

Chapter 15 presents the design of the community site for supporting multiple motor-skill development. The authors have presented the design and implementation of a web-community system that integrates different skill-communities to interact with each other. A number of trials are conducted to validate the approach.

Chapter 16 is on community size estimation of internet forum by posted article distribution. A number of experiments are conducted to validate the proposed approach.

Chapter 17 presents the design of lightweight reprogramming for wireless sensor networks. The scheme avoids the need of reprogramming the sensor network in case, there is change in environment. This aspect makes the system efficient with respect to the service availability and energy consumption.

Chapter 18 proposes the design of an adaptive traffic signal controller based on the expected traffic congestion. By using simulations, it is demonstrated that the adaptive traffic controller reduces the travelling time of the vehicle and thus helps in reducing the road congestion.

The final chapter presents a comparative study on communication protocols in disaster areas with virtual disaster simulation systems. The virtual disaster areas are constructed using hazard maps for predicting damage of disaster. Using experiments, it is demonstrated that the proposed system is superior than the systems reported in the literature.

3 Conclusion

This chapter presents an introduction into recent *Advances in Information Processing Paradigms.* It take the reader on an abbreviated journey into many of the paradigms discussed in this book. We discussed the basic concepts of AIP, knowledge representation, decision support systems and AI in decision making before introducing the most recent topics by many experts in their domain.

References

[1] Ackley, H., Hinton, E., Sejnowski, J.: A learning algorithm for boltzmann machines. Cognitive Science, 147–169 (1985)

[2] Balbo, F., Pinson, S.: A transportation decision support system in agent-based environment. Intelligent Decision Technologies 1(3), 97–115 (2007)

[3] Beyer, H.: The Theory of Evolutionary Strategies. Springer, Berlin (2001)

[4] Bichindaritz, I., Marling, C.: Case-based reasoning in the health sciences: What's next? Artificial Intelligence in Medicine 36, 127–135 (2006)

[5] Bigus, J.P., Bigus, J.: Constructing Intelligent Agents Using Java: Professional Developer's Guide, 2nd edn. Wiley, New York (2001)

[6] Bourg, D.M., Seeman, G.: AI for Game Developers. O'Reilly Media, Sebastopol (2004)

[7] Bratman, M.E.: What is intention? In: Cohen, P.R., Morgan, I.L., Pollock, M.E. (eds.) Communications, pp. 15–32. MIT Press, Cambridge (1990)

[8] Callan, R.: Artificial Intelligence. Palgrave MacMillan, Hampshire (2003)

[9] Campbell, M., Jr., A.J.H., Hsu, F.: Deep blue. Artificial Intelligence 134(1-2), 57–83 (2002)

[10] Carpenter, G., Grossberg, S.: Art 2: Self-organization of stable category recognition codes for analog input patterns. Applied Optics 26(23), 4919–4930 (1987)

[11] Cerrito, P.B.: Choice of antibiotic in open heart surgery. Intelligent Decision Technologies 1(1-2), 63–69 (2007)

[12] Chan, C.: An expert decision support system for monitoring and diagnosis of petroleum production and separation processes. Expert Systems with Applications 29, 127–135 (2005)

[13] Chauhan, B.K., Hanmandlu, M.: Load forecasting using wavelet fuzzy neural network. International Journal of Knowledge-Based and Intelligent Engineering Systems 13(4), 1327–2314 (2010)

[14] Coram, R.: Boyd: The Fighter Pilot who changed the Art of War. Little Brown and Company, Boston (2002)

[15] Crispin, A., Rankov, V.: Evolutionary algorithm for pcb inspection. International Journal of Knowledge-Based and Intelligent Engineering Systems 13(4), 1327–2314 (2009)

[16] De Paz, J.F., Rodríguez, S., Bajo, J., Corchado, J.M.: Case-based reasoning as a decision support system for cancer diagnosis: A case study. Int. J. Hybrid Intell. Syst. 6(2), 97–110 (2009)

[17] Dhar, V., Stein, R.: Intelligent decision support methods: the science of knowledge work. Prentice-Hall, Inc., Upper Saddle River (1997)

[18] d'Inverno, M., Kinny, D., Luck, M., Wooldridge, M.: A formal specification of dMARS. In: Agent Theories, Architectures, and Languages, pp. 155–176 (1997)

[19] Dodson, C.S., Johnson, M.K., Schooler, J.W.: The verbal overshadowing effect: why descriptions impair face recognition. Mem. Cognit. 25(2), 129–139 (1997)

[20] Emden, M.H.V., Kowalski, R.A.: The semantics of predicate logic as a programming language. J. ACM 23(4), 733–742 (1976)

[21] Fazlollahi, B., Vahidov, R.: Multi-agent decision support system incorporating fuzzy logic. In: 19th International Conference of the North American Fuzzy Information Processing Society, NAFIPS, pp. 246–250 (2000)

[22] Feigenbaum, E., McCorduck, P., Nii, H.P.: The Rise of the Expert Company. Times Books, New York (1988)

[23] Franklin, S., Graesser, A.: Is it an agent, or just a program?: A taxonomy for autonomous agents. In: Proceedings of the Third International Workshop on Agent Theories, Architectures and Languages, Budapest, Hungary, pp. 193–206 (1996)
[24] Freeman, E., Freeman, E.: Head First: Design Patterns. O'Rielly, Sebastopol (2004)
[25] Frize, M., Yang, L., Walker, R., O'Connor, A.: Conceptual framework of knowledge management for ethical decision-making support in neonatal intensive care. IEEE Transactions on Information Technology in Biomedicine 9, 205–215 (2005)
[26] Grevier, D.: AI – The Tumultuous History of the Search for Artificial Intelligence. Basic Books, New York (1993)
[27] Grossberg, S.: Competitive learning: From interactive activation to adaptive resonance. Cognitive Science 11, 23–63 (1987)
[28] Hammond, G.T.: The Mind of War: John Boyd and American Security. Smithsonian Institution Press, Washington (2004)
[29] Haugeland, J.: Artificial Intelligence: The Very Idea. MIT Press, Cambridge (1985)
[30] Holland, J.: Adaptation in Natural and Artificial Systems: An Introductory Analysis with Applications to Biology, Control and Artificial Intelligence. MIT Press, Cambridge (1975)
[31] Hopfield, J.: Neurons with graded responses have collective computational properties like those of two-state neurons. Proceedings of the National Academy of Sciences (USA) 81, 3088–3092 (1984)
[32] Iliadis, L.: A decision support system applying an integrated fuzzy model for long-term forest fire risk estimation. Environmental Modelling and Software 20, 613–621 (2005)
[33] Jackson, P.: Introduction to Expert Systems, 3rd edn. Addison-Wesley, Reading (1999)
[34] Jain, A., Jain, A., Jain, S., Jain, L. (eds.): Artificial Intelligence Techniques in Breast Cancer Diagnosis and Prognosis. Machine Perception and Artificial Intelligence, vol. 39. World Scientific Publishing, Hackensack (2000)
[35] Jennings, N., Wooldridge, M.: Software agents. IEE Review, The Institution of Engineering and Technology (IET) 42(1), 17–20 (1996)
[36] Jiang, E.: Detecting spam email by radial basis function networks. International Journal of Knowledge-Based and Intelligent Engineering Systems 11(6), 409–418 (2007)
[37] Jones, M.T.: AI Application Programming. Charles River Media, Inc., Hingham (2003)
[38] Kanoh, H.: Dynamic route planning for car navigation systems using virus genetic algorithms. International Journal of Knowledge-Based and Intelligent Engineering Systems 11(1), 65–78 (2007)
[39] Kenneth, I.C., Bowen, A., Buettner, A., Turk, A.K.: The design and implementation of a high-speed incremental portable prolog compiler. In: Shapiro, E. (ed.) ICLP 1986. LNCS, vol. 225, pp. 650–656. Springer, Heidelberg (1986)
[40] Khazab, M., Tweedale, J., Jain, L.: Web-based multi-agent system architecture in a dynamic environment. International Journal of Knowledge-Based and Intelligent Engineering Systems 14(4), 217–227 (2010)
[41] Kirkpatrick, S., Gelatt, C.D., Vecchi, M.P.: Optimization by simulated annealing. Science 220(4598), 671–680 (1983)
[42] Klien, G.: Sources of Power. MIT Press, Cambridge (1998)
[43] Klien, G.A.: Recognition-primed decisions. In: Rouse, W.B. (ed.) Advances in Man Machine System Research, vol. 5, pp. 47–92. JAI Press, Greenwich (1989)

[44] Klien, G.A.: A recognition-primed decision (RPD) model of rapid decision making. In: Klien, G.A., Orasanu, J., Calderwood, R., Zsambok, C.E. (eds.) Decision Making in Action: Models and Methods, pp. 138–147. Aplex Publishing Coroporation (1993)
[45] Klien, G.A., Calderwood, R., McGregor, D.: Critical decision method of eliciting knowledge. IEEE Transactions on Systems, Man and Cybernetics 19, 462–472 (1989)
[46] Knuth, D.: The art of computer programming. In: Grace Murray Hopper Award. Association for Computing Machinery, ACM Press, New York (1971)
[47] Kotsiantis, S., Tzelepis, D., Koumanakos, E., Tampakas, V.: Selective costing voting for bankruptcy prediction. International Journal of Knowledge-Based and Intelligent Engineering Systems 11(2), 409–418 (2007)
[48] Koza, J.: Genetic Programming: On the Programming of Computers by Means of Natural Selection. MIT Press, Cambridge (1992)
[49] Kubota, N., Kamijima, S.: Intelligent control for vision-based soccer robots. International Journal of Knowledge-Based and Intelligent Engineering Systems 10(1), 83–91 (2007)
[50] Lambert, D., Scholz, J.: Ubiquitous command and control. Intelligent Decision Technologies 1(3), 157–173 (2007)
[51] Linsker, R.: Self-organization in a perceptual network. Computer 21(3), 105–117 (1988)
[52] Liu, T.-I., Khan, W.H., Oh, C.T.: A knowledge-based system of high speed machining for the manufacturing of products. International Journal of Knowledge-Based and Intelligent Engineering Systems 14(4), 185–199 (2010)
[53] Mackworth, A.: The coevolution of AI and AAAI. AI Magazine 26, 51–52 (2005)
[54] McCarthy, J.: Programs with common sense. In: Symposium on Mechanization of Thought Processes, National Physical Laboratory, Teddington (1958)
[55] McCorduck, P.: Machines who think, pp. 1–375. Freeman, San Francisco (1979)
[56] Minsky, M.: Society of Mind. Simon and Schuster, Pymble (1985)
[57] Minsky, M.L., Papert, S.A.: Perceptrons. MIT Press, Cambridge (1969)
[58] Nadathur, G., Miller, D.: Higher-order horn clauses. J. ACM 37(4), 777–814 (1990)
[59] Negoita, M.G., Arslan, T.: Adaptive hardware/evolvable hardware – the state of the art and the prospectus for future development. International Journal of Knowledge-Based and Intelligent Engineering Systems 12(3), 183–185 (2009)
[60] Nilsson, N.: Artificial Intelligence: A New Synthesis. Morgan Kaufmann Publishers, San Francisco (1998)
[61] Nolfi, S., Elman, J.L., Parisi, D.: Learning and evolution in neural networks. Technical report, Technical Report 9019, Center for Research in Language. University of California, San Diego (1990)
[62] Phillips-Wren, G., Jain, L.: Preface. In: Phillips-Wren, G., Jain, L. (eds.) Intelligent Decision Support Systems in Agent-Mediated Environments. Frontiers in Articial Intelligence and Applications, vol. 115, pp. vii – ix. IOS Press, Amsterdam (2005)
[63] Poole, D., Mackworth, A., Goebel, R.: Computational Intelligence: A Logical Approach. Oxford University Press, New York (1998)
[64] Rao, A., Georgeff, M.: BDI Agents: From theory to practice. In: Proceedings for the 1st International Conference on Multi-Agent Systems (ICMAS 1995), pp. 312–319. AAAI Press, California (1995)
[65] Rasmussen, J.: Outlines of a hybrid model of the process plant operator. In: Sheridan, T.B., Johannsen, G. (eds.) Monitoring Behaviour and Supervisory Control. Plenum Press, New York (1976)

[66] Rasmussen, J.: Skills, signs, and symbols, and other distinctions in human performance models. System Design for Human Interaction 13(3), 291–300, IEEE Transactions on Systems, Man, and Cybernetics, Piscataway, NJ, USA (1983)
[67] Rasmussen, J.: Information Processing and Human-Machine Interaction: An Approach to Cognitive Engineering. Elsevier Science Inc., Amsterdam (1986)
[68] Rasmussen, J.: Diagnostic reasoning in action. IEEE Transactions on Systems, Man, and Cybernetics 23(4), 981–992 (1993)
[69] Rasmussen, J., Pejtersen, A.M., Goodstein, L.P.: Cognitive Systems Engineering. John Wiley & Sons Inc., Chichester (1994)
[70] Rasmussen, J., Pejtersen, A.M., Schmidt, K.: Taxonomy for cognitive work analysis. Technical report, Risø National Laboratory (September 1990)
[71] Ren, X., Thompson, H.A., Fleming, P.J.: An agent-based system for distributed fault diagnosis. International Journal of Knowledge-Based and Intelligent Engineering Systems 10(5), 319–335 (2006)
[72] Rich, E., Knight, K.: Artificial Intelligence. McGraw-Hill College, New York (1991)
[73] Ross, T.J.: Fuzzy Logic with Engineering Application, 3rd edn. Wiley, Chichester (2010)
[74] Rumelhart, D., Hinton, G., Williams, R.: Learning internal representations by error propagation. In: Rumelhart, D., McClelland, J. (eds.) Parallel Distributed Processing: Explorations in the Microstructure of Cognition, vol. 1, MIT Press, Cambridge (1986)
[75] Russell, S.J., Norvig, P.: Artificial Intelligence: A Modern Approach, 2nd edn. Prentice Hall, Pearson Education, Inc., Upper Saddle River, NJ, USA (2003)
[76] Schütz, H.: Generating minimal herbrand models step by step. In: Murray, N.V. (ed.) TABLEAUX 1999. LNCS (LNAI), vol. 1617, pp. 263–277. Springer, Heidelberg (1999)
[77] Stanley, K.O., Bryant, B.D., Miikkulainen, R.: Evolving neural network agents in the nero video game. In: Proceedings of the IEEE 2005 Symposium on Computational Intelligence and Games (CIG 2005). IEEE, Piscataway (2005)
[78] Stanley, K.O., Bryant, B.D., Miikkulainen, R.: Real-time neuroevolution in the nero video game. IEEE Transactions on Evolutionary Computation 9, 653–668 (2005)
[79] Stanley, K.O., Miikkulainen, R.: Evolving neural networks through augmenting topologies. Technical Report AI2001-290, Department of Computer Sciences. The University of Texas at Austin (2002)
[80] Sutton, R.S.: Learning to predict by the methods of temporal differences. Machine Learning 3, 9–44 (1988)
[81] Sutton, R.S., Barto, A.G.: Reinforcement Learning: An Introduction. MIT Press, Cambridge (1998)
[82] Taghezout, N., Zaraté, P.: Supporting a multi-criterion decision making and multi-agent negotiation in manufacturing systems. Intelligent Decision Technologies 3(3), 139–155 (2009)
[83] Thagard, P.R.: Computational Philiosphy of Science. MIT Press, Cambridge (1993)
[84] Turing, A.: Intelligent machinery. In: Meltzer, D. (ed.) Machine Intelligence, vol 5, pp. 3 – 23. Edinburgh University Press, Orginally, a National Physics Laboratory Report (1948)
[85] Turing, A.: Computing machinery and intelligence. Mind 59(236), 433–460 (1950) (unpublished until 1968)

[86] Urlings, P.J.M., Spijkervet, A.L.: Expert systems for decision support in military aircraft. In: Murthy, T.K.S., Münch, R.E. (eds.) Computational Mechanics Publications, pp. 153–173. Springer, Heidelberg (1987)
[87] Valluri, A., Croson, D.: Agent learning in supplier selection models. Decision Support Systems 39, 219–240 (2005)
[88] Vincente, K.J.: CongnitiveWork Analysis: Towards Safe, Productive and Healthy Computer-Based Work. Lawrence Erlbaum Associatesn, Mahwah (1999)
[89] Watkins, C.J., Dayan, P.: Technical note: Q-learning. Machine Learning 8(3), 279–292 (1992)
[90] Wirth, N.: Hardware architectures for programming languages and programming languages for hardware architectures. SIGARCH Comput. Archit. News 15(5), 2–8 (1987)
[91] Wooldridge, M., Muller, J., Tambe, M.: Agent theories, architectures, and languages: A bibliography. In: Intelligent Agents II Agent Theories, Architectures, and Languages, pp. 408–431. Springer, Heidelberg (1996)
[92] Zadeh, L.A.: Fuzzy sets. Information and Control 8(3), 338–353 (1965)

Chapter 2
The Extraction of Figure-Related Sentences to Effectively Understand Figures

Ryo Takeshima and Toyohide Watanabe

Department of Systems and Social Informatics,
Graduate School of Information Science, Nagoya University
Furo-cho, Chikusa-ku, Nagoya 464-8603, Japan
{takeshima,watanabe}@watanabe.ss.is.nagoya-u.ac.jp

Abstract. In research the related activities, such as searching, reading, and managing papers are important parts of the investigation process in both the pre-stage and post-stage of research. The number of academic papers, related in some way to a research topic, is large. It is difficult to read them completely from beginning to end. There are various types of comprehension by which we understand papers, so as to be appropriate to the research objective. In one case, it may be useful even if the abstractly summarized story should be grasped; and in the other case it may be necessary to understand them in detail.

Here, we propose an automatic extraction process of sentences which are related to figures effectively since the sentences explain the corresponding figures. This method is based on our experience. In many cases figures serve important roles to explain papers successfully. Our research objective is to introduce a weight propagation mechanism which is then applied to words and sentences between repeatedly processes such as "estimation of word importance" and "update of sentence weight."

Keywords: Figure Explanation, Weight Propagation, Reading of Paper.

1 Introduction

We can now obtain much information easily and rapidly from the Internet. This phenomena may also be observed in the research and development fields. Scientific and technical papers play an important role for both researchers and investigators. The initial step is to grasp both the limits and the motivation for the research field; the progress step is to understand the research objective, the approach and method, and the experimental results and discussions from interesting paper with a view to determining its partient research viewpoint. In the final step, the attempt is to classify related papers into the citation-oriented references. This may be done in order to prepare the research paper.

Extraction of figure-related explanatory sentences may be considered to be a kind of summary composition. Many methods for composing automatically

T. Watanabe and L.C. Jain (Eds.): Innovations in Intell. Machines – 2, SCI 376, pp. 19–31.
springerlink.com

summary are based on the sentence-selective extraction methods [1,2,3]. Such methods calculate the importance of individual sentences from the information, and then compose summaries by using the importance [4,5,6]. Some studies extract important sentences using machine learning [7,8]. Kupiec, and others considered the extraction of important sentences as a form of a statistical classification problem. They introduced a function that calculates the importance ratio of a sentence by the use of training data analyzed on the basis of Bayesian classifier [9]. Lin composed summaries by using decision tree learning [10]. Machine learning requires much training data. Hirao, et al. extracted important sentences by using a Support Vector Machine that has a good generalization ability [11]. These methods generate summaries or extract important sentences statically. They are not effective for sentences supplied when required by the user. Our aim is to extract figure-related explanatory sentences, which is not always attainable by these summary-oriented approaches.

The understanding of research papers is very strongly dependent on information management in the research phase. This includes fast reading in order to extract the important features. Careful reading is necessary to know the description contents in detail. Pin-point reading is used to reconfirm the already-determined content. Our objective in this paper is to develop a smart reading function. The idea of the research is to support the figure-related explanation means for figures used often in scientific and technical papers. This may partly explain illustratively important concepts, procedural methods, experimental environments and results. It may also be regarded as a fundamental resources related directly to the paper, corresponding to the interests of the researchers and developers. Traditionally, some methods such as Summary Composition, Topic Extraction, for example, have been investigated as intelligent support for paper understanding. These research subjects provide useful effects to help natural language processing. The topic extraction or the extraction of important words is one basic technical method for summary composition [12,13]. These extraction methods of topics, important words or co-related words take an important role which identifies a set of sentences related to the figures. In particular, the extraction of co-related words is useful. This is because the figure-related explanatory sentences should be selectively recognized from successive sentences. The idea is to propagate the weights of important words, included in directly referred sentences, to the succesive sentences and predecessor sentences which have a relationship for important words, one by one.

2 Approach

Our objective is to extract well-expressed sentences which are related strongly to the figures, with a view to grasping them effectively. The central idea here is to focus on the dependency between the sentence and the word. The concept of this dependency was proposed by Karov, et al. in order to get rid of ambiguity attendant on the word meanings [14]. Here is a dependency between word and sentence. This means that similar words should be included in related sentences.

Also similar sentences must contain correspondingly related words [15,16,17]. Generally, the explanatory sentences related to a figure in academic and technical papers should be located after the figure was first referred to. This observation is however not universally else where true. In some cases, the detail explanations or related explanations exist. These may be different from the mutual reference.

The use is made of stepwise propagation of weights, which are assigned to important words, for the surrounding sentences. The weight assigned to important words is useful when selecting meaningfully related sentences from other sentences in the paper. The weight propagation, based on the dependency between the sentence and the word, is suitable when choosing appropriate explanatory sentences. These must be meaningfully co-related with sentences. It is firstly referred to figures in the logical structure.

The weight propagation process is composed of 2 steps. One is the calculation of word importance; and the other is the update of the sentence weight. Until today, many investigations have been reported, which are to extract most important keywords on the basis of this calculated word. Edmundson proposed a keyword extraction method, using the access words [18]. Kimoto investigated a method to exclude noises from the extracted keywords on the basis of the meaningful relationship between keywords, which are derived from the sentence structure, access words or the thesaurus [19]. It is difficult to apply these conventional methods to our objective. This is because our application fields are not fixed to special research scopes with a predefined forms. The amount of data to be preset becomes too large if we wish to manage all fields. Luhn proposed another method of keyword extraction based on the frequency of word occurrences [20]. The frequency-based extraction method is likely to choose general words; it is necessary to exclude these general words by using *tf-idf*, with a view to making the extraction ratio high [21]. This method is useful to distinguish individual important words since the extraction scope is limited to application-specific fields. In our objective, it is necessary to develop some advanced methods or approaches in the use of these traditional methods. Weight propagation cannot be completed by only one trial, but must be repeated. We apply the word frequency to estimate word importance. It is possible to make the importance of general words lower even when we are not use *tf-idf* or others. This is because we can look upon the important words whose weights are low and whose frequencies are high, as general words. From this viewpoint, we suggest a method to estimate the word importance from the word frequency.

Propagation of Weight: We use weight which is assigned to each sentence. This is done to select appropriate figure-related explanatory sentences. It is not sufficient to only extract figure-related explanatory sentences by the use of positional relationships, derived from the corresponding paragraphs. Explanatory sentences do not always appear close to reference sentences. This is shown in Figure 1. To improve this insufficient process, we introduce the weight as an evaluation factor. This has a computable value, and can distinguish useful sentences on the basis of semantic relationships between sentences after having propagated

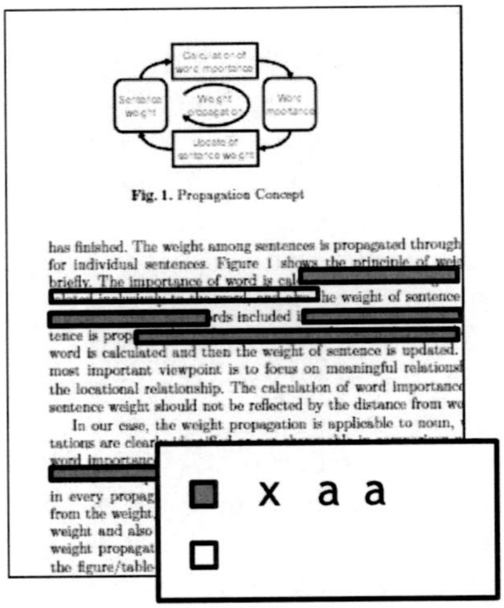

Fig. 1. Explanation and reference sentences

the weight mutually over the sentences. Our weight indicates that the suitability of its sentence was evaluated for a figure-related explanatory sentence, when the weight propagation process was finished. The weight in the sentences is propagated through common words for individual sentences. Figure 2 briefly shows the principle of weight propagation. The importance of a word is obtained by using the weights of all the sentences related to that word, and the weight of the sentence is from the importance of all words in the sentences. Thus, the weight of the sentence is propagated to other sentences one by one; when the importance of word is calculated the weight of sentence is updated. The most important viewpoint in this idea is to focus on semantic relationship between the word and the sentence. It is not on locational relationship. The calculation of the word importance and updating of the sentence weight are not affected by the distance from word or the sentence.

Here, the weight propagation is applicable only to nouns. The representations are clearly identified or not changed in comparison to the others. The word importance is calculated using all nouns in the target paper. This importance is only a temporary value used in the weight propagation. It is initialized at every propagation step. In this weight definition, the importance of word is dependent on the weight. The word is counted many times in a large weight sentence and is also a small weight sentence. We assume that the weight propagation assigns a large weight value to the sentence which is the figure-related explanatory sentence. Here, the initial value is applied to the sentences which are referred to as the focused figures. Generally, the locations where figures are explained are those as the first reference.

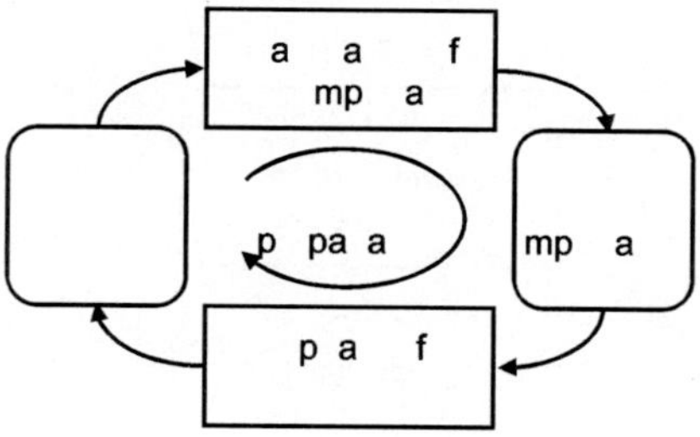

Fig. 2. Propagation concept

3 Method

The processing how used for extracting the figure-related explanatory sentences is shown in Figure 3. Firstly, the system calculates the initial values of the weights in each sentence which is dependent on the position in the sentences. Next, the calculation of the word importance and the update of the sentence weight are repeated until the user-specified repeating number is satisfied. The importance of the word is calculated using the weights of the sentences, and the weight of the sentence is calculated from the importance of the word. The weight is propagated from one sentence to another sentence. The weights of appropriate sentences for figure-related explanation become greater by the repeating use of the propagation of weight. The system ranks the individual sentences in an order based on the weight and extracts them from the longer ones.

3.1 Calculation of Initial Weight

The initial value for sentence weight is set using the relative positions of the sentences. This is based on the idea that the sentences describing a figure are located near a figure reference sentence. Firstly, the system looks for figure reference sentences. Next, for each sentence s_l, the initial value of weight $Weight_0(s_l)$ is calculated using the following formula. The initial weights of sentences are calculated based on the distances from the reference sentences as shown in Figure 4.

$$Weight_0(s_l) = \alpha \sum_{r \in R_f} \frac{1}{\sqrt{2\pi}} \exp\left\{-\frac{(l-r)^2}{2}\right\} \quad (1)$$

Equation 1 contains a normal distribution formula whose average is an index of figure reference sentence r and whose standard deviation is 1, where l represents the index of the sentence, and s_l is the sentence. If there are multiple figure

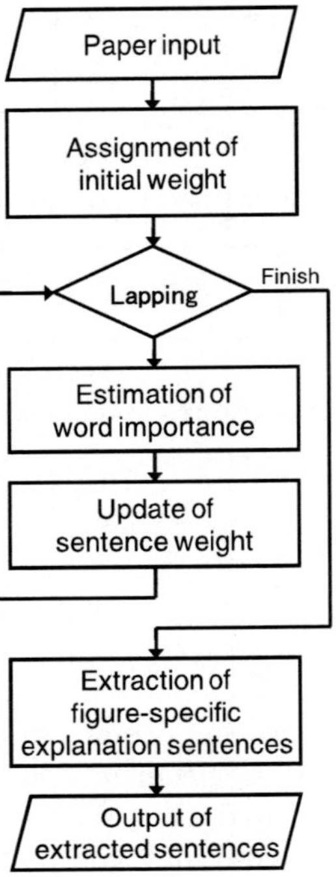

Fig. 3. Processing flow

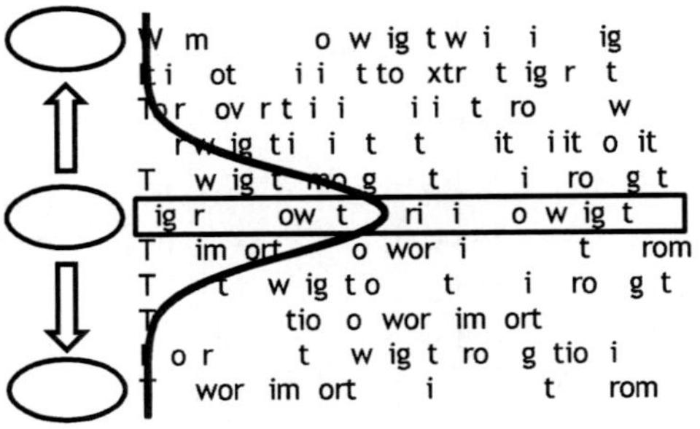

Fig. 4. Weight initialization

reference sentences, R_f has multiple elements and the weight is then summed up. Here, α is a normalization factor and is defined as follows.

$$\alpha = \frac{1}{\sum_{l \in L_s} \sum_{r \in R_f} \frac{1}{\sqrt{2\pi}} \exp -\frac{(l-r)^2}{2}} \tag{2}$$

L_s is the set of indices of all sentences in the paper. α is the inverse of the sum of the weights.

3.2 Calculation of Word Importance

The importance of a word is calculated using the weights of sentences which include the word. For each word w_l, the importance is defined using $Importance_p(w_l)$.

$$Importance_p(w_l) = \frac{1}{|S_{w_l}|} \sum_{s \in S_{w_l}} Weight_{p-1}(s), \tag{3}$$

where S_{w_l} is the set of sentences containing w_l. p represents the number of propagation steps. The sum of the weights of sentences in S_{w_l} is divided by the number of sentences in S_{w_l}. In this way, the importance of the words in the paper is restricted.

3.3 Update of Sentence Weight

The weight of sentence is updated using the idea that semantically similar statements will share many words. The weight of a figure-related explanation is increased by the sentences which are composed of important words and also the sentences whose weights are larger. The weight of a sentence $Weight_p(s_l)$ is updated by the following definition.

$$Weight_p(s_l) = \beta \left\{ Weight_{p-1}(s_l) + \gamma \sum_{w \in W_{s_l}} Importance_p(w) \right\} \tag{4}$$

Where W_{s_l} is the set of the words that compose the sentence s_l. $Weight_p(s_l)$ is the sum of the importances of words composing the sentence s_l. The sentence weight in the previous iteration is also a component. γ is a coefficient which adjusts the speed of propagation. β is the normalization factor which is defined as follows.

$$\beta = \frac{1}{\sum_{l \in L_s} \left(Weight_{p-1}(s_l) + \gamma \sum_{w \in W_{s_l}} Importance_p(w) \right)} \tag{5}$$

3.4 Extraction of Figure-Related Explanation Sentences

After the propagation phase, the sentences that have higher weights are extracted as figure-related sentences. The procedure for extracting figure-related sentences is illustrated in Algorithm 1. This algorithm takes the set of all sentences in research papers/articles S as an input and returns a set of extracted sentences E as an output. First, the sentences in S are sorted in descending order by their weights. Then, the sentences are added to E from the top of S while the condition $l < l_{min}$ is true. Here, l is the total length of sentences in E and l_{min} is the predefined minimum value of the total length of extracted sentences. $Length(S[i])$ is a function that returns the length of the sentence $S[i]$.

In this algorithm, the number of extracted sentences is decided according to the total length of sentences. This is because the amount of information contained in a sentence varies according to its length. Even though the numbers of extracted sentences are the same, the amounts of information contained in extracted sentences are different according to the total length of sentences.

Here, we explain briefly the processing in Figure 5. First, sentences are set in Figure 5(a) and keywords are extracted in Figure 5(b). Words are extracted from sentences before initialization step. Next, the initial weight for propagation is assigned to every sentence in Figure 5(c). Then, weight propagation is repeated. For example, a keyword "importance" is included in two sentences; so the word importance of "importance" is 3 as an average of those weights in Figure 5(d). In the same way, the importances of other words are estimated in Figure 5(e). Next, sentence weights are updated using word importance. A new weight is calculated by adding the importance of all words included in the sentence to the previous weight in Figures 5(f) and 5(g). Finally, figure-related explanatory sentences are extracted in Figure 5(h).

Algorithm 1. Extraction algorithm

$Sort(S)$;
$l \leftarrow 0$;
$i \leftarrow 0$;
$E \leftarrow \phi$;
while $l < l_{min}$ *and* $i < |S|$ **do**
  $E \leftarrow E \cup S[i]$;
  $l \leftarrow l + Length(S[i])$;
  $i \leftarrow i + 1$;
end while
$Express(E)$;

4 Prototype System

We implemented a prototype system for extracting figure-related explanatory sentences and supporting that a user understands a paper. When the user points out a figure with a view to understanding, the system first calculates the weights of sentences and then extracts figure-related explanatory sentences. Figures 6 and 7 show the interface windows of the system. The system consists of two windows. Figure 6 is the main window. The paper that the user must read is shown in this window. The user can indicate their intention using this window directly. When a paper is the input, marks showing the positions of figures are displayed. The user can change the propagation count using an up-down control system. Figure 7 is the window which displays the extracted explanatory sentences. Here, these sentences contain individually the "attainable region" or the "moving distance".

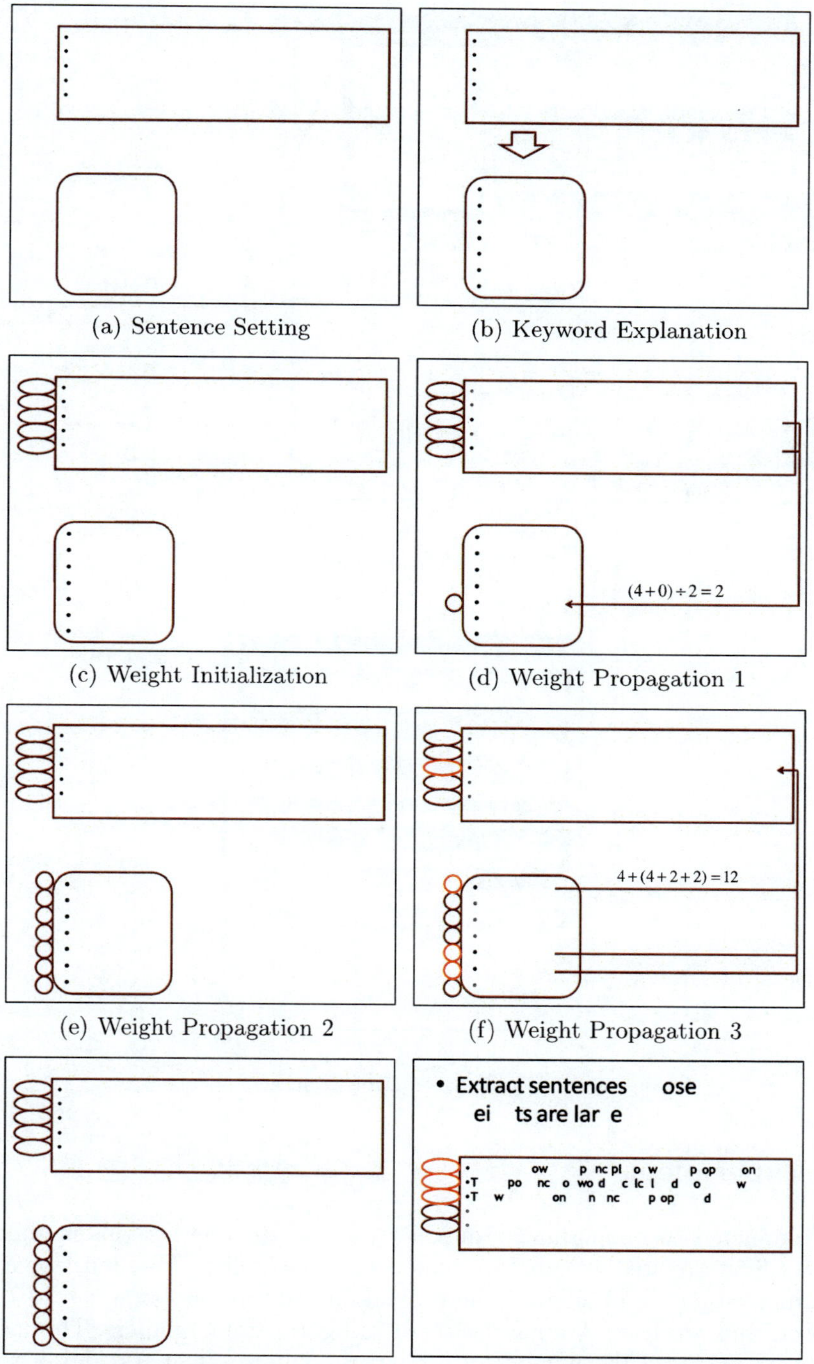

(a) Sentence Setting
(b) Keyword Explanation
(c) Weight Initialization
(d) Weight Propagation 1
(e) Weight Propagation 2
(f) Weight Propagation 3
(g) Weight Propagation 4
(h) Extraction

Fig. 5. Processing

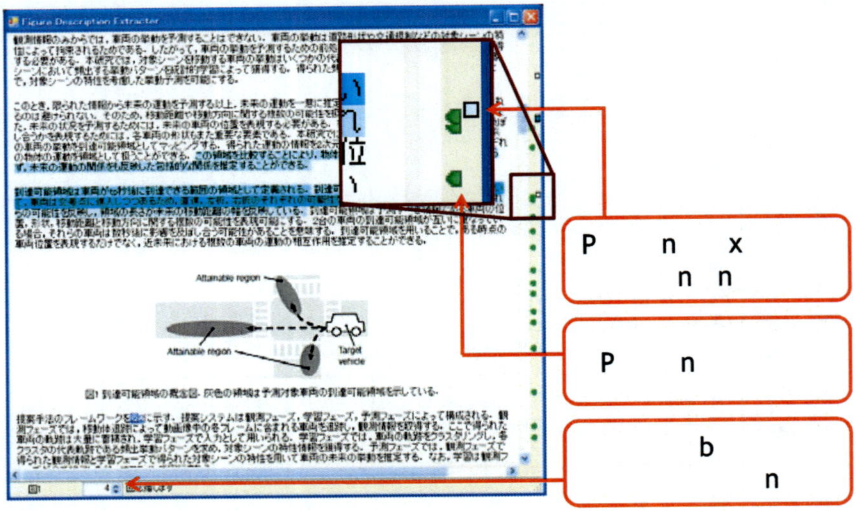

Fig. 6. Main window

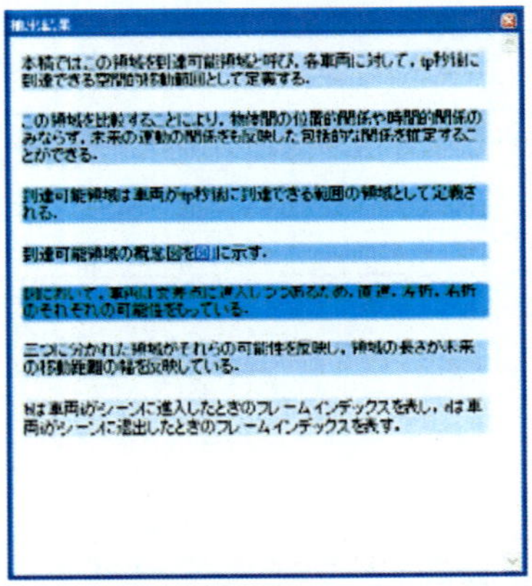

Fig. 7. Extraction window

5 Experiment

We conducted two experiments in order to evaluate whether this method can extract figure-related explanatory sentences successfully. These must be consistent understanding objective. The evaluation criterion is *extraction of correct answers* and *precision*. We have selected 24 figures from 4 papers. The speed of propagation γ and the number of propagations were set at 0.1 and 4, respectively. We evaluated the proposed method as to whether it can efficiently extract data from the appropriate sentences which can aid figure understanding.

Table 1. Experimental result obtained when extracting the correct answers.

Number of correctly extracted sentences	0	1	2	3
Number of cases	0	5	11	8

Table 2. Experimental result on precision

No.	Extracted sentences	Correct sentences	Percentage of correct sentences
1	3	2	66.7%
2	5	4	80.0%
3	6	6	100.0%
4	7	5	71.4%
5	5	3	60.0%
6	5	4	80.0%
7	6	3	50.0%
8	6	5	83.3%
9	6	5	83.3%
10	4	2	50.0%
11	7	6	85.7%
12	7	6	85.7%
13	7	6	85.7%
14	7	6	85.7%
15	7	6	85.7%
16	8	7	87.5%
17	7	4	57.1%
18	9	8	88.9%
19	8	3	37.5%
20	6	6	100.0%
21	6	4	66.7%
22	7	6	85.7%
23	6	4	66.7%
24	6	5	83.3%
Total	151	116	76.8%

Extraction of Correct Answers: We noted that 3 sentences are required for each figure, and whether we can regard the sentences as providing correct answers. We examined how many correct sentences are extracted by the system. Some experimental results are shown in Table 1.

Precision: We also investigated how many sentences related to the focused figure were extracted by using our method. We consider sentences, which include contents relevant to the figures, as correct sentences. Experimental results are shown in Table 2. The ratio of correct sentences was 76.8%. Some of the extracted sentences were not helpful to understand the figures, but many sentences had contents which were relevant to figures.

6 Conclusion

In this paper, a method for extracting figure-related explanatory sentences has been proposed. Generally speaking, the figures show the important contents of the papers. It is desirable to understand the figure-related explanatory sentences well and with respect to the paper reading. We proposed a weight propagation method successfully to extract the meaning explanatory sentences.

In this method, the result is extracted as a set of sentences. It is difficult to understand the extracted sentences by only reading sentences in the set, users may need to read the relevant sentences. It is necessary to improve the method used for calculating weights and presenting results. Also, the definition of propagation must be reconsidered, and additional parameters for calculating the weight of sentence need to be introduced.

From the experimental results, it is confirmed that our method can extract figure-related explanatory sentences contained in a paper. Since figures represent the important contents in papers, understanding figures gives an understanding of the whole paper. However, it is not clear just how a user will understand the paper, when they read figure-related explanatory sentences which have been extracted by the use of our method. In order to clarify the method, we need to conduct more experimentation and then to compare our method with other methods for efficiency.

References

1. Mani, I.: Automatic Summarization. John Benjamins Pub. Co., Amsterdam (2001)
2. Radev, D.R., Hovy, E., McKeown, K.: Introduction to the Special Issue on Summarization. Computational Linguistics 28(4), 399–408 (2002)
3. Hahn, U., Mani, I.: The challenges of automatic summarization. Computer 22(11), 29–36 (2000)
4. Knight, K., Marcu, D.: Summarization beyond sentence extraction: A probabilistic approach to sentence compression. Artificial Intelligence 139(1), 91–107 (2002)
5. Ko, Y., Kim, K., Seo, J.: Topic keyword identification for text summarization using lexical clustering. IEICE Trans. on Inf. & Syst. E86D(9), 1695–1701 (2003)
6. Yu, L., Ma, J., Ren, F., Kuroiwa, S.: Automatic Text Summarization Based on Lexical Chains and Structural Features. In: Proceedings of the 8th ACIS International Conference on Software Engineering, Artificial Intelligence, Networking, and Parallel/Distributed Computing, vol. 2 (2007)
7. Manning, C.D., Schütze, H.: Foundations of Statistical Natural Language Processing. MIT Press, Cambridge (1999)
8. Pradhan, S., Hacioglu, K., Krugler, V., Ward, W., Martin, J.H., Jurafsky, D.: Support vector learning for semantic argument classification. Journal of Machine Learning 60(1-3), 11–39 (2005)
9. Kupiec, J., Pedersen, J., Chen, F.: A Trainable Document Summarizer. In: Proceedings of the 18th Annual International ACM SIGIR Conference on Research and Development in Information Retrieval, pp. 68–73 (1995)
10. Lin, C.Y.: Training a Selection Function for Extraction. In: Proceedings of the 8th International Conference on Information and Knowledge Management, pp. 55–62 (1999)

11. Hirao, T., Takeuchi, K., Isozaki, H., Sasaki, Y., Maeda, E.: SVM-Based Multi-Document Summarization Integrating Sentence Extraction with Bunsetsu Elimination. IEICE Trans. on Inf. & Syst. E86-D(9), 399–408 (2003)
12. Sebastiani, F.: Machine Learning in Automated Text Categorization. The Journal of ACM Computing Surveys 20(1), 19–62 (2005)
13. Hotho, A., Nürnberger, A., Paaß, G.: A Brief Survey of Text Mining. LDV Forum-GLDV Journal for Computational Linguistics and Language Technology 20(1), 19–62 (2005)
14. Karov, Y., Edelman, S.: Similarity-based Word Sense Disambiguation. Computational Linguistics 24(1), 41–59 (1998)
15. Barzilay, R., McKeown, K.: Sentence Fusion for Multidocument News Summarization. Computational Linguistics 31(3) (2005)
16. Daumé III, H., Marcu, D.: Induction of Word and Phrase Alignments for Automatic Document Summarization. Computational Linguistics 31(4) (2005)
17. Dorr, B., Gaasterland, T.: Exploiting aspectual features and connecting words for summarization-inspired temporal-relation extraction. Information Processing & Management 43(6), 1681–1704 (2007)
18. Edmundson, H.P.: New Methods in Automatic Extracting. Journal of the ACM 16(2), 264–285 (1969)
19. Kimoto, H.: Automatic Indexing and Evaluation of Keywords for Japanese Newspapers. The Transactions of the Institute of Electronics, Information and Communication Engineers 74, 556–566 (1991)
20. Luhn, H.P.: Statistical Approach to Mechanized Encoding and Searching of Literary Information. IBM Journal of Research and Development 1(4), 309–317 (1957)
21. Amati, G., Carpineto, C., Romano, G., Bordoni, F.U.: FUB at TREC-10 Web Track: A probabilistic framework for topic relevance term weighting. In: Proceeding of 10th Text Retrieval Conference, NIST online publication (2001)

Chapter 3
Alignment-Based Translation Unit for Simultaneous Japanese-English Spoken Dialogue Translation

Koichiro Ryu[1], Shigeki Matsubara[2], and Yasuyoshi Inagaki[3]

[1] Graduate School of International Development, Nagoya University, Furo-cho, Chikusa-ku, Nagoya, 464-8601, Japan
[2] Graduate School of Information Science, Nagoya University, Furo-cho, Chikusa-ku, Nagoya, 464-8601, Japan
[3] Toyohashi University of Technology, 1-1 Hibarigaoka, Tempaku-cho, Toyohashi, Aichi-ken, 441-8580, Japan

Abstract. Recently, the development of simultaneous translation systems has been desired. However, no previous study has proposed the appropriate translation unit for a simultaneous translation system. In this paper, we propose a translation unit for simultaneous Japanese-English spoken dialogue translation. The proposed unit is defined based on the word alignment between a source sentence and its translation. The advantage of using the proposed unit is that a translation system can independently translate an input based on each translation unit. To confirm that such translation unit is effective for simultaneous translation, we evaluated the translation unit from the viewpoints of the length and the detectability.

Keywords: speech translation, translation unit, simultaneous interpretation, sentence segmentation.

1 Introduction

Recently, speech-to-speech translation systems have become important tools for supporting communication between different languages. With the advancement of natural language processing and speech processing, several speech-to-speech translation services have been developed. For example, InterACT has developed Jibbigo [5], an iPhone application of speech-to-speech translatoin, and NTT DoCoMo has released mobile phones on which a program of speech-to-speech translation is installed [17].

Most of the existing studies on machine translation give priority to high-quality translation [3, 4, 9, 15, 16]. However, it is not necessarily enough to provide high-quality translation for the users in a smooth cross-lingual communication. Most of the current machine translation systems, because of their sentence-by-sentence fashions, cannot start to translate a sentence until it has been fully spoken. Then, the following problems would arise in cross-lingual communication:

- It takes the same length of time as the speaker's utterance time to translate it, which decreases the efficiency of communication because it takes twice as long as normal communication.

T. Watanabe and L.C. Jain (Eds.): Innovations in Intell. Machines – 2, SCI 376, pp. 33–44.
springerlink.com

- The speaker has to wait for the response of the listener because the difference between the beginning time of the speaker's utterance and that of its translation is increased in such systems.

These problems are likely to cause some awkwardness in conversations [18]. One effective method of resolving them is that a translation system begins to translate without waiting for the end of the speaker's utterance. Therefore, the purpose of our research is to develop a simultaneous translation system.

A simultaneous translation system has to use a translation unit which is smaller than a sentence [1, 2, 6, 7, 8]. We have proposed a framework for simultaneous translation of incrementally executing detection, translation and generation processing for a translation unit. However, an appropriate translation unit for simultaneous translation is not obvious.

In this paper, we propose a translation unit for simultaneous translation, which can be translated independently and immediately. To acquire the translation units, we indicate our approach of segmenting a source sentence into translation units using its translation. In our approach, we segment a source sentence into translation units based on the word alignment between the source sentence and its translation[1]. However, in our translation method, a simultaneous translation system needs to segment a source sentence into translation units without using its translation. In this paper, we indicate that the translation units can be detected with moderately high precision without using the translation information.

The rest of the paper is organized as follows: In the next section, we discuss an appropriate unit for simultaneous translation. In Section 3, we propose a translation unit for simultaneous translation and describe the construction of a corpus annotated with translation units. In Section 4, we evaluate the incremental detectability of the translation units.

2 Translation Unit for Simultaneous Translation System

Conventional speech-to-speech translation systems employ a sentence as a translation unit, namely they translate an input in a sentence-by-sentence way. However, a simultaneous translation system has to use translation units smaller than a sentence. In this section, we examine an appropriate translation unit for simultaneous translation.

2.1 Simultaneous Translation Unit

The advantage of using a sentence as a translation unit is that a translation system can translate an input independently and immediately. In this study, we propose a translation unit which is shorter than a sentence and can be translated independently and immediately.

[1] The task of word alignment is to find correspondences between the words of a source sentence and a target sentence.

In Fig. 1, we show a flow of translating a Japanese sentence (J1) “ のと 予定 が が遅れ が の ". First, the Japanese sentence is segmented into translation units : “ のと " “予定 が", “ が", “遅れ が の " and “ ". Each of these units can be translated into an English phrase : "For now", "it’s on time", "the departure", "might be delayed" and "please understand it". The English phrases are generated incrementally.

The proposed unit does not necessarily correspond to linguistic units such as words, phrases and clauses. For example, when we segment the same sentence into clause units, it is segmented into “ のと 予定 が", “ が遅れ ", “ が の " and “ ". These units do not correspond to the proposed units as above. The translations of an adverb phrase such as “ のと " and a subject such as “ が", which generally appear at the beginning of a sentence in Japanese, also appear at the beginning of a sentence in English. Therefore, an adverb phrase and a subject can become a translation unit. The proposed unit is defined by the relation between a source language and a target language. In this paper, we describe a method of acquiring the proposed units by using a parallel corpus.

2.2 Comparing with Linguistic Unit

Conventional simultaneous translation systems have used linguistic units, such as words, phrases or clauses, as translation units. These translation systems execute parsing, transfer and generation processing unit by unit [14, 19]. In studies of simultaneous translation that used clauses as translation units, translation systems detect clauses using CBAP [12] and control the output timing. However, a word or a phrase do not satisfy

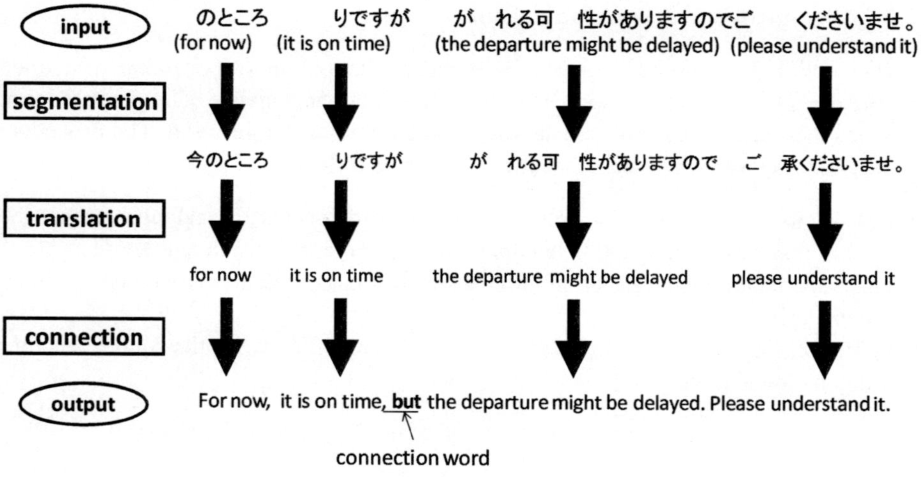

Fig. 1. Simultaneous translation model

the independence of translation and immediate translation because they are not semantically enough coherent units. On the other hand, a clause satisfies the independence of translation because it is a semantically enough coherent unit. However, a clause does not satisfy immediate translation because the appearance order of clauses is different between English and Japanese.

There are four properties that should be satisfied by a translation unit for simultaneous translation. The first is that a translation unit is shorter than a sentence. The second is that translation units are detected incrementally. The third is that a translation unit is translated independently. The fouth is that a translation unit is translated immediately. Table 1 shows the properties of each unit. We proposed a translation unit which satisfies the independence of translation and immediate translation. However, we have to confirm that each translation unit is smaller than a sentence and can be detected incrementally. In this paper, we evaluate whether the proposed unit satisfies these properties.

Table 1. Properties of each unit

	length shorter than a sentence	incremental detection	independence of translation	immediate translation
word	○	○	×	×
phrase	○	○	×	×
clause	○	○	○	×
sentence	×	○	○	○
proposed unit	?	?	○	○

3 Alignment-Based Translation Unit and Its Analysis

3.1 Alignment-Based Translation Unit

In this section, we propose an alignment-based translation unit (ATU), which is defined by the alignment between a source sentence and a target sentence. The proposed unit satisfies the independence of translation and the immediate translation. The procedure for segmenting a Japanese sentence into ATUs is as follows:

Step1: Translating a Japanese sentence into a word-for-word translation so that the word order of the translation becomes similar to that of its source utterance.

Step2: Segmenting the translation into the smallest units which can be translated independently.

Step3: Merging the units into an ATU by the alignment between the source sentence and its translation.

Fig. 2 shows an example of the segmentation of the Japanese sentence (J1).

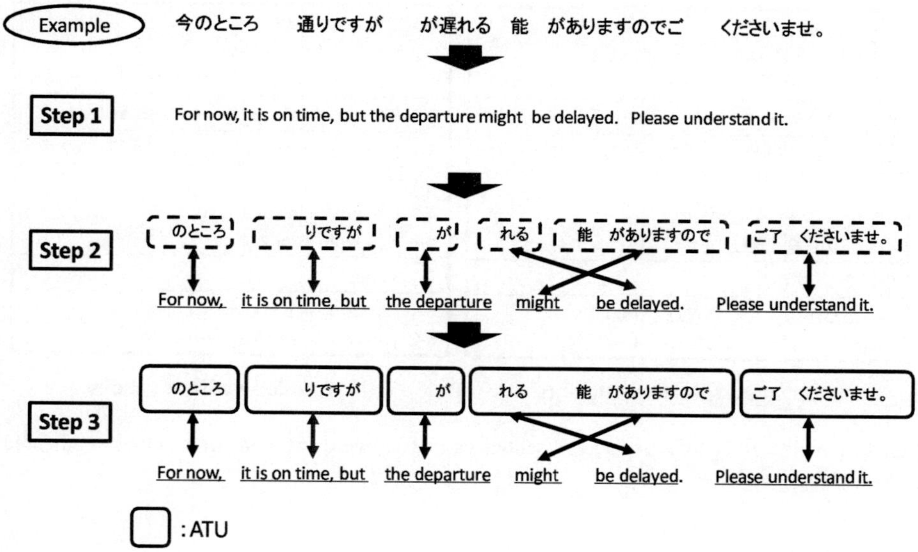

Fig. 2. Acquisition of ATU

Table 2. Statistics of ATU corpus

item	number
dialogues	216
sentences	8736
ATUs	4701
morphemes	57016

3.2 Construction of the ATU Corpus

We constructed an ATU corpus to confirm that each ATU is shorter than a sentence and can be detected incrementally. Japanese speaker's utterances in the simultaneous interpretation database (SIDB) [20] were used to construct the ATU corpus. In the SIDB, Japanese speaker's utterances are annotated with the information of the utterance units. Utterance unit boaders are set at 200ms-or-longer pauses in the speech of speakers. In the SIDB, language tags are also added onto fillers, hesitations and corrections. We removed all these fillers, hesitations and corrections from the SIDB before our analysis.

Fig. 3(a) shows the sample data of Japanese speaker's utterances segmented into ATUs. Fig. 3(b) shows the sample of the word-for-word translations of Japanese speaker's utterances. The numbers at the left side in Figs. 3(a) and (b) indicate the sentence IDs and ATU IDs.

Table 2 shows the statistical information of the data of the proposed units made by the procedure indicated in Sec. 3.1.

1-1	お は
1-2	道 いではないんですけれども
1-3	ビルの にあります。
2-1	ビルはすぐ つけていただけると います。
3-1	テレビ という きなタワーのすぐ ですので。
4-1	
4-2	と いますので
4-3	からお きします。
5-1	そうですね。
6-1	せっかくお しいただいてるので
6-2	城を られたらと いますね。

(a) Japanese speaker's utterances

1-1	The restaurant
1-2	is not on the street but
1-3	It's on the second floor of Hayashi building.
2-1	You can find Hayashi building easily.
3-1	It's just next to a tall tower called TV tower.
4-1	Perhaps
4-2	it is written in Japanese.
4-3	So I write it for you.
5-1	I see.
6-1	As you took a trouble to come here,
6-2	you should see Nagoya castle.

(b) Word-for-word translations

Fig. 3. Samples of Japanese speaker's sentences and its word-for-word translations segmented into ATUs

3.3 Length of ATU

To confirm that the ATU can be enough shorter than a sentence, we examined the length of ATUs. The average length of an ATU is 4.22 morphemes. The average length of a sentence is 6.53. This indicates that the ATU can be enough shorter than a sentence. Fig. 4 shows the distribution of the lengths of sentences and ATUs composed of more than 14 morphemes. According to Fig. 4, most long sentences are segmented into two or more ATUs.

4 Detection of ATUs

The ATU is defined by using the alignment information. However, our simultaneous translation method has three modules, detection, translation and generation, and these modules are independent from each other. So the detection module must detect an ATU without using the alignment information.

In this section, we examine that ATUs can be detected without using the alignment information. The information of the previous or next words and the boundaries of utterance units can be parsed incrementally. Therefore, we examine the relation between ATU boundaries and the information at first. Next, we propose a method of automatically detecting ATUs and show the results of an experiment for ATU detection using the method.

4.1 Analysis of ATUs

We analyzed the relation between an ATU boundary and its previous or next words and between ATU boundaries and utterance unit boundaries. We used 180 dialogues in the ATU copurs for the analysis.

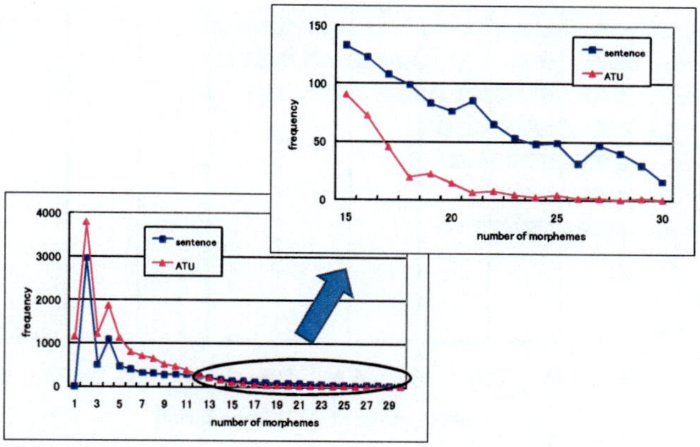

Fig. 4. Distribution of the lengths of sentences and ATUs

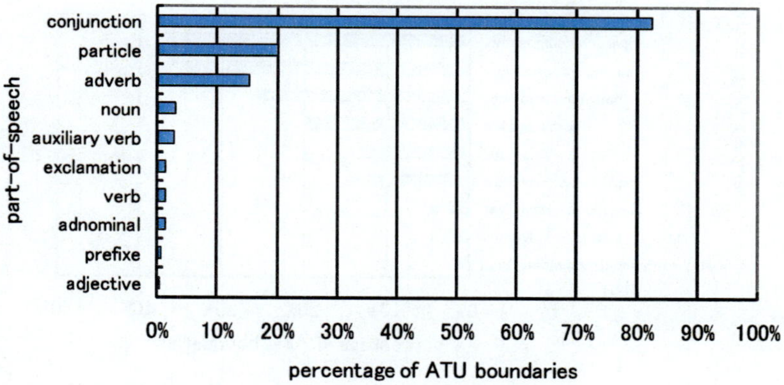

Fig. 5. Part-of-speeches of morphemes before ATU boundaries

We analyzed the morphemes around ATU boundaries. The Chasen [13] was used for morphological analysis. 8.0 % of all the morpheme boundaries coincided with ATU boundaries. Fig. 5 shows the distribution of parts of speech (POSs) of the previous words. The percentage for conjunctions was extremely high. The percentages of particles and adverbs were 19.9% and 15.4%, respectively. The rest were less than 5.0%.

Fig. 6 shows the distribution of POSs of the next words. The percentages of particles, auxiliary verbs and verbs were less than 5.0%.

Fig. 7 shows the distribution of sub parts of speech of the previous words. The percentages of particles-conjunctive, particles-dependency were more than 30.0%. On the other hand, the percentages of particles-adnominalizer, particles-adverbial and particles-adverbial/conjuctive/final were less than 5.0%.

Fig. 8 shows the distribution of surface forms of the previous mophemes (particles) that morpheme boundaries coincide with ATU boundaries. The prcentages of partciles-case "de" and "ga" were more than 40.0%. On the other hand, the prcentages of partciles-case "to", "ni", "wo" and "toiu" were less than 10.0%.

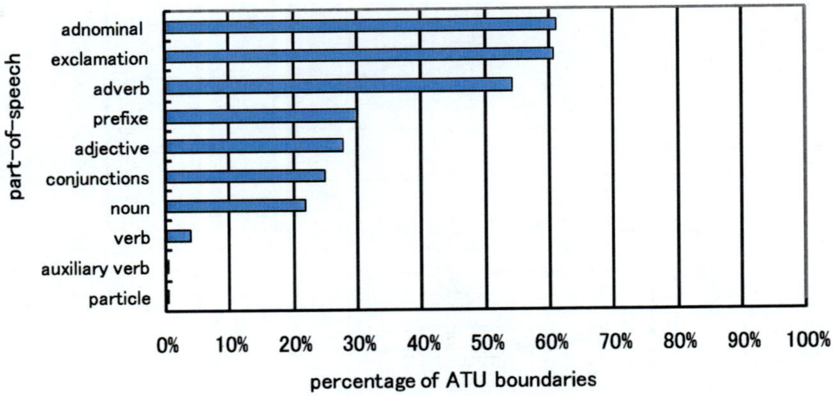

Fig. 6. Part-of-speeches of morphemes after ATU boundaries

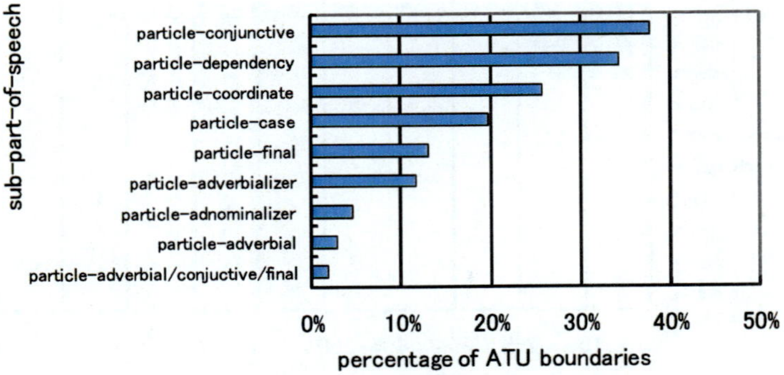

Fig. 7. Sub-part-of-speeches of morphemes before ATU boundaries

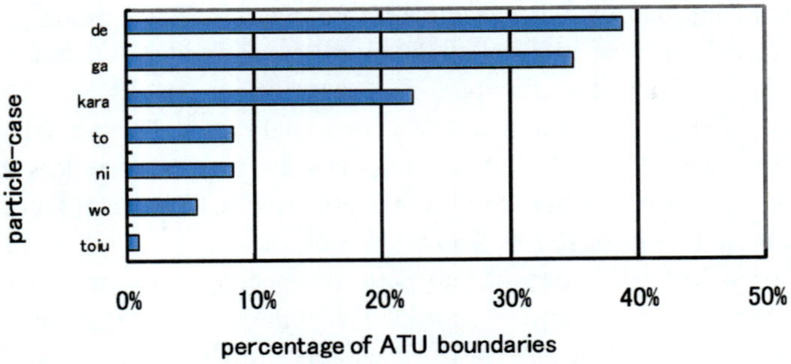

Fig. 8. Particle-case before ATU boundaries

Table 3. Experimental result

method	precision	recall	F-value
our method	80.8%(329/407)	74.3%(329/443)	77.4

We examined the relation between ATU boundaries and utterance units. The boundaries of utterance units in a sentence were examined. The number of the boundaries of utterance units in a sentence was 3252. 53.9 %(1754/3252) of the utterance units coincide with ATUs. 44.4 %(1754/3950) of the ATUs coincide with the utterance units.

4.2 Method of Detecting

To detect ATUs, we calculate the probability that each bunsetsu[2] boundary coincides with an ATU boundary . The probability is calculated using the following function.

$$p(x = 1|y) \tag{1}$$

The function denotes the conditional probability of predicting an outcome x on seeing the context y. The $x = 1$ means that the bunsetsu boundary is the ATU boundary. According to our analysis in Sec. 4.1, the previous or next words, and utterance unit boundaries were used as context y. If the probability of $p(x = 1|y)$ is over the threshold, then we judge that the bunsetsu boundary is the ATU boundary. We used a maximum entropy model as the probability model.

4.3 Experiment

We had an experiment on segmenting Japanese sentences into the proposed units by using the method described in the previous section.

We used 216 dialogues introduced in Sec. 3.2 as experimental data. Among them, 189 were used as training set. 18 were used as the data set for training the features. 18 were used as a test set.

We used the maximum entropy modeling toolkit [10] to train maximum entropy models. This tool needs to train the parameters of the model repeatedly. The number of training steps was 50[3]. Moreover, we adopted the L-BFGS [11] to estimate parameters. If the probability of being the boundary of the proposed unit is more than 50%, then the boundary was judged as the boundary of the proposed unit. As the features of the maximum entropy model, we selected the previous and next three words, and whether a proposed-unit boundary is an utterance unit boundary or not.

Table 3 shows the experimental results. The precision is 80.8% and the recall is 74.3%. The results indicate that it is possible to detect ATUs incrementally by using only a source sentence.

To analyze the detection error, we show the performance for each type of boundary in Fig. 9 . In the figure, the word "correct" means the correct boundary number detected by

[2] A bunsetsu is a basic linguistic unit in Japanese.

[3] The number of times was decided by considering the result of the preliminary experiments.

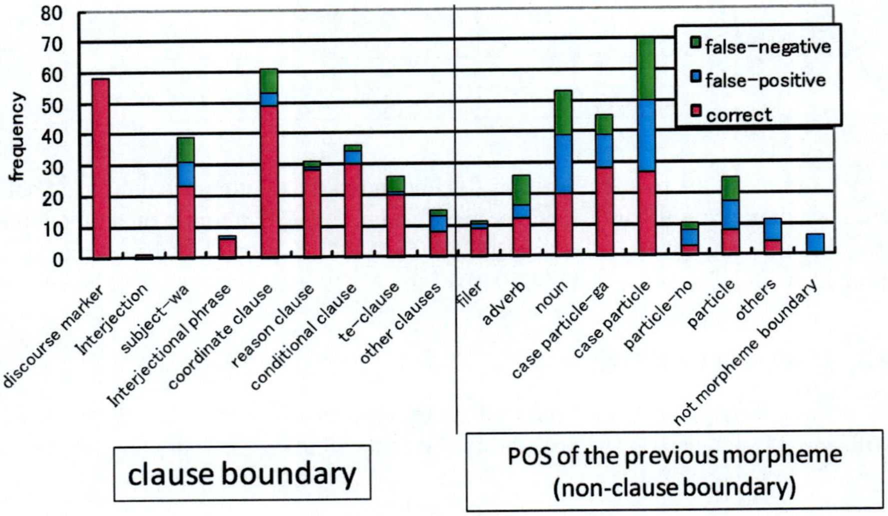

Fig. 9. Performance for each type of boundary

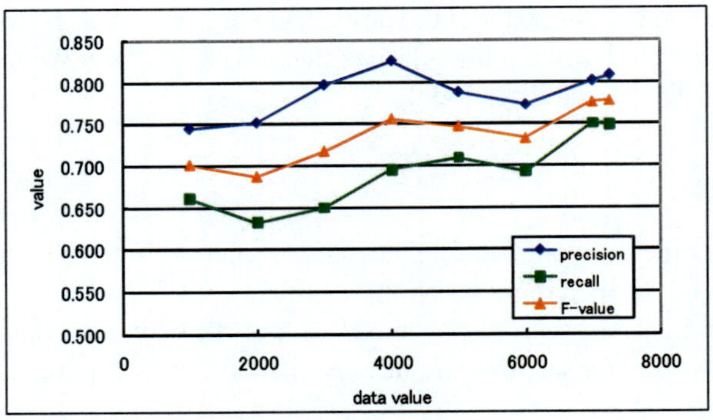

Fig. 10. Effect of different amounts of training data

using our method. The false-negative means that it did not detect a boundary of ATU as a boundary of ATU. The false-positive means that it detected a non-ATU boundary as a ATU boundary. The boundaries of ATUs that are also clause boundaries can be detected with about 90% precision and recall. On the other hand, the boundaries of ATUs that are not clause boundaries can be detected with about 60% precision and recall.

Fig. 10 shows the effect of different amounts of training data. The results in the figure indicate that we can not further improve the performance by increasing training data.

5 Conclusion

In this paper, we proposed ATUs as translation units for simultaneous translation and gave the method of acquiring the proposed translation units. We confirmed that each of the proposed translation units is enough shorter than a sentence and can be detected incrementally.

For future work, we aim at improving the precision and recall of detecting ATUs using the features that we did not use this time. We will try to use syntactic features and acoustic features to train the maximum entropy model.

Acknowledgement. This research was partially supported by the Grant-in-Aids for Scientific Research (B) (No. 20300058) and for Young Scientists (B) (No. 22720154) of JSPS, and Artificial Intelligence Research Promotion Foundation.

References

[1] Amtrup, J.W.: Chart-based incremental transfer in machine translation. In: Proceedings of the 6th International Conference of Theoretical and Methodological Issues in Machine Translation, pp. 188–195 (1995)

[2] Casacuberta, F., Vidal, E., Vilar, J.M.: Architectures for speech-to-speech. In: Proceedings of the Workshop on Speech-to-Speech Translation: Algorithms and System, pp. 39–44 (2002)

[3] Fuhua, L., Yuqing, G., Liang, G., Michael, P.: Noise robustness in speech to speech translation. IBM Tech. Report RC22874 (2003)

[4] Isotani, R., Yamada, K., Ando, S., Hanazawa, K., Ishikawa, S., Iso, K.: Speech-to-speech translation software on PDAs for travel conversation. NEC Research and Development 42(2), 197–202 (2003)

[5] Jibbigo, http://www.jibbigo.com/website/index.php.

[6] Kashioka, H., Maruyama, T.: Segmentation of Semantic Unit in Japanese Monologu. In: Proceedings of International Conference on Speech Database and Assessments, pp. 87–92 (2004)

[7] Kitano, H.: PhiDMDIALOG:A speech-to-speech dialogue translation system. Machine Translation 5(4), 301–338 (1990)

[8] Kolss, M., Wolfel, M., Kraft, F., Niehues, J., Paulik, M., Waibel, A.: Simultaneous German-English Lecture Translation. In: Proceedings of the 5rd International Workshop on Spoken Language Translation, pp. 175–181 (2008)

[9] Lazzari, G.: TC-STAR: a Speech to Speech Translation Project. In: Proceedings of the 3rd International Workshop on Spoken Language Translation, pp. 14–15 (2006)

[10] Le, Z.: Maximum entropy modeling toolkit for Python and C++ (2004)

[11] Liu, D.C., Nocedal, J.: On the limited memory BFGS method for large scale optimization. Math. Programming 45, 503–528 (1989)

[12] Maruyama, T., Kashioka, H., Kumano, T., Tanaka, H.: Development and evaluation of Japanese clause boundaries annotation of general text. Journal of Natural Language Processing 11(3), 39–68 (2004) (in Japanese)

[13] Matsumoto, Y., Kitauchi, A., Yamashita, T., Hirano, Y., Matsuda, H., Takaoka, K., Asahara, M.: ChaSen morphological analyzer version 2.4.0 user's manual. Nara Institute of Science and Technology (2007), http://chasen-legacy.sourceforge.jp/.

[14] Mima, H., Iida, H., Furuse, O.: Simultaneous interpretation utilizing example-based incremental transfer. In: Proceedings of the 17th International Conference on Computational Linguistics and the 36th Annual Meeting of the Association for Computational Linguistics, pp. 955–961 (1998)

[15] Nakamura, S., Markov, K., Nakaiwa, H., Kikui, G., Kawai, H., Jitsuhiro, T., Zhang, J., Yamamoto, H., Sumita, E., Yamamoto, S.: The ATR multilingual speech-to-speech translation. IEEE Transactions on Audio, Speech and Language Processing 14(2), 365–376 (2006)

[16] Ney, H., Och, J.F., Vogel, S.: The RWTH System for Statistical Translation of Spoken Dialogues. In: Proceedings of the 1st International Conference on Human Language Technology Research, pp. 1–7 (2001)

[17] NTT DoCoMo Press Release Article, `http://www.nttdocomo.com/pr/2007/001372.html`.

[18] Ohara, M., Ryu, K., Matsubara, S., Kawaguchi, N., Inagaki, Y.: Temporal Features of Cross-Lingual Communication Mediated by Simultaneous Interpreting: An Analysis of Parallel Translation Corpus in Comparison to Consecutive Interpreting. The Journal of the Japan Association for Interpretation Studies, 35–53 (2003) (in Japanese)

[19] Ryu, K., Matsubara, S., Inagaki, Y.: Simultaneous English-Japanese spoken language translation based on incremental dependency parsing and transfer. In: Proceedings of the 21st International Conference on Computational Linguistics and the 44th Annual Meeting of the Association for Computational Linguistics, pp. 683–690 (2006)

[20] Tohyama, H., Matsubara, S., Kawaguchi, N., Inagaki, Y.: Construction and utilization of bilingual speech corpus for simultaneous machine interpretation research. In: Proceedings of 9th European Conference on Speech Communication and Technology (Eurospeech 2005), pp. 1585–1588 (2005), `http://slp.el.itc.nagoya-u.ac.jp/sidb/`

Chapter 4
Automatic Collection of Useful Phrases for English Academic Writing

Shunsuke Kozawa, Yuta Sakai, Kenji Sugiki, and Shigeki Matsubara

Graduate School of Information Science, Nagoya University, Furo-cho, Chikusa-ku, 464-8601, Japan
kozawa@el.itc.nagoya-u.ac.jp, sakai@el.itc.nagoya-u.ac.jp, sugiki@el.itc.nagoya-u.ac.jp, matubara@nagoya-u.jp

Abstract. English academic writing is indispensable for researchers to present their own research achievement. It is hard for non-native researchers to write research papers in English. They often refer to phrase dictionaries for academic writing to know useful expressions in academic writing. However, lexica available in the market do not have enough expressions and example sentences to serve the purpose since the lexica are created by hand. In order to respond to the demand for the better lexica, this paper proposes a method for extracting useful expressions automatically from English research papers. The expressions are extracted from research papers based on four characteristics of the expressions. The extracted expressions are classified into five classes; "introduction", "related work", "proposed method", "experiment", and "conclusion". In our experiment using 1,232 research papers, our proposed method achieved 57.5% in precision and 51.9% in recall. The f-measure was higher than those of the baselines, and therefore, we confirmed the validity of our method. We developed a phrase search system using extracted phrasal expressions to support English academic writing.

1 Introduction

The aim of our research is to support English academic writing because it is not an easy task for non-native researchers although English academic writing is indispensable for researchers to present their own research achievement. The researchers often consult bilingual dictionaries to translate source language words into English words, refer to lexica of phrases on English research papers to know useful expressions in academic writing, or use search engines to learn English grammar and usage.

Some studies for supporting English writing have been conducted by focusing on English grammar and usage. Search systems for example sentences [1,7,11,12,14,22] and automatic correction systems [5,13] have been developed to assist confirmation of English grammar and usage. In contrast, no study focusing on useful expressions in academic writing has been conducted. Researchers use lexica available in the market (e.g. [17,19]) to find the expressions. The lexica are useful because researchers can use expressions in them without any modification. However, the lexica do not have enough expressions and example sentences because they are produced manually. When they

T. Watanabe and L.C. Jain (Eds.): Innovations in Intell. Machines – 2, SCI 376, pp. 45–59.
springerlink.com

Table 1. Examples of phrasal expression

In this paper, we propose ···
To the best of our knowledge,
The rest of this paper is organized as follows.
In addition to,
With the exception of ···
with respect to ···
as we have seen
as discussed in ···
It is interesting to note that
It must be noted that

cannot find suitable expressions in lexica, researchers have to refer to research papers in their fields to search for the expressions.

If a lexicon of expressions useful for academic writing could be automatically generated, it would help researchers to write research papers. Recently, a considerable number of research papers have been published electrically [8]. This allows us to create a lexicon of expressions useful for academic writing. However, the expressions useful for academic writing could not be acquired by using the conventional methods for extracting collocations or idioms [2,3,4,9,20,21].

This paper proposes a method for extracting useful expressions automatically from English research papers. We call the useful expression *phrasal expression*. We analyze a lexicon available in the market to capture the characteristics of phrasal expressions. The phrasal expressions, which include idioms, idiomatic phrases, and collocations are extracted from research papers using statistical and syntactic information. Then, the extracted phrasal expressions are classified into five classes such as "introduction" and "experiment" to make their usages clear. By using the extracted phrasal expressions, we developed a phrase search system for supporting writing research papers in English.

The remainder of this paper is organized as follows: Section 2 shows the characteristics of phrasal expressions. Section 3 presents a method for acquiring phrasal expressions automatically from research papers. Section 4 presents a method for classifying the extracted phrasal expressions. In Section 5, we report the experimental results. In Section 6, we introduce a phrase search system which uses the extracted phrasal expressions. In Section 7, we draw our conclusions and present the future works.

2 Characteristics of Phrasal Expression

Phrasal expressions are expressions useful for academic writing which include idioms, idiomatic phrases and collocations. Table 1 shows some examples of phrasal expressions. In order to capture the characteristics of phrasal expressions, we analyzed the expressions appeared in the book [17], which is one of the most popular books to refer to when writing research papers in English. By analyzing 1,119 expressions appeared in the book, we found four characteristics of the phrasal expressions; a unit of phrasal expressions, phrasal signs (see Sec. 2.2), statistical characteristics, and syntactic constraints. The following subsections describe these.

2.1 Unit of Phrasal Expression

The expressions appeared in the book represents a coherent semantic unit as seen from examples “in the early part of the paper” or “As a beginning, we will examine”. In other words, the expressions which do not represent a coherent semantic unit such as “in the early part of the” and “As a beginning, we will” are not considered as full phrasal expressions.

We analyzed the expressions based on a base-phrase. A base-phrase is a phrase that does not dominate another phrase [18][1]. Each expression was checked whether it was a sequence of base-phrases or not by using JTextPro [15] for base-phrase chunking. Consequently, out of 1,119 expressions, 1,082 (96.7%) expressions were constituted of base-phrases. Thus, we assume that a base phrase is a minimum unit of a phrasal expression.

2.2 Phrasal Sign

The ellipsis symbol “...” which represents the omission of phrases or clauses is frequently used in the book (e.g. “With the exception of ...”). 859 expressions (76.8%) appeared in the book contained the symbol. The symbol is a useful means to present fixed expressions with an open slot. We use <NP> and <CL> (we call them *phrasal signs*) instead of the symbol to represent the slot. They are a noun phrase and a clause, respectively.

2.3 Statistical Characteristics

We found the following statistical characteristics by analyzing the book:

- **It occurs frequently**
  The expressions appeared in the book are frequently used in academic writing.
- **The length is not too short**
  The expressions composed of one or two base-phrases account for only 6.9% of all expressions appeared in the book.
- **The preceding/succeeding words are various**
  The phrasal expressions are used in various contexts. Thus, phrasal expressions can be preceded/succeeded by many kinds of base-phrases. Let us consider the expressions “in spite” (not phrasal expression) and “in spite of” (phrasal expression). As for “in spite”, term frequency was 36 and succeeding base-phrase was only “of” in the research papers used in our experiments. On the other hand, as for “in spite of”, term frequency was same as “in spite.” However, the frequency of kinds of succeeding base-phrases was 36 (e.g. “their inability”, “the noise”, “the significant error rate”, etc.). This shows that phrasal expressions have a tendency to precede/succeed various base-phrases.

[1] For example, the sentence “In this paper, we propose a new method.” is converted into a sequence of base-phrases “[PP In] [NP this paper] , [NP we] [VP propose] [NP a new method] .”. Here, parenthetical parts are base-phrases.

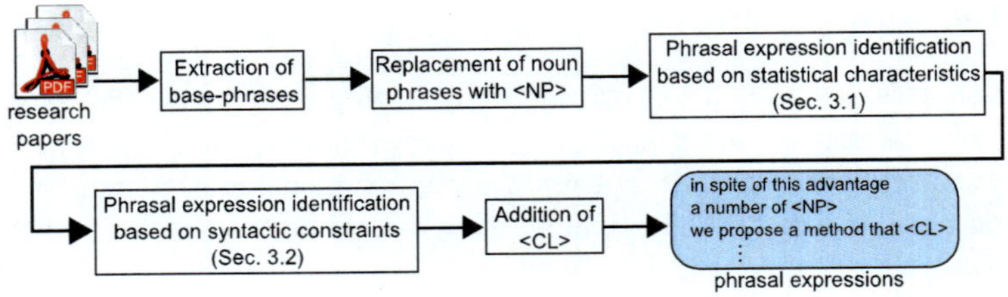

Fig. 1. Flow for acquiring phrasal expressions

2.4 Syntactic Constraints

We found some syntactic constraints, that is, some syntactic patterns were not used for writing research papers. For example, "stem from" appeared in the book. On the other hand, "stem in" and "stem with" did not appear. This means, syntactic patterns are an important factor in determining whether a given expression is a phrasal expression or not. In addition, the book contains not only general expressions such as "in other words" but also specialized expressions for writing research papers such as "The purpose of this paper is to" and "The result of the experiment was that". This shows that the specialty of expressions provides a clue to identify expressions as phrasal expressions.

3 Acquisition of Phrasal Expression

Phrasal expressions are extracted from research papers based on the characteristics shown in Section 2. The processing flow is shown in Figure 1. First, sequences of base-phrases are extracted from research papers. Secondly, the noun phrases in them are replaced by <NP>. Note that sequences of base-phrases which contain three or more <NP> are not generated. Thirdly, sequences of base-phrases satisfying statistical characteristics are acquired from them. Then, sequences of base-phrases which do not satisfy syntactical constraints are eliminated from them. Finally, the sequences of base-phrases which contains a complementizer phrase (e.g. "that", "which", "so that") as the last base-phrase are postfixed with <CL>. The following subsections describe our method for acquiring phrasal expressions using statistical characteristics and syntactic constraints.

3.1 Phrasal Expression Identification Based on Statistical Characteristics

Candidates for phrasal expressions are extracted from sequences of base-phrases using statistical information. Note that we do not acquire sequences of base-phrases which meet conditions that relative document frequency is less than 1% or the number of base-phrases is one.

We used the scoring functions *Lscore* and *Rscore* based on Ikeno et al's method [6] in order to identify whether the given sequence of base-phrases has statistical characteristics. The functions are described as follows:

$$Lscore(E) = \log(tf(E)) \times length(E) \times H_l(E).$$

$$Rscore(E) = \log(tf(E)) \times length(E) \times H_r(E).$$

Here, E is a sequence of base-phrases. $length(E)$ denotes the number of base-phrases contained in E. $tf(E)$ represents the term frequency of E in target research papers. $H_l(E)$ and $H_r(E)$ denote the entropies of probability distributions of the preceding and succeeding base-phrase, respectively. The scores are higher if the frequency of the kinds of preceding and succeeding base-phrases are high and their frequency shows uniformity. $H_l(E)$ and $H_r(E)$ are formulated by the following equations, respectively:

$$H_l(E) = -\sum_i Pl_i(E) \log Pl_i(E).$$

$$H_r(E) = -\sum_i Pr_i(E) \log Pr_i(E).$$

Pl_i/Pr_i is a probability that E is preceded/succeeded by a base-phrase Xi.

$$Pl_i(E) = P(X_iE|E) = \frac{P(X_iE)}{P(E)} \approx \frac{tf(X_iE)}{tf(E)}.$$

$$Pr_i(E) = P(EX_i|E) = \frac{P(EX_i)}{P(E)} \approx \frac{tf(EX_i)}{tf(E)}.$$

The first, second and third terms in *Lscore* and *Rscore* represent the length, the term frequency and the type of preceding and succeeding base-phrases, respectively. That is to say, the more the sequence reflects statistical characteristics described in Sec. 2.3, the higher the score is.

Our method considers E as candidate for a phrasal expression if E satisfies the following inequations:

$$Lscore(E) > Lscore(XE)$$

$$Rscore(E) > Rscore(EX)$$

Here, X is a preceding/succeeding base-phrase. This means that EX/XE has more base-phrases than E. If E satisfies the above two equations, E is extracted.

3.2 Phrasal Expression Identification Based on Syntactic Constraints

Phrasal expressions have syntactic characteristics described in Sec. 2.4. However, since the characteristics are too various, it is difficult to identify whether a target expression has them. Therefore, in our method, sequences which do not have any of the syntactic characteristics are eliminated by a rule-based approach.

In order to generate a rule, 809 sequences of base-phrases were extracted at random from the candidates of phrasal expressions and judged whether a given sequence is a phrasal expression or not. We generated the rule composed of 25 patterns based on grammatical information according to the analysis of the sequences. The generated rule is shown in Table 2. NP, VP, PP, ADVP, ADJP and VBG represent a noun phrase, verb phrase, prepositional phrase, adverbial phrase, adjective phrase and gerund, respectively. Note that <NP> is different from NP. The rule is not applied if a given

Table 2. Rule based on grammatical constraints

<table>
<tr><th>Pattern</th></tr>
<tr><td>A sequence does not include interrogatives, adjective phrases, noun phrases which do not consists only of pronouns, and verb phrases which do not consists only of copulas</td></tr>
<tr><td>The first or last word of a sequence is “and”.</td></tr>
<tr><td>A sequence has complementizer phrase which are neither the first nor last base-phrase.</td></tr>
<tr><td>A sequence is ended with “[complementizer|;|:|,] <NP>” or “PP”.</td></tr>
<tr><td>The last word of a sequence is a nominative pronoun (“we”, “I”, “he”, “she”, “they”).</td></tr>
<tr><td>A sequence is begun with to-infinitive, a complementizer “that” or “PP <NP>”</td></tr>
<tr><td>A sequence contains to-infinitive and does not contain an infinitive verb.</td></tr>
<tr><td>A sequence contain “<NP> [of|in|,|and] <NP>” or “<NP> “(” <NP> “)””.</td></tr>
<tr><td>NP of NP</td></tr>
<tr><td>NP [of|in] <NP>the threshold of <NP></td></tr>
<tr><td><NP> PP NP (PP <NP>)</td></tr>
<tr><td>PP NP</td></tr>
<tr><td>PP VBG (PP is not “without”)</td></tr>
<tr><td>PP VBG (<NP>)</td></tr>
<tr><td>NP ADVP</td></tr>
<tr><td>NP interrogative</td></tr>
<tr><td>interrogative VP</td></tr>
<tr><td>pronoun (ADVP) VP (PP) (<NP>)</td></tr>
<tr><td>(NP| <NP>) copula (ADVP) (NP|ADJP) (PP)</td></tr>
</table>

sequence appears in existing dictionaries or has specialized nouns or verbs in academic writing.

Specialized nouns and verbs are acquired by comparing relative term frequency in the target research papers with the frequency in general documents such as newspapers and Web. Given a word w, it is identified as specialized words if it satisfies the following conditions:

- Relative document frequency in the target research papers is larger than or equal to α %.
- Relative term frequency in the target research papers is more than β times more frequent than relative term frequency in general documents.

The thresholds α and β are set empirically.

4 Classification of Phrasal Expressions

In this section, we describe a method for classifying the phrasal expressions utilizing the composition of research papers. The phrasal expressions in the book [19] are categorized by section name which they are frequently used (e.g. expressions appeared such as in “introduction” or “conclusion”). This categorization helps users find and use appropriate phrasal expressions efficiently. In this research, we assume that research papers are composed of five sections, namely, “introduction”, “related work”, “proposed

Table 3. Clue expressions of each section class

section class	clue expression
introduction	"introduction"
related work	"past work", "related work", "previous work", "recent work"
experiment	"result", "experiment", "evaluation", "discussion"
conclusion	"conclusion", "future work", "summary"

method", "experiment", and "conclusion". They are selected by taking into account the field of computer science. Phrasal expressions are classified into the above five classes (*section classes*) by using frequencies of their appearance in each section class.

4.1 Structuring Research Papers

In order to classify phrasal expressions according to the structure of the research papers, research papers are structuralized since research papers in pdf format do not have the structure. To structure a research paper, the section titles are identified since a research paper are divided into sections. Each section title is described in a same form in a research paper even though the forms of section titles are little different from each other, depending on the difference of the type of the research papers.

The title of section 1 is identified by using the following regular expression (Perl specification). Other section titles are identified by using the matched pattern.

- /^1(\.?)\s+[A-Z].{2,}/

4.2 Section Class Identification

The sections in the research papers are classified into five section classes to learn which phrasal expression frequently appears in which section class. The section titles are classified into the section classes by using clue expressions since section titles contain common words in many research papers in same section class. Table 3 shows clue expressions of each section class. The section S is classified under the section class C if the title of the section S contains the clue expressions of a section class C. The section S is classified as the section class "proposed method" if the title of the section S do not contain the clue expressions.

We carried out a preliminary experiment to evaluate our method for classifying the sections into the section classes. We randomly selected 100 papers from 1,232 papers in proceedings of ACL[2] from 2001 to 2008 as evaluation data and classified 753 sections into the five section classes. Consequently, we achieved 91.4% (688/753) in accuracy. The experimental result shows our method is valid in identification of section classes.

4.3 Phrasal Expression Classification Based on Locality

The phrasal expressions are classified into five section classes based on *locality* calculated by using the frequency of the phrasal expressions in each section class. The

[2] The Conference of Association for Computational Linguistics.

Table 4. Statistics of experimental data

papers	sentences	base-phrases	words
1,232	204,788	2,683,773	5,516,612

locality represents how frequently a phrasal expression appears in a section class and is calculated by the following formula:

$$locality(E,c) = \frac{ndf_{E,c}}{\sum_{c_k \in C} ndf_{E,c_k}}$$

$$ndf_{E,c} = \frac{df_{E,c}}{N_c}$$

Here, E, c, and C represent a phrasal expression, a section class, and the set of section classes, respectively. $ndf_{E,c}$ is the ratio of the number of papers containing the section identified as the class c ($df_{E,c}$) to the number of papers containing the section which is identified as the class c and contains the phrasal expression E (N_c). We used the number of papers instead of the frequency of phrasal expressions so that we could avoid the influence of the phrasal expression frequently used in a particular research paper. Moreover, $ndf(E,c)$ is used to avoid the effect of the difference in the number of sections which were classified under each section class. The phrasal expression E falls into the section class c if the locality is greater than or equal to the threshold γ. If the locality is smaller than γ for each section classes, the phrasal expression E is not classified and is considered as the expression which appears anywhere.

5 Experiments

5.1 Experiment on Phrasal Expression Acquisition

5.1.1 Experimental Settings

As for our experimental data set, we used the proceedings of the ACL from 2001 to 2008. Table 4 shows statistical information of the set. We evaluated our method which extracted 4,945 phrasal expressions from experimental data. We selected Eijiro 4th Edition [16] as a dictionary used in Sec. 3.2. Specialized nouns and verbs were extracted by comparing the experimental data set with the Wall Street Journal data from the Penn Treebank [10]. The thresholds α, β for nouns and β for verbs were manually set to 1, 4 and 2, respectively, by comparing the Wall Street Journal data with the proceedings of COLING[3]2000, COLING2002 and COLING2004. We extracted 1,119 nouns and 226 verbs with these thresholds. We used xpdf[4] to convert pdf to plain text and JTextPro [15] for base-phrase chunking.

As for our evaluation data, 500 sequences of base-phrases were extracted from the experimental data at random and they were judged by one of the authors who is familiar with academic writing. We evaluated our method based on precision (the ratio of

[3] The International Conference on Computational Linguistics.

[4] http://www.foolabs.com/xpdf/

Table 5. Experimental result

	precision (%)	recall (%)	f-measure
Baseline	16.20 (81/500)	100.00 (81/81)	27.88
Statistical	23.51 (59/251)	72.84 (59/81)	35.54
Syntactic	44.07 (52/118)	64.20 (52/81)	52.26
Proposed	57.53 (42/73)	51.85 (42/81)	54.55

successfully extracted phrasal expressions to the total number of the extracted phrasal expressions) and recall (the ratio of successfully extracted phrasal expressions to the total number of the correct phrasal expressions). We compared the following four methods to evaluate our method:

Baseline : phrasal expressions were acquired at random.
Statistical : phrasal expressions were acquired using only statistical information.
Syntactic : phrasal expressions were acquired using only syntactic information.
Proposed : phrasal expressions were acquired using both statistical and syntactic information.

5.1.2 Experimental Result

The experimental results are shown in Table 5. Out of 500 base-phrases in the evaluation data, 81 was correct phrasal expressions. Our proposed method achieved 57.53% in precision and 51.85% in recall. In comparison with random extraction, the methods using both or either statistical and syntactic information improved in f-measure. The results show that the use of both statistical and syntactic information is available for acquiring phrasal expressions. Therefore, we have confirmed the validity of our method.

Table 6 shows the examples of phrasal expressions acquired successfully. Expressions appeared in dictionaries such as "As a result," or "adding <NP> to <NP>" were acquired. Furthermore, some useful expressions which do not appear in dictionaries such as "In this paper, we propose" and "<NP> divided by the total number of <NP>" could be acquired.

5.1.3 Discussion

We investigated why the recall decreased in the method using statistical information. Out of 22 correct phrasal expressions which were not extracted, seven (31.8%) were frequently succeeded by "of" (e.g., "we have performed <NP> (of)" and "we also show <NP> (of)"). They were eliminated because their *Rscore* were lower than *Rscore* for "we have performed <NP> of" and "we also show <NP> of" since noun phrases are frequently succeeded by "of" and "of" is succeeded by various noun phrases. This problem will be solved by replacing "<NP> of <NP>" with "<NP>".

We investigated why the precision was not significantly improved with statistical information. Out of 192 incorrect phrasal expressions extracted by using statistical information, 29 (15.1%) were base-phrases which lack a nominative noun phrase (e.g., "(<NP>) is treated as <NP>" and "(<NP>) is created for <NP>"). The correct phrasal expressions were "<NP> is treated as <NP>" and "<NP> is created for <NP>". However, *Lscore* for "<NP> is treated as <NP>" was not larger than *Lscore*

Table 6. Examples of successfully acquired phrasal expressions

phrasal expression
<NP> is set to <NP>
<NP> is shown in Figure <digit>.
<NP> leads to <NP>,
<NP> depends on <NP>,
<NP> attached to <NP>
<NP> applied to <NP>
<NP> divided by the total number of <NP>
<NP> is not statistically significant.
<NP> is consistent with <NP>
Using <NP> as <NP>
As a result,
extracting <NP> from <NP>,
adding <NP> to <NP>.
the results obtained with <NP>
the occurrence of <NP>
N is the total number of <NP>
when <NP> are used.

Table 7. Experimental result of phrasal expression classification

section class	correct	incorrect
introduction	32	18
related work	35	15
proposed method	18	32
experiment	16	34
conclusion	24	3
total	125	102

for “is treated as <NP>” since “is treated as <NP>” was preceded by various noun phrase and “<NP> is treated as <NP>” was frequently preceded by a preposition “on”. We will have to reconsider the formula by taking into account the nominative noun phrases.

5.2 Experiment on Phrasal Expression Classification

We selected frequently appeared 50 phrasal expressions for each section class and evaluated whether they were correctly classified. Note, however, that only 27 phrasal expressions fall into the section class ”conclusion” and therefore available for this purpose. We set the threshold value of the locality for classification of phrasal expression to 0.5.

Experimental results are shown in Table 7. We achieved 55% (125/227) in accuracy. Some of the phrasal expressions which were classified under “introduction” are shown

Table 8. Example of phrasal expressions classified into "introduction"

phrasal expression	*locality*
in Section <digit>, <NP> present	1.00
*is funded by <NP>	1.00
<NP> of this paper is organized as follows	1.00
<NP> conclude in <NP>	1.00
*we would like to thank <NP>, <NP>,	1.00
*we would like to thank <NP>,	1.00
the rest of this paper is organized as follows	1.00
*we would like to thank <NP> for <NP>	1.00
section <digit> introduces	0.97
in section <digit>, <NP> describe <NP>	0.93
section <digit> discusses	0.93
section <digit> describes <NP>,	0.92
<NP> <digit> reports on <NP>	0.92
section <digit> summarizes	0.92
in Section <digit> we describe <NP>	0.91
in this paper we present	0.90
in section <digit>, <NP> describe	0.89
we then present	0.88
*<NP> then present	0.88
in Section <digit>, we present	0.86
*<NP> in Section <digit>	0.82
in this paper, <NP> introduce	0.82
in this paper <NP> describe	0.81
in this paper we describe	0.81
in this paper, we introduce	0.81
in this paper, <NP> present	0.80
in this paper , <NP> propose	0.80
in Section <digit> we describe	0.80
in Section <digit> <NP> describe	0.80
this paper describes	0.79

in Table 8. Expressions marked with an asterisk "*" in the Table 8 are incorrectly classified.

In order to learn causes of the errors, we investigated 102 phrasal expressions which were incorrectly classified. Consequently, we found that 88.2% (90/102) are expressions which appears in any section class. In addition, the locality of them are close to the threshold value. Figure 2 shows the accuracy when the threshold value of the locality is changed from 0.5 to 1.0. The higher the threshold value was, the better the accuracy was. This indicates the locality has effect on classification of phrasal expressions. Note that the accuracy is decreased when the threshold value was 1.0. This is because the expressions to be classified under "acknowledgment" were incorrectly classified under "introduction". This problem will be solved by adding "acknowledgment" class to the section classes.

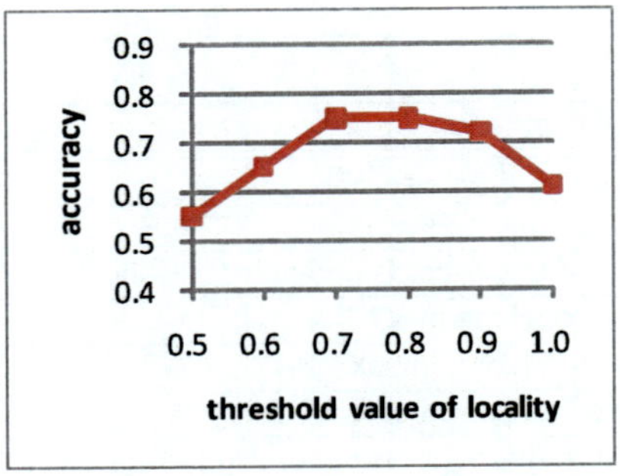

Fig. 2. Relation between the threshold value of the *locality* and accuracy

6 Phrasal Expression Search System

The aim of our research is to support English writing. To achieve the aim, we developed SCOPE (System for Consulting Phrasal Expressions) by using extracted phrasal expressions as index. SCOPE provides phrasal expressions and the example expressions using them. In addition, SCOPE can provide the phrasal expressions which are classified under the section class by selecting any of the five section classes (introduction, related work, proposed method, experiment and conclusion).

SCOPE first receives one or more English words and the type of section class as queries, it then retrieves phrasal expressions which contain the input English words and appear in the input section class, and finally it provides the phrasal expressions ranked by the frequency. If the section class is not selected, phrasal expressions containing the input English words will be provided. SCOPE offers example sentences of the phrasal expression when a particular phrasal expression is clicked. SCOPE has the following functions:

- Searching for phrasal expressions containing query keywords
- Searching for phrasal expressions according to the type of section class
- Showing the frequency and many examples of searched phrasal expressions

SCOPE was implemented using Perl. Tokyo Cabinet[5] was used as the database for searching for phrasal expressions. We used 7,769 phrasal expressions extracted from the proceedings of ACL from 2001 to 2008 and the proceedings of COLING2000, 2002, 2004 and 2008.

Let us consider the situation where a user who wants to write experimental results searches SCOPE for phrasal expressions by the keyword "result" and by selecting "experiment" from the section classes. Figure 3 shows the result using "result" as a query and "experiment" as a section class. Here, the value of the <digit> embedded in the phrasal expressions can be any number. The user will learn that his experimental results can be described by using the phrasal expressions such as "Table <digit> shows the results of <NP>" and "we present the results <NP>". The user will also know that two most frequently used expressions are "<NP> shows the results of <NP>" and "Table

[5] http://1978th.net/

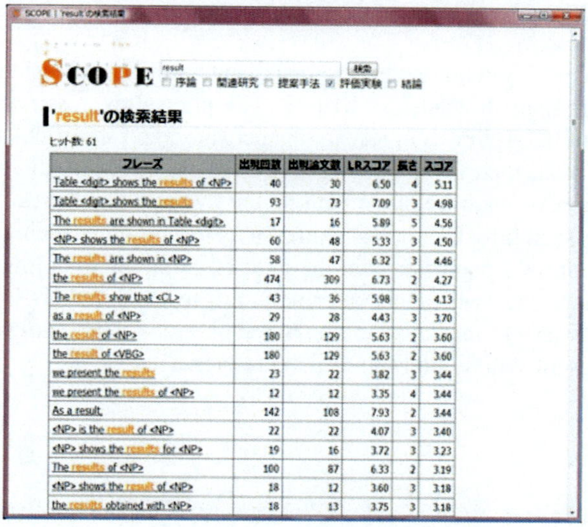

Fig. 3. Search result using "result" as a query

Fig. 4. Detail information of "Table <digit> shows the results of <NP>"

<digit> shows the results of <NP>" by referring to the frequencies of them. In addition, the example sentences are available as shown in Figure 4 by clicking the phrasal expression "Table <digit> show the results of <NP>". The user can find expressions suitable to his need by referring to the examples.

SCOPE has been in operation at the following Web site:

http://scope.itc.nagoya-u.ac.jp/

7 Conclusion

In this paper, we proposed the method for acquiring phrasal expressions from research papers to support English academic writing. The phrasal expressions were extracted from sequences of base-phrases in research papers based on statistical and syntactic information found by analyzing the existing lexicon of phrasal expressions in this method. The extracted expressions were classified into the five section classes. Then, we developed SCOPE for searching for phrasal expressions to support academic writing.

In this paper, phrasal expressions in the field of computational linguistics were acquired. In the future, we will apply our method to research papers in other fields. In that case, we will have to improve the rule based on grammatical information. We also would like to present synonymous phrasal expressions.

References

1. Ando, K., Tsunashima, Y., Okada, M.: A Writing Support Tool for Learners of English and/or Japanese as a Second Language. In: Proceedings of World Conference on Educational Multimedia, Hypermedia and Telecommunications 2008, pp. 5921–5927 (2008)
2. Bouma, G., Villada, B.: Corpus-based acquisition of collocational prepositional phrases. Language and Computers 45(1), 23–37 (2002)
3. Cook, P., Fazly, A., Stevenson, S.: Pulling their weight: exploiting syntactic forms for the automatic identification of idiomatic expressions in context. In: Proceedings of the Workshop on A Broader Perspective on Multiword Expressions, pp. 41–48 (2007)
4. Fazly, A., Stevenson, S.: Automatically constructing a lexicon of verb phrase idiomatic combinations. In: Proceedings of the 11th Conference of the European Chapter of the Association for Computational Linguistics, pp. 337–344 (2006)
5. Ge, S.L., Song, R.: Automated Error Detection of Vocabulary Usage in College English Writing. In: Proceedings of 2010 IEEE/WIC/ACM International Conference on Web Intelligence and Intelligent Agent Technology, pp. 178–181 (2010)
6. Ikeno, A., Hamaguchi, Y., Yamamoto, E., Isahara, H.: Techinical term acquisition from web document collection. Transactions of Information Processing Society of Japan 47(6), 1717–1727 (2006)
7. Kato, Y., Egawa, S., Matsubara, S., Inagaki, Y.: English sentence retrieval system based on dependency structure and its evaluation. In: Proceedings of 3rd International Conference on Information Digital Management, pp. 279–285 (2008)
8. Lawrence, S., Lee Giles, C., Bollacker, K.: Digital libraries and autonomous citation indexing. IEEE Computer 32(6), 67–71 (1999)
9. Lin, D.: Automatic identification of non-compositional phrases. In: Proceedings of the 37th Annual Meeting of the Association for Computational Linguistics on Computational Linguistics, pp. 317–324 (1999)
10. Marcus, M.P., Santorini, B., Marcinkiewicz, M.A.: Building a large annotated corpus of English: the Penn Treebank. Computational Linguistics 19(4), 313–330 (1993)
11. Miyoshi, Y., Ochi, Y., Kanenishi, K., Okamoto, R., Yano, Y.: An illustrative-sentences search tool using phrase structure "SOUP". In: Proceedings of 2004 World Conference on Educational Multimedia, Hypermedia and Telecommunications, pp. 1193–1199 (2004)
12. Narita, M., Kurokawa, K., Utsuro, T.: Web-based English abstract writing tool using a tagged E-J parallel corpus. In: Proceedings of 3rd International Conference on Language Resources and Evaluation, pp. 2115–2119 (2002)

13. Nishimura, N., Meiseki, K., Yasumura, M.: Development and evaluation of system for automatic correction of English composition. Transactions of Information Processing Society of Japan 40(12), 4388–4395 (1999) (in Japanese)
14. Oshika, H., Sato, M., Ando, S., Yamana, H.: A translation support system using search engines. IEICE Technical Report. Data Engineering 2004(72), 585–591 (2004) (in Japanese)
15. Phan, X.H.: JTextPro: A Java-based text processing toolkit (2006), `http://jtextpro.sourceforge.net/`
16. Eppstein, D.: Project. Eijiro, 4th edn. ALC Press Inc. (2008)
17. Sakimura, K.: Useful expressions for research papers in English. Sogen-sha (1991) (in Japanese)
18. Sang, E.F.T.K., Buchholz, S.: Introduction to the CoNLL-2000 shared task: Chunking. In: Proceedings of 4th Conference on Computational Natural Language Learning and of the 2nd Learning Language in Logic Workshop, vol. cs.CL/0009008, pp. 127-132 (2000)
19. Sugino, T., Ito, F.: How to write a better English thesis. Natsume-sha (2008) (in Japanese)
20. Van de Cruys, T., Moirón, B.V.: Semantics-based multiword expression extraction. In: Proceedings of the Workshop on A Broader Perspective on Multiword Expressions, pp. 25–32 (2007)
21. Widdows, D., Dorow, B.: Automatic extraction of idioms using graph analysis and asymmetric lexicosyntactic patterns. In: Proceedings of the ACL-SIGLEX Workshop on Deep Lexical Acquisition, pp. 48–56 (2005)
22. Yamanoue, T., Minami, T., Ruxton, I., Sakurai, W.: Learning usage of English KWICLY with WebLEAP/DSR. In: Proceedings of 2nd International Conference on Information Technology and Applications (2004)

Chapter 5
An Effectively Focused Crawling System

Yuki Uemura, Tsuyoshi Itokawa, Teruaki Kitasuka, and Masayoshi Aritsugi

Computer Science and Electrical Engineering,
Graduate School of Science and Technology,
Kumamoto University, Japan
{uemura@dbms., itokawa@dbms.,
kitasuka@, aritsugi@}cs.kumamoto-u.ac.jp

Abstract. In this article, we illustrate design and implementation of a focused crawling system for effectively collecting webpages concerning specific topics. An algorithm for deciding where to crawl next is developed by exploiting not only anchor texts but also the concept of PageRank. Given a topic to be focused on, our system attempts to collect webpages concerning the topic by crawling webpages that are expected to have not only close similarities to the topic but also high rank. Experimental results using many topics are reported and investigated in this article.

1 Introduction

WWW has provided an enormous amount of data these days, and it is useful for innovative and creative activities of human beings to collect necessary information from WWW effectively and efficiently. In this article, we illustrate design and implementation of a focused crawling system for collecting webpages concerning specific topics.

Although there are several products of general purpose WWW retrieval systems based on crawlers such as Google and Yahoo!, we develop a focused crawler because it needs less resources including storage capacity and network bandwidth than a general purpose WWW crawler. As a result, we can run it on our own machine, thereby not only preserving our privacy but also keeping collected webpages up-to-date more easily.

There are mainly three problems developing a focused crawler that can run on a personal machine (Fig. 1). One is how to extract specific topics from interests of a user. Since interests of a user must be different from those of another user, it is important to extract appropriately specific topics of a user. Another is how to crawl webpages concerning the topics efficiently. The other is how to manage resources including network bandwidth, computing power and disk space according to the characteristics of a personal environment. For example, crawling for user A in Fig. 1 must be optimized because the user has small resources, while crawling for user B may be processed in parallel. In this article, we adapt the concept of PageRank [2] to contribute the second problem.

We proposed a prioritization algorithm of webpages for deciding where to crawl next for focused crawlers [24]. In the algorithm, we attempt to integrate the concept of PageRank into the decision. PageRank has been applied to general purpose crawlers so

T. Watanabe and L.C. Jain (Eds.): Innovations in Intell. Machines – 2, SCI 376, pp. 61–76.
springerlink.com

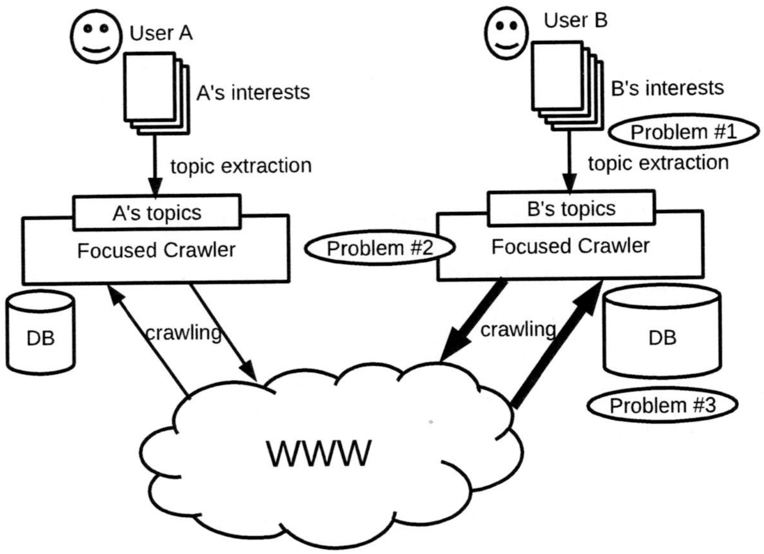

Fig. 1. Three problems developing a focused crawler

far. To our knowledge, one of our contribution is to consider how to apply the concept of PageRank to focused crawling. Concretely, our algorithm is based on personalized PageRank [13] and a lower bound of PageRank [7], and integrate them with similarity measurement to specific topics. In this article, we illustrate design and implementation details of our system, and discuss some places of our current implementation to be improved. In addition, results of experiments with many topics different from [24] are reported for discussing the effectiveness and points to be improved of our system.

A focused crawler was firstly proposed in [3], and many studies of focused crawlers have followed (e.g., [8,10]). Chakrabarti et al. [3] attempted to identify hubs for specific topics for focused crawling. In [3], topics are specified by using exemplary documents. On the other hand, in this study topics are modeled by a simple way in which feature words extracted from given webpages as seeds are used, and we focus on the strategy of deciding where to crawl next. Diligenti et al. [8] developed context focused crawler, in which context graphs were generated as compact context representations for modeling context hierarchies. Ester et al. [10] introduced a unique focused crawler that attempted to select websites instead of webpages. Shchekotykhin et al. [22] proposed a focused crawling method that exploited existing navigational structures such as index pages, hierarchical category structures, menus, and site maps derived from Kleinberg's HITS [14] based algorithm, implemented a random restart strategy, and also included a query generation algorithm for exploiting public search engines. While these studies use graph-based approaches constructed in their own ideas, we exploit the concept of PageRank. This idea was inspired by [7].

There have been many studies of web crawling ordering [6,15,1,7]. Cho and Schonfeld [7] discussed crawler coverage guarantee and crawler efficiency. They defined RankMass as the sum of PageRank values of crawled webpages, and developed a set of algorithms using RankMass for providing a theoretical coverage guarantee. In [7], they also defined a lower bound of PageRank, and we exploit it in this study. Main different points of our work from them are to focus on focused crawlers instead of general purpose crawlers and to use precision and target recall [19,23,20] in evaluation.

The remainder of this article is organized as follows. Section 2 proposes personalized PageRank for focusing on a specific topic. Section 3 describes how to prioritize webpages. Section 4 illustrates our crawling algorithm. Section 5 reports some experimental results to evaluate our system, and Section 6 concludes this article.

2 Personalized PageRank for Focusing on a Topic

Our algorithm is to integrate the concept of PageRank [2] into a focused crawling system. In this section, we propose a simple way of calculating rank of each webpage for our system.

PageRank is based on the random surfer model. The importance of each webpage is calculated as a probability that the webpage is accessed in the model. In other words, the higher the PageRank value of a webpage is, the more important the webpage is supposed to be in WWW.

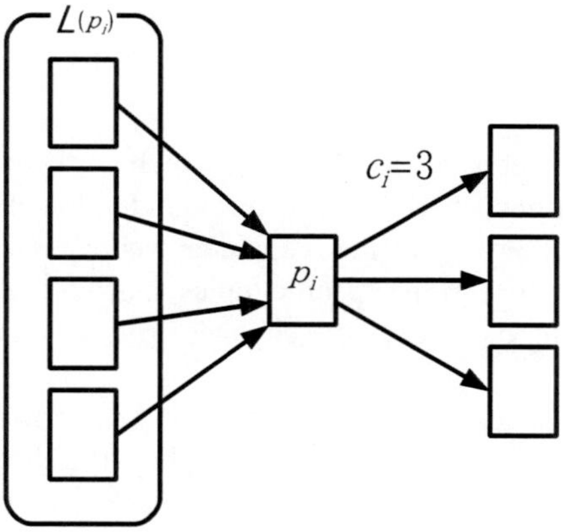

Fig. 2. Example of links of webpage p_i

In [13], the original PageRank is refined as personalized PageRank. Let $L(p_i)$ be the set of webpages that have at least a link to webpage p_i, and c_i be the number of out-links that webpage p_i has. An example of links concerning webpage p_i is shown in Fig. 2 where the numbers of element webpages in $L(p_i)$ and out-links of webpage p_i are four and three, respectively. Then, personalized PageRank of webpage p_i, which is expressed as r_i, is defined as follows:

$$r_i = d\left[\sum_{p_j \in L(p_i)} \frac{r_j}{c_j}\right] + (1-d)t_i. \tag{1}$$

In Equation (1), d is a constant value called a damping factor and is often set to 0.85 [12], and t_i is the trust score of webpage p_i. If a webpage is supposed to be trusted, then its trust score is a non-zero value and $\sum_i t_i = 1$.

We assume a variation of the random surfer model, in which a Web surfer who attempts to access only webpages that are related to a specific topic, and propose another PageRank based on the model. For simplicity, a specific topic is supposed to be modeled with feature words extracted from webpages that are given by a user in this study. Let T be a specific topic modeled with feature words, the probability that a Web surfer, who wants to access webpages concerning T, accesses webpage p_i is defined as follows:

$$r_i = d \left[\sum_{p_j \in L(p_i)} \frac{sim(a_{ij}, T) \times r_j}{\sum_{a_{kj} \in A_j} sim(a_{kj}, T)} \right] + (1-d)t_i, \tag{2}$$

where A_j is the set of anchor texts that webpage p_j has, a_{ij} is an anchor text for the link to webpage p_i in webpage p_j, and $sim(a_{kj}, T)$ is the cosine similarity between anchor text a_{kj} and topic T. By taking the similarity into account, ranks calculated with Equation (2) can be used in prioritizing webpages, as will be described in the following sections.

3 Prioritization

We prioritize unvisited webpages for deciding which of them we should crawl next. Our prioritization algorithm is based on a lower bound of PageRank proposed in [7]. In this article, we focus on how to prioritize candidate webpages to be crawled next in initial collecting webpages concerning specific topics; consideration on multiple iteration crawling [11] for keeping the collected webpages up-to-date is included in our future work.

A candidate way of deciding a webpage to crawl next is to calculate PageRanks of all webpages and to select a webpage with the highest PageRank. Note, however, that it is naturally impossible to calculate precise PageRanks of webpages that have never been accessed yet. Instead of calculating precise each PageRank, Cho and Schonfeld [7] proposed calculating a lower bound of it based on visited webpages.

In this study, we attempt to integrate the idea of a lower bound of PageRank into focused crawlers. Assume that there is a path from webpage p_j to p_i in WWW. According to Equation (2), we can say that webpage p_j has $(1-d)t_j$ as a lower bound of its PageRank, regardless of link structures around the webpages. Let w_{ji} and W_{ji} be a path and the set of all paths from p_j to p_i, respectively, and $|w_{ji}|$ be the number of clicks to get webpage p_i along path w_{ji}. Then, the probability of being on webpage p_i from webpage p_j along path w_{ji} without being interrupted can be expressed as $d^{|w_{ji}|}$. Let p_k be a webpage on path w_{ji}, and S_k and s_k be the sum of similarities between a specific topic and all anchor texts webpage p_k holds and the similarity between the topic and the anchor text on webpage p_k that the surfer clicks, respectively. Then, the probability that the surfer gets webpage p_i from webpage p_j along path w_{ji}, expressed as $PP(w_{ji})$, is calculated as follows:

$$PP(w_{ji}) = \left(\prod_{p_k \in w_{ji}} \frac{s_k}{S_k} \right) (1-d) t_j d^{|w_{ji}|}. \tag{3}$$

Let D_c be the set of webpages crawled already. Then, we can calculate the probability that the surfer accesses webpage p_i, or a lower bound of its PageRank, as follows:

$$r_i \geq \sum_{p_j \in Dc} \sum_{w_{ji} \in W_{ji}} PP(w_{ji}). \tag{4}$$

In Equation (4), a lower bound of PageRank of a webpage is propagated to those linked by the webpage. As a result, we can calculate a lower bound of PageRank of each webpage linked from a webpage by means of Equation (4). This equation is calculated during crawling, and we can decide where to crawl next by selecting the webpage with the highest value of this lower bound.

4 Crawling Algorithm

Figure 3 summarizes our crawling algorithm in pseudo code, where we omit a sleeping process not for accessing a website too frequently. Webpages expressing a specific topic are supposed to be given first in this study as SeedSet. Also, we express the topic in a simple manner with feature words which are extracted from the webpages. Then, each webpage in SeedSet is assigned to its score equally. A database storing urls, their out-links with similarities to the topic, and two queues, one for urls with scores and the other for crawled webpages, are used in the algorithm. Line 27 corresponds to Equation (4).

Figure 4 shows a flow of processing our crawling algorithm on an example where there are five webpages. In the example, we set d to 0.85. Two webpages are used as seeds, and thus they have the same score (line 4) as 0.075 and the others have 0 at the first step. One of the two webpage is selected (line 7) and crawled (line 15) at step 2. Then, the score 0.075 is propagated to two outlinks of the crawled webpage with taking into account of similarities as 0.0213 and 0.0425 as shown in step 3 (line 27). After that, the score of the crawled webpage is set to 0 (step 4). The webpage with the highest score at step 4 is crawled next, and the score is propagated to its outlinks in the same manner and then set to 0 (line 29), as shown in step 5. The webpage with the highest score at step 5 is then crawled next and its score is set to 0 (step 6). Note that the score of webpages crawled can be propagated to other crawled webpages. At step 6, the webpage with the highest score is one already crawled. In the example, the webpage is selected and extracted from DB (line 10) and same processes are performed.

Input: SeedSet
Output: a collection of webpages

```
FeatureWords = extractFeature(SeedSet)
                  // extract feature words from seeds
foreach u in SeedSet
   u.score = (1 − d) / |SeedSet|
                  // score of each webpage is set equally
   enqueue(UrlQueue, u)
end foreach
while()
   url = dequeue(UrlQueue)
   SimTotal = 0
   if url ∈ CrawledPages then
      UrlList = extractFromDB(url)
      foreach u in UrlList
         SimTotal = SimTotal + u.sim
      end foreach
   else
      Webpage = crawlPage(url)
                  // crawl the webpage
      enqueue(CrawledPages, url)
      UrlList = extractUrls(Webpage)
                  // anchors and outlinks of url are extracted from the webpage
      foreach u in UrlList
         u.sim = similarity(u.anchor, FeatureWords)
                  // calculate similarity between anchor and feature words
         SimTotal = SimTotal + u.sim
         if u ∉ UrlQueue then
            enqueue(UrlQueue, u)
         end if
      end foreach
   end if
   foreach i in UrlList
      i.score = i.score + ((d × i.sim × url.score) / SimTotal)
   end foreach
   url.score = 0
   updateDB(url, UrlList)
   reorderQueue(UrlQueue)
                  // reorder UrlQueue with scores
```

Fig. 3. Crawling algorithm

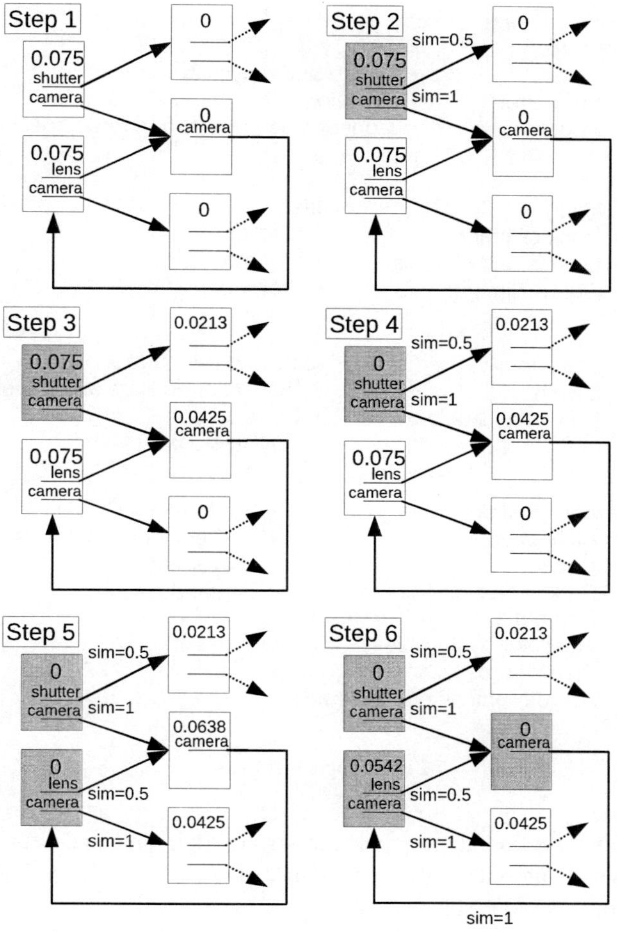

Fig. 4. Example of results of our crawling algorithm

5 Experiments

We report results of experiments with many topics different from [24]. In the evaluation we used precision and target recall [19,23,20] as metrics.

5.1 Environment

The data we used in the experiments come from the Open Directory Project (ODP) [17], which is a human-edited webpage directory. Webpages are categorized into topics and the topics construct topic hierarchies in ODP. In the experiments, we assumed that the topics and topic hierarchies of webpages in ODP were correct.

The experiments were conducted in Japanese. To evaluate the effectiveness of our proposal, we randomly selected eight topics, namely building types, environment,

- Building types
  - Castles
  - Ruins
  - Structures
  - Houses
- Environment
  - Energy
  - Waste and recycling
  - Pollution
  - Conservation and endangered species

- Climbing
  - Rock climbing
  - Guides and schools
  - Gorge climbing
- Glass
  - Stained
  - Lampworking
  - Blowing

- Theme parks
  - Disney
  - Universal Studio
- Trains and railroads
  - Discontinued lines and travels
  - Directories
  - Rolling stock

- Neurological disorders
  - Alzheimer's
  - Stroke
  - Parkinson's disease
  - Dementia
  - Muscular dystrophies
  - Spinal cord
  - Amyotrophic lateral sclerosis
- Gardening
  - Residential gardens
  - Vegetables
  - Landscaping

Fig. 5. Topic hierarchies

climbing, gardening, glass, theme parks, trains and railroads, and neurological disorders, from those having relatively large number of webpages. Figure 5 shows the hierarchies of the topics used in the experiments.

We used 20 webpages in each topic as its seeds, and the rest of the webpages were used as its targets. The 20 webpages were randomly selected from each directory, and the proportion of them were set to the proportion of webpages in each topic hierarchy. For example, assume topic A has only one child topic B and the numbers of webpages categorized in A and B are 20 and 30, respectively. In this case the numbers of seeds from topics A and B are 8 and 12, respectively. We extracted feature words from the 20 webpages by using tf-idf method, and each main topic was modeled with the feature vector consisting of the words. We calculated the cosine similarity between the feature vector and anchor texts in crawling.

In the experiments, we implemented three crawlers, namely our proposal, a focused crawler using anchor texts only, and a crawler based on breadth-first crawling. In the anchor texts only strategy, as in [19,21], the score of each linked webpage was estimated as follows:

$$score = \beta \times page\_score + (1 - \beta) \times context\_score, \tag{5}$$

where *page_score* stands for the cosine similarity between the feature vector constructed with words extracted from all seeds using tf-idf method and the feature vector constructed with words extracted from the crawled webpage using tf-idf method, and $context\_score$ stands for the cosine similarity between the feature vector constructed with words extracted from all seeds using tf-idf method and the feature vector constructed with anchor texts from the crawled webpage using tf-idf method. We set $\beta = 0.25$, which come from [19,21], in the experiments.

To evaluate the three crawlers, we decided to use precision and target recall [19,23,20] as metrics. After crawling N webpages, the precision is defined as follows:

$$precision = \frac{1}{N}\sum_{i=1}^{N} sim(T, p_i), \tag{6}$$

where $sim(T, p_i)$ is the cosine similarity between topic T and webpage p_i. Let T_t be the set of targets of topic t, and C_t^N be the set of crawled N webpages according to the topic. Then, the target recall is defined as follows:

$$target\ recall = \frac{|T_t \cap C_t^N|}{|T_t|}. \tag{7}$$

In the following, we report not only values of target recall but also those of average target recall, which is calculated by dividing the sum of the values of target recall by the number of crawled webpages. We think this metric is significant for focused crawlers running on personal computers with poor computing resources because the results can tell us how fast a crawler can collect targets.

5.2 Results

We run the three crawlers with identical seeds of each of the eight main topics shown in Fig. 5 for collecting 10,000 webpages. Figures 6 to 13 show the performances of precision and average target recall of the three crawlers on the eight topics. Table 1 reports the values of target recall of the crawlers after crawling the 10,000 webpages.

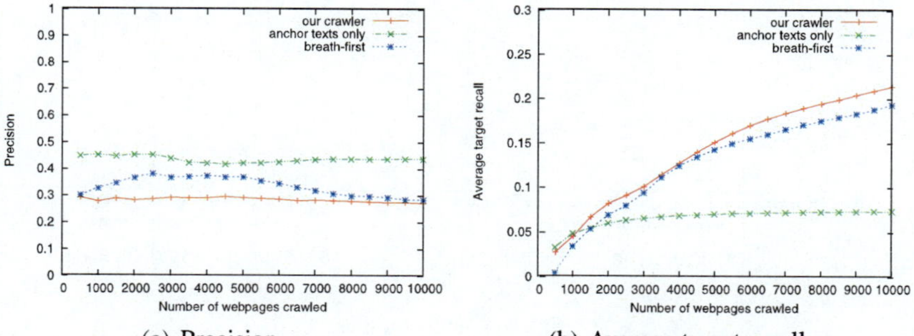

(a) Precision (b) Average target recall

Fig. 6. Results on building types

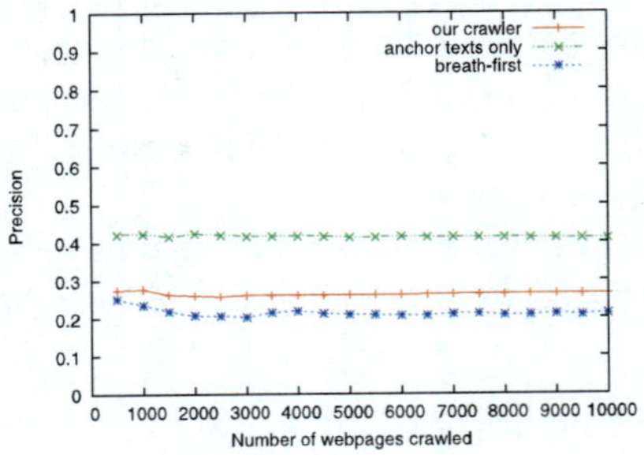

(a) Precision

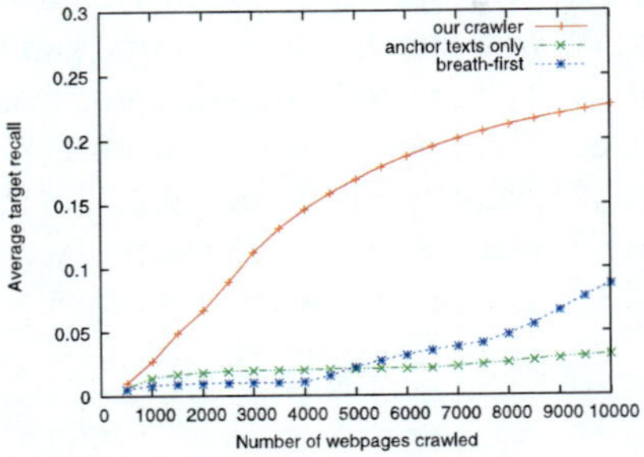

(b) Average target recall

Fig. 7. Results on environment

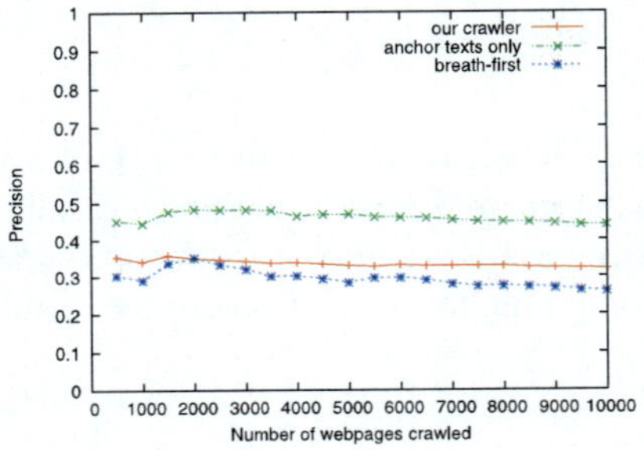

(a) Precision

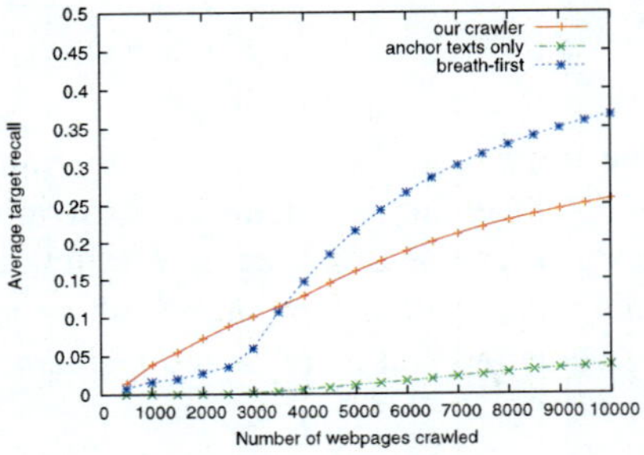

(b) Average target recall

Fig. 8. Results on climbing

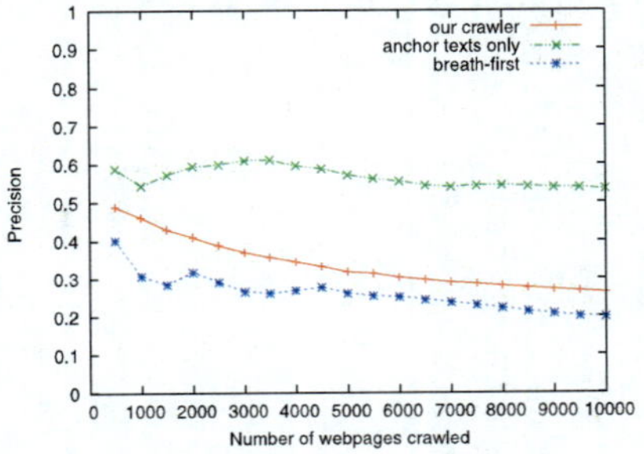

(a) Precision

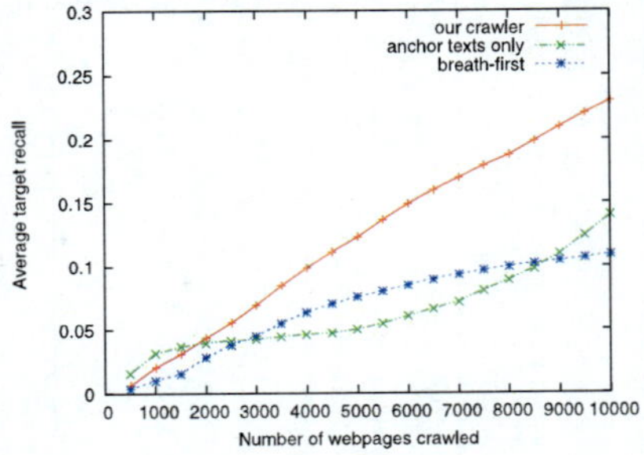

(b) Average target recall

Fig. 9. Results on glass

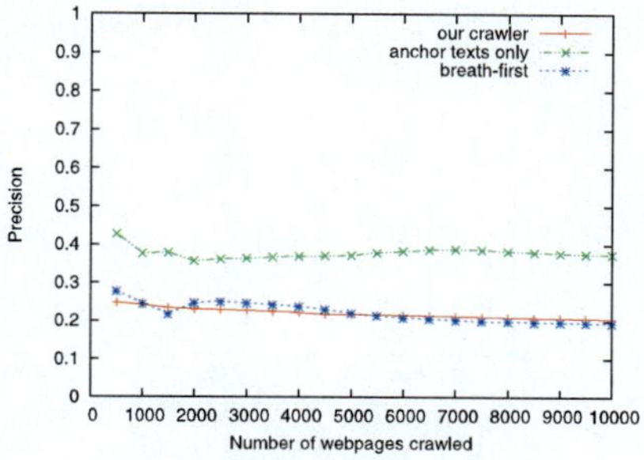

(a) Precision

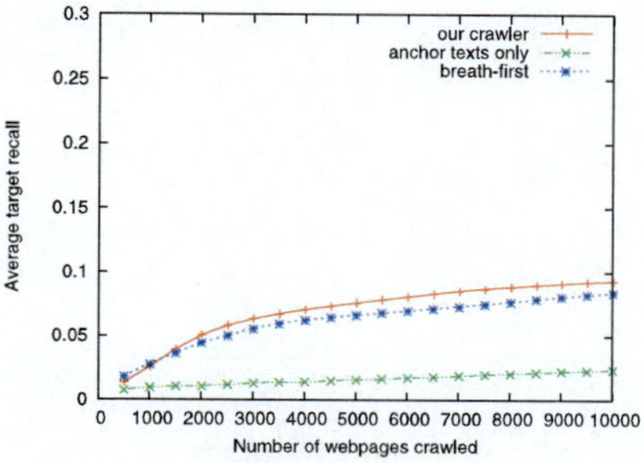

(b) Average target recall

Fig. 10. Results on theme parks

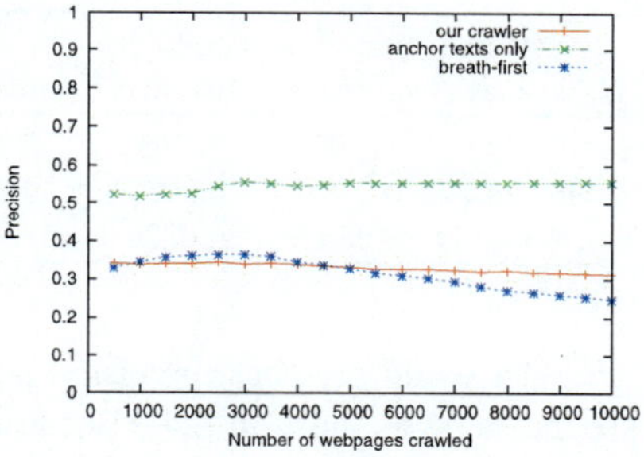

(a) Precision

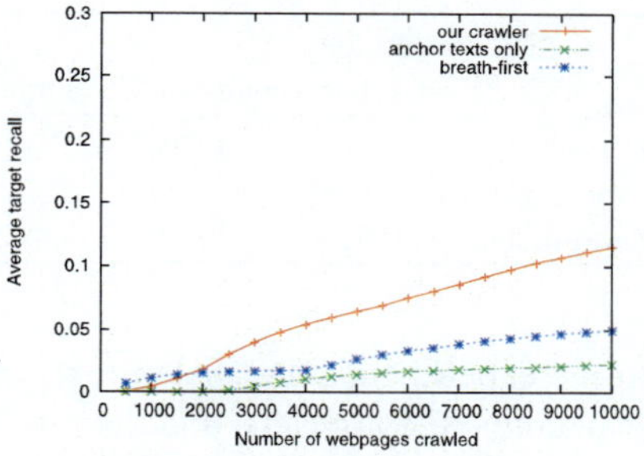

(b) Average target recall

Fig. 11. Results on trains and railroads

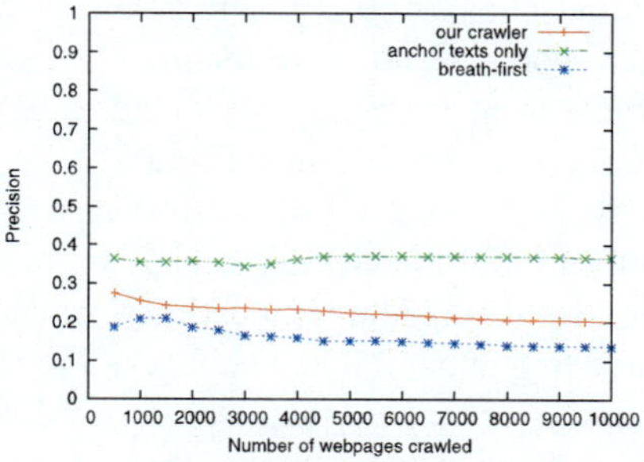

(a) Precision

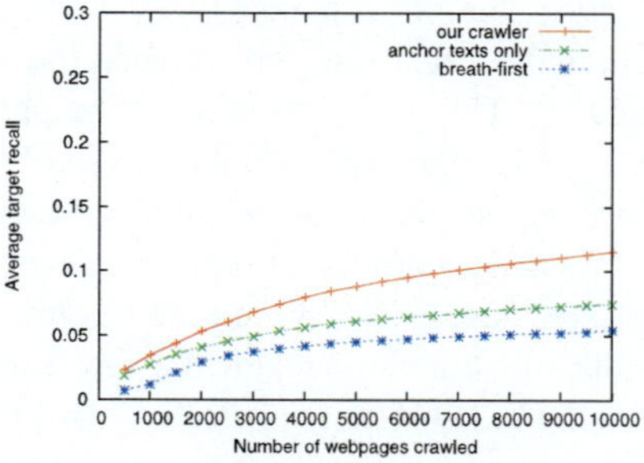

(b) Average target recall

Fig. 12. Results on neurological disorders

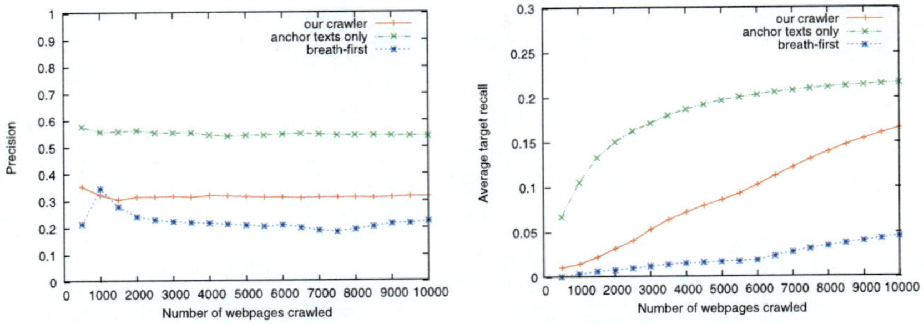

(a) Precision (b) Average target recall

Fig. 13. Results on gardening

Table 1. Target recalls after crawling 10,000 webpages

	building types	environment	climbing	glass	theme parks	trains and railroads	neurological disorders	gardening
our crawler	0.304	0.295	0.388	0.410	0.112	0.190	0.165	0.282
anchor texts only	0.0759	0.0682	0.100	0.476	0.140	0.0381	0.0909	0.235
breadth-first	0.316	0.273	0.525	0.152	0.0789	0.0857	0.0826	0.0941

In the experiments, the anchor texts only crawler could give best precision performance among the crawlers, and the results are the same as those in [24]. Because the precision performance is calculated with Equation (6), the results can be reasonable. Our proposal could give second best precision performance in almost all main topics except for building types. As shown in Fig. 6(a), the breadth-first crawler could give better precision especially in the range between the beginning of the crawling and the time when crawling about 6,000 webpages. However, at the time when crawling 10,000 webpages, the performance of the two crawling schemes became almost the same. To improve the precision performance too much would result in collecting only similar webpages. This situation may not be preferable in many cases. We intend to develop how to assess the precision more suitably for focused crawlers in the future.

In contrast, our crawler could give best average target recall performance among the crawlers in almost all main topics except for climbing and gardening, as shown in Figs. 8(b) and 13(b). In the case of climbing, the average target recall performance of breadth-first crawler became the best after crawling about 3,000 webpages. The reason is that the breadth-first crawler accessed portal site webpages around the time. Although our crawler also accessed the same portal site webpages, many anchor texts in the webpages consisted of personal names or names of stores and, as a result, the performance of our crawler could not change as good as that of breadth-first crawler. In fact, we found three portal site webpages having 118, 192, and 187 outlinks, respectively, accessed in the experiments. The breadth-first crawler naturally crawled the three webpages and 497 webpages linked from them, while our crawler crawled the three webpages and 70, 4, and 39 webpages linked from them, respectively.

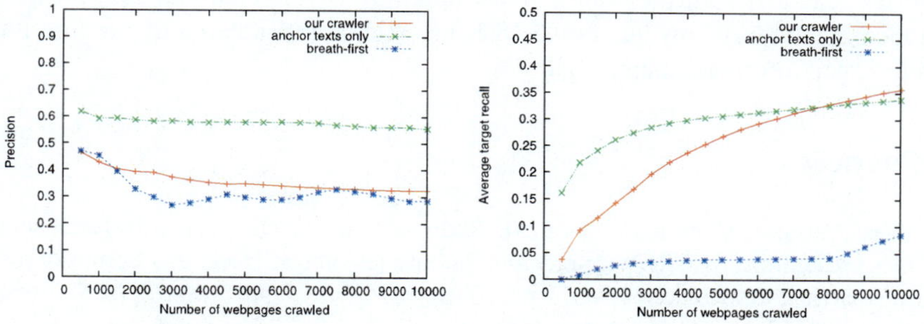

(a) Precision (b) Average target recall

Fig. 14. Results on vegetables

In the case of gardening, the average target recall performance of anchor texts only crawler became the best. To study this case, we did the same experiments using topic vegetables, which was a child topic of gardening as shown in Fig. 5. Figure 14 shows the results. Observing Figs. 13 and 14, the average target recall performances on topic vegetables were slightly better than those on topic gardening. In addition, our crawler could give best average target recall performance after crawling 10,000 webpages in the results on vegetables. Because topic vegetables is narrower than topic gardening, we think the topic expression in the case of topic vegetables is better than that of topic gardening. In other words, if we can express a specific topic appropriately, our crawler can give good performance.

As shown in Table 1, our crawler could give good performance of target recall after crawling 10,000 webpages independent of topics compared with the two crawlers. In fact, our crawler could give the best target recall performance in four out of the eight main topics and the second best in the rest main topics. On the other hand, the performance of the other two crawlers depended on topics.

To summarize, the performance of our crawler was good in many topics, several places to be improved were also found, though. We intend to improve our crawler in terms of how to exploit useful portal site webpages and how to express topics in the future. Shchekotykhin et al. [22] develop xCrawl in which portal sites have higher priority to be crawled and thus will help improve our proposal.

6 Conclusion

In this article, we have illustrated design and implementation of a focused crawling system for effectively collecting webpages concerning specific topics. Our proposal is based on personalized PageRank and a lower bound of PageRank, and integrates them with similarity measurement to specific topics. We have reported and investigated some results of experimental evaluation with many topics different from [24]. According to the results, our proposal can give good target recall performance regardless of topics on which the crawler system focuses. Also, we have found the future directions that the current implementation should be improved in terms of how to exploit useful portal site webpages and how to express topics more appropriately. In addition, multiple iteration and incremental web crawlers [4,9,5,18,16] should be integrated into our algorithm in order to keep information up-to-date.

References

1. Baeza-Yates, R., Castillo, C., Marin, M., Rodriguez, A.: Crawling a country: better strategies than breadth-first for Web page ordering. In: Special Interest Tracks and Posters of the 14th International Conference on World Wide Web, WWW 2005, pp. 864–872. ACM Press, New York (2005), `http://doi.acm.org/10.1145/1062745.1062768`, doi:10.1145/1062745.1062768
2. Brin, S., Page, L.: The anatomy of a large-scale hypertextual Web search engine. Computer Networks and ISDN Systems 30(1-7), 107–117 (1998), `http://www.sciencedirect.com/science/article/B6TYT-3WRC342-2N/2/63e7d8fb6a64027a0c15e6ae3e402889`, doi:10.1016/S0169-7552(98)00110-X; Proceedings of the Seventh International World Wide Web Conference
3. Chakrabarti, S., van den Berg, M., Dom, B.: Focused crawling: a new approach to topic-specific Web resource discovery. Computer Networks 31(11-16), 1623–1640 (1999), `http://www.sciencedirect.com/science/article/B6VRG-405TDWC-1F/2/f049016cf8fefd114f056306b5ae4a86`, doi:10.1016/S1389-1286(99)00052-3
4. Cho, J., Garcia-Molina, H.: The evolution of the web and implications for an incremental crawler. In: Abbadi, A.E., Brodie, M.L., Chakravarthy, S., Dayal, U., Kamel, N., Schlageter, G., Whang, K.Y. (eds.) Proceedings of 26th International Conference on Very Large Data Bases, VLDB 2000, pp. 200–209. Morgan Kaufmann, San Francisco (2000), `http://www.vldb.org/conf/2000/P200.pdf`
5. Cho, J., Garcia-Molina, H.: Effective page refresh policies for Web crawlers. ACM Trans. Database Syst. 28(4), 390–426 (2003), `http://doi.acm.org/10.1145/958942.958945`, doi:10.1145/958942.958945
6. Cho, J., Garcia-Molina, H., Page, L.: Efficient crawling through URL ordering. Computer Networks and ISDN Systems 30(1-7), 161–172 (1998), `http://www.sciencedirect.com/science/article/B6TYT-3WRC342-2G/2/122be31915c6e16c444898fb12cfdf87`, doi:10.1016/S0169-7552(98)00108-1; Proceedings of the Seventh International World Wide Web Conference
7. Cho, J., Schonfeld, U.: RankMass crawler: a crawler with high personalized PageRank coverage guarantee. In: Proceedings of the 33rd International Conference on Very Large Data Bases, VLDB 2007, pp. 375–386. VLDB Endowment (2007), `http://www.vldb.org/conf/2007/papers/research/p375-cho.pdf`
8. Diligenti, M., Coetzee, F., Lawrence, S., Giles, C.L., Gori, M.: Focused crawling using context graphs. In: Proceedings of the 26th International Conference on Very Large Data Bases, VLDB 2000, pp. 527–534. Morgan Kaufmann Publishers Inc., San Francisco (2000), `http://www.vldb.org/conf/2000/P527.pdf`

9. Edwards, J., McCurley, K., Tomlin, J.: An adaptive model for optimizing performance of an incremental web crawler. In: Proceedings of the 10th International Conference on World Wide Web, WWW 2001, pp. 106–113. ACM Press, New York (2001), http://doi.acm.org/10.1145/371920.371960, doi:10.1145/371920.371960
10. Ester, M., Kriegel, H.P., Schubert, M.: Accurate and efficient crawling for relevant websites. In: Proceedings of the Thirtieth International Conference on Very Large Data Bases, VLDB 2004, vol. 30, pp. 396–407. VLDB Endowment (2004), http://www.vldb.org/conf/2004/RS10P3.PDF
11. Fetterly, D., Craswell, N., Vinay, V.: The impact of crawl policy on web search effectiveness. In: Proceedings of the 32nd International ACM SIGIR Conference on Research and Development in Information Retrieval, SIGIR 2009, pp. 580–587. ACM Press, New York (2009), http://doi.acm.org/10.1145/1571941.1572041, doi:10.1145/1571941.1572041
12. Haveliwala, T.H.: Topic-sensitive PageRank: A context-sensitive ranking algorithm for Web search. IEEE Transactions on Knowledge and Data Engineering 15(4), 784–796 (2003), http://doi.ieeecomputersociety.org/10.1109/TKDE.2003.1208999, doi:10.1109/TKDE.2003.1208999
13. Jeh, G., Widom, J.: Scaling personalized web search. In: Proceedings of the 12th International Conference on World Wide Web, WWW 2003, pp. 271–279. ACM Press, New York (2003), http://doi.acm.org/10.1145/775152.775191, doi:10.1145/775152.775191
14. Kleinberg, J.M.: Authoritative sources in a hyperlinked environment. J. ACM 46(5), 604–632 (1999), http://doi.acm.org/10.1145/324133.324140, doi:10.1145/324133.324140
15. Najork, M., Wiener, J.L.: Breadth-first crawling yields high-quality pages. In: Proceedings of the 10th International Conference on World Wide Web, WWW 2001, pp. 114–118. ACM Press, New York (2001), http://doi.acm.org/10.1145/371920.371965, doi:10.1145/371920.371965
16. Olston, C., Pandey, S.: Recrawl scheduling based on information longevity. In: Proceeding of the 17th International Conference on World Wide Web, WWW 2008, pp. 437–446. ACM Press, New York (2008), http://doi.acm.org/10.1145/1367497.1367557, doi:10.1145/1367497.1367557
17. Open Directory Project, http://www.dmoz.org/
18. Pandey, S., Olston, C.: User-centric Web crawling. In: Proceedings of the 14th International Conference on World Wide Web, WWW 2005, pp. 401–411. ACM Press, New York (2005), http://doi.acm.org/10.1145/1060745.1060805, doi:10.1145/1060745.1060805
19. Pant, G., Menczer, F.: Topical crawling for business intelligence. In: Koch, T., Sølvberg, I.T. (eds.) ECDL 2003. LNCS, vol. 2769, pp. 233–244. Springer, Heidelberg (2003), http://www.springerlink.com/content/p0n6lh04f4j7y26u, doi:10.1007/978-3-540-45175-4_22
20. Pant, G., Srinivasan, P.: Learning to crawl: Comparing classification schemes. ACM Trans. Inf. Syst. 23(4), 430–462 (2005), http://doi.acm.org/10.1145/1095872.1095875, doi:10.1145/1095872.1095875

21. Pant, G., Srinivasan, P.: Link contexts in classifier-guided topical crawlers. IEEE Transactions on Knowledge and Data Engineering 18(1), 107–122 (2006), http://doi.ieeecomputersociety.org/10.1109/TKDE.2006.12, doi:10.1109/TKDE.2006.12
22. Shchekotykhin, K., Jannach, D., Friedrich, G.: xCrawl: a high-recall crawling method for Web mining. Knowledge and Information Systems 25(2), 303–326 (2010), http://dx.doi.org/10.1007/s10115-009-0266-3, doi:10.1007/s10115-009-0266-3
23. Srinivasan, P., Menczer, F., Pant, G.: A general evaluation framework for topical crawlers. Information Retrieval 8(3), 417–447 (2005), http://dx.doi.org/10.1007/s10791-005-6993-5, doi:10.1007/s10791-005-6993-5
24. Uemura, Y., Itokawa, T., Kitasuka, T., Aritsugi, M.: Where to crawl next for focused crawlers. In: Setchi, R., Jordanov, I., Howlett, R.J., Jain, L.C. (eds.) KES 2010. LNCS, vol. 6279, pp. 220–229. Springer, Heidelberg (2010), http://dx.doi.org/10.1007/978-3-642-15384-6_24, doi:10.1007/978-3-642-15384-6_24

Chapter 6
Web-Pages Re-ranking, Based on Relevant/Irrelevant Feedback Information

Toyohide Watanabe and Kenji Matsuoka

Department of Systems and Social Informatics,
Graduate School of Information Science, Nagoya University
Furo-cho, Chikusa-ku, Nagoya 464-8603, Japan
watanabe@is.nagoya-u.ac.jp

Abstract. A keyword-based retrieval engine, which is most usable recently, extracts appropriate Web-pages by means of keywords in user-specified queries. However, it is not always easy to extract the user-preferred Web-pages correctly, because the user-specified keywords have several meanings in many cases. In such case, we must find out relevant Web-pages and exclude irrelevant Web-pages. Also, in case that we cannot retrieve the desirable Web-pages, we must retry after modifying the original query. In this paper, we propose an advanced Web-page retrieval method to find out user-preferred Web-pages in case that relevant pages could not be extracted. The idea is to make use of user's unconscious reactions to judge which pages are relevant or not, when the retrieved results were listed up. Our method is to infer user-preference on the basis of relevant or irrelevant indications for the page and reflect the inferred preference into the next retrieval query with a view to improving the retrieved results.

1 Introduction

Even if we knew retrieval keywords which identify target pages in Web successfully it is difficult to extract the target pages effectively [1, 2]. Of course, it is not easy to retrieve appropriate pages if we did not know powerful retrieval words. In many cases, the pages which are not always adjusted to target pages appropriately are unnecessarily selected without any avoidance means [3, 4]: users must individually distinguish meaningful pages by their own operations from retrieved results, and this work is trivial for the users, but heavy. Also, the retrieval process must be repeated, but it is not always easy to modify the query directly based on the retrieved results and reference features; and it is important to be able to refer to the results individually so as to judge whether they are relevant or not, and then modify the query repeatedly, if necessary [5, 6]. Under these situations we focused on the evaluation process in which users can estimate whether the retrieved results are acceptable or not. If users can infer the page features to be accepted or rejected on the basis of user judgments for retrieved results, we can extract the most adjusted pages and also support the query modification process for specifying the retrieval conditions with the better retrieval terms or words.

T. Watanabe and L.C. Jain (Eds.): Innovations in Intell. Machines – 2, SCI 376, pp. 77–90.
springerlink.com

In this paper, two ideas are introduced to attain our retrieval process successively: user-preference with relevance/irrelevance specification, and re-search based on query modification. From these two viewpoints, we look upon different features between retrieved pages, which are suitable to the retrieval purpose and the features of irrelevant pages, as classification factors for distinguishing irrelevant pages from relevant pages; and we propose a re-ranking method of retrieved results so as to strengthen the difference through computation of all pages. Additionally, we propose the query generation means by adding the words, which represent relevant features with large difference, and deleting the words, which include irrelevant effects. Thus, users can select relevant pages suitable to their own preferences through their unconscious evaluation for individual pages in retrieved results, and also can take out better retrieval results by its modified/refined query.

2 Approach

Until today, many researches have been investigated to make the accuracy of retrieval pages high, and can be in general classified into two types, depending on whether the feedback mechanism is applied to the query processing or not [7-9]. In the feedback-independent approaches, one is the research which infers the retrieval preference of user from input query, and another is the research based on the user profile, arranged from the retrieval histories. It is, however, not easy to infer the retrieval preference automatically though these researches have an advantage which is not to impose additional loads on users. On the other hand, the feedback-dependent approaches such as selection of keywords, feature-oriented classification of pages, permutation of pages, etc. make it possible to estimate the retrieval preference and enable to apply the estimated result to the query composition. In this paper, we address the feedback-dependent approach, based on the interaction between system and user with respect to the estimation of user preference and query modification. Generally, the feedback-dependent approach is called relevant feedback.

Yamamoto, et al. proposed a method to indicate user intention by deleting/adding any keywords from/to the title snippets [10]. Also, in this case they focused on the effective re-ranking means by looking up frequently occurred keywords as a set such as the tag clouds. Also, Karube, et al. proposed a method to re-rank the retrieved results by selecting document segments, corresponding to target pages, from all retrieved documents [11]. They chose partial documents because the retrieval means based on keywords cannot specify fully the information set and also the means based on pages includes much un-meaning information [12].

The difference between these approaches and our approach is dependent on whether the user or the system must selectively control the preference process. Although the keyword-based re-ranking method makes it possible to propagate the retrieval preference without noises, keywords to be included in the target pages must be always accurately inferred. However, it is difficult for this method to represent complex retrieval conditions. The method based on partial documents can propagate the preference flexibly without noises, but must find out partial sentences applicable to target pages. While, our method decreases hopefully the decision-making loads, which user must select individually results, in comparison with the existing methods because our system function infers successfully relevant pages on the basis of page features, contained respectively in the desirable pages or useless pages.

Our processing framework consists of two different successive processes: pre-processing and feedback. Figure 1 shows the processing flow. The pre-processing procedure extracts identically the features of words contained observably in retrieved pages by using mainly various kinds of techniques to be useful in natural language processing. While, in the feedback procedure the practical ranking mechanism is effective on retrieved pages under the interaction: evaluation of relevance/irrelevance, re-ranking, query modification, etc.

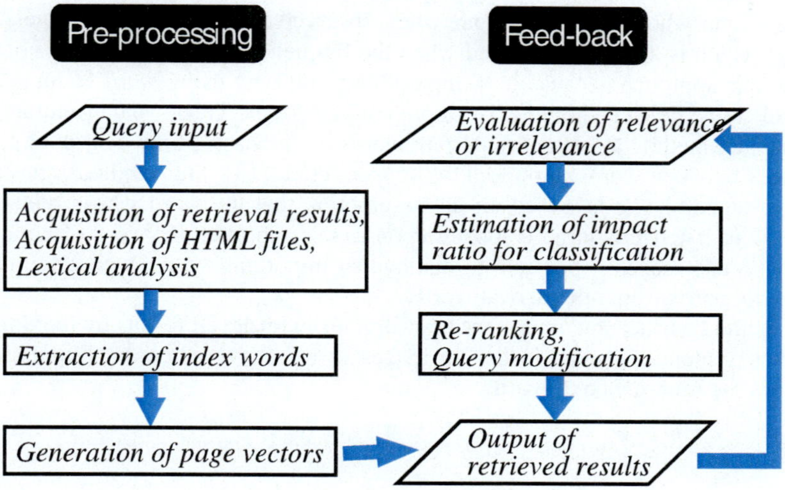

Fig. 1. Processing flow

3 Re-ranking Based on Feedback

3.1 Retrieved Results and Lexical Analysis

In our research, we use Google as the basic retrieval engine. The main procedure is as follows:

1) retrieve appropriate pages by using input query;
2) select title, snippet and URL related individually to each page from such retrieved results;
3) extract sentences, excluded HTML tags from them.
4) analyze these extracted sentences lexically with a view to looking upon a semantic unit as the word.

Under this lexical analysis, individual words are lexically distinguished, and categorized into functional words and content words. Functional words are auxiliary verb (jyodoshi), jyoshi and grammatically-specified words, while content words include noun, verb, adjective, adverb and so on as glossarial words. In our approach, we do not focus on the sentence structure, but are interesting in the fact of whether individual words present the page content or not. Thus, the content words are selectively used and extracted in the lexical analysis process, and moreover in our case only nouns are manipulated because the nouns generally are combined with verbs or adjectives and take main roles to compose sentences.

3.2 Extraction of Index Keywords

The method based on the features of words which are included in the corresponding pages has been commonly investigated to catch up page features: ***tf.idf*** is typical. ***tf.idf*** is a standard criteria for assigning high scores to words whose frequencies are too many, and whose occurrences are often observed in only particular pages. However, ***idf*** which is adjustably defined when the frequency of pages is less is not applicable to our approach because it is impossible to rank by using word features unless the words to be deleted should exist throughout all pages. Thus, we concentrate on an experimental method to extract important words on the basis of the distribution of co-existences between words, proposed by Matsuo, et al. [13]. This method can estimate the importance of words, based on the assumption that the word whose relationship for word coexistence is large is looked upon as the important keyword of high possibility, and also makes it possible to distinguish important words, whose frequencies are not too many, from other trivial words.

Our method extracts the important keywords from retrieved results by looking upon individual sentences ranked in the top pages as effective sentences. Expression (1) calculates the importance of word:

$$X^2(w) = \Sigma_g \in_G (freq(w, g) - n_w \cdot p_g)^2 / n_w \cdot p_g \tag{1}$$

Here, $freq(w, g)$ is a summation of co-existing numbers between sentence-lines for word w and those for word g. w is a word for which we calculate the importance, and g is a word whose frequency is many. n_w is a summation of word numbers in sentences which include word w. p_g indicates the number of words g for the number of words in all sentences. Namely, in this expression (1) $(freq(w, g) - n_w \cdot p_g)$ represents the difference of the number of the expected co-occurrence in the word g from the number of co-occurrence between words w and g. $(freq(w, g) - n_w \cdot p_g)^2 / n_w \cdot p_g$ indicates the variation when the co-occurrence between words w and g is compared with average co-occurrence between words g and other words. Thus, $X^2(w)$ computes the degree of variation for co-occurrence of words w with respect to a set G of all frequently occurred words.

Our method regards a set of sentences included in all pages as only one sentence, and calculates the importance of words in retrieved results. Also, we compare X^2-based ranking with frequency-based ranking, and make use of most changeable words in ranking process as indexes. An example is shown in Figure 2. In Fig.2, the ordering transition from X^2-based ranking to frequency-based ranking is denoted. The area surrounded by broken-line segments is a large part of order deviation. In this example, we can choose words w_1, w_2 and w_3 from higher 3 words of X^2-based ranking as indexes. Using MeCab [14] as a Japanese lexical analyzer we selected nouns, which are classified by IPA (Information-technology Promotion Agency, Japan) in parts of speech system.

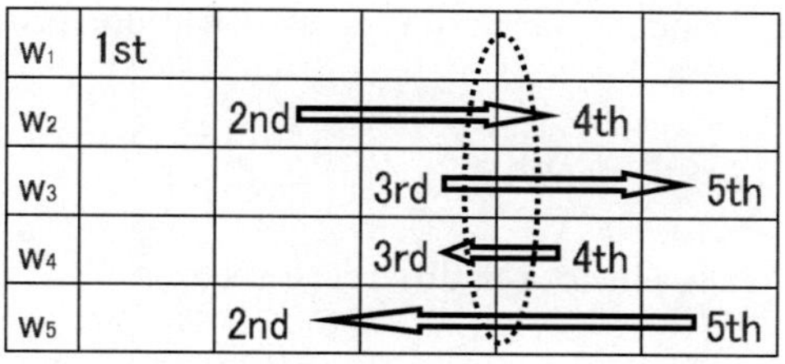

Fig. 2. Transition of word ranking

3.3 Feature Vector in Page

Generally, some words are frequently occurred in a page *A* and also are so in other pages as well as the page *A*. The information acquired effectively from the word usage is not always valuable. On the other hand, we may acquire more information in case that the corresponding words are not included in the page. Our method looks upon the differential information about how many the important words are included in some pages, as the page feature. We show our procedural steps for calculating the feature vector of page:

Step1: Set the number of index words contained in each page as this value of each dimension p_j in page vector $\boldsymbol{P}_i$.

$$\boldsymbol{P}_i = (Count(w_1, i), Count(w_2, i), \ldots, Count(w_n, i)) \tag{2}$$

where $Count(w_n, i)$ is the number of index words w_n included in the i-th page.

Step2: Calculate the ratio of individual index words for each page:

$$\boldsymbol{P}_i = \boldsymbol{P}_i / \Sigma_{j=1,N}\, p_j \tag{3}$$

Thus, we can compare them without depending on the difference among number of words.

Step3: Estimate the ranking of each page in index words, and then compute individual dimension values of page vector according to the ranking order:

$$\boldsymbol{P}_i = (PageCount\text{-}Rank(p_1, w_1), PageCount\text{-}Rank(p_2, w_2), \ldots, PageCount\text{-}Rank(p_n, w_n)) \tag{4}$$

PageCount is the total number of pages, and $Rank(p_i, w_i)$ is the value order of p_i in ranking for frequency ratio of index word w_i. Thus, we can suppress the occurrence.

Step4: Compute the difference vector $\boldsymbol{D}_i$ between the page vector $\boldsymbol{P}_i$ and the average vector $\boldsymbol{M}$:

$$\boldsymbol{D}_i = \boldsymbol{P}_i - \boldsymbol{M} \tag{5}$$

Step5: Compute the standard deviation $\boldsymbol{\sigma}_j$ of each dimensional value in page vector:

$$\boldsymbol{\sigma}_j = \sqrt{(1/N)\Sigma_{i=1,N}\,(d_{i,j} \cdot d_{i,j})} \tag{6}$$

$d_{i,j}$ is j-th dimensional value of i-th difference vector.

Step6: Regard the value in each dimension for individual difference vectors $\boldsymbol{D}_i$ which is divided by the corresponding standard deviation, as the page vector.

$$\boldsymbol{P}_i = (d_{i,1}/\boldsymbol{\sigma}_1,\ d_{i,2}/\boldsymbol{\sigma}_2,\ \ldots,\ d_{i,n}/\boldsymbol{\sigma}_n) \tag{7}$$

Thus, it is possible to represent the difference of feature for other pages.

Step7: Normalize the norm of page vector as 1.

Table 1 shows the computed values of page vector for Homepage in Nagoya University, selected from the top-200 retrieved results by query "Nagoya University".

Table 1. Example of page vector (from Homepage in Nagoya Univ.)

#	(word)	value
1	cooperation	.190
2	ｼﾝﾎﾟｼﾞｳﾑ(symposium)	.188
3	新 (news paper)	.186
4	応(correspondence)	.172
5	(academic)	.172
6	(system)	.172
7	果(result)	.168
8	講 (section)	.163
9	(co-related)	.163
10	請 (request)	.156

3.4 Calculation of Evaluation Criterion

We can execute a re-ranking process after having computed individually the criteria for the relevant pages based on acceptable evaluation and the irrelevant page based on unacceptable evaluation. So, we prepare two different criterion for relevant and irrelevant pages, respectively. This is in general because all page features which were pointed out as the irrelevance are not always completely unnecessary page features. For example, consider the case that a page A is relevant and a page B is irrelevant. If the similarity between pages A and B is lower, the feature of page A is a target page and the feature of page B is an unnecessary page for this preference. However, if the similarity between pages A and B is higher, the feature of page B may be consistent to that of target page. In this case, the feature of unnecessary page for page B is far from that for the page *A*. Thus, we address the re-ranking algorithm, which can change the criterion calculation process flexibly, with respect to the difference for the similarity between relevant and irrelevant pages.

Our evaluation criteria ***EC*** is represented, using the summation of difference between average vector of relevant pages and each irrelevant page vector.

$$\boldsymbol{EC} = \Sigma_{j=1,N} \left(1/N^{-} \cdot \Sigma_{i=1,N^{+}} (\boldsymbol{D}_i^{+}) - \boldsymbol{D}_j^{-} \right) \qquad (8)$$

N is the total number of pages, N^+ or N^- is the number of relevant pages or the number of irrelevant pages, $\boldsymbol{D}_i^+$ or $\boldsymbol{D}_j^-$ is a relevant page vector or an irrelevant page vector, respectively. Table 2 is an example of evaluation criterion vector, which were computed after having evaluated by 4 times with query "Apple" and retrieval intension "maker".

Table 2. Example of evaluation criterion vector (query: Apple, intention: maker)

#	(word)	value
1	STORE	.878
2	(purchase)	.834
3	ﾘﾘｰｽ(release)	.729
4	(product)	.727
5	(consultation)	.696
6	採 (adoption)	.637
7	(usage)	.631
8	ﾌﾟﾗｲﾊﾞｼｰﾎﾟﾘｼｰ (privacy policy)	.597
9	APPLE	.568
10	合 (lodging)	-.566

3.5 Score Computation and Re-ranking of Retrieved Result

We evaluate individually these cases such as relevance/irrelevance, relevance and irrelevance:

1) Evaluation of relevance/irrelevance

According to relevant/irrelevant evaluation criterion, compute the similarity for average vector from the features of relevant pages, as a positive score. Also, compute the maximum similarity for irrelevant pages on the basis of evaluation criterion, as a negative score:

$$SCORE(i) = \boldsymbol{P}_i \cdot (1/N^{+}) \cdot \Sigma_{j=1,N^{+}} (\boldsymbol{D}_j^{+} \cdot |\boldsymbol{EC}|) - \boldsymbol{max}_j (\boldsymbol{P}_i \cdot \boldsymbol{D}_j^{-} \cdot |\boldsymbol{EC}|) \qquad (9)$$

2) Evaluation of relevance

Compute the similarity for average vector of relevant page vectors, as positive score:

$$SCORE(i) = \boldsymbol{P}_i \cdot (1/N^{+}) \cdot \Sigma_{j=1,N^{+}} \boldsymbol{D}_j^{+} \qquad (10)$$

3) Evaluation of irrelevance

Compute the summation of similarities for all irrelevant pages so as to make the score of page vectors high, which are located in most partial space to be far from irrelevant page vector over vector space:

$$SCORE(i) = - \Sigma_{j=1,N^-} (\mathbf{P}_i \cdot \mathbf{D}_j^-) \qquad (11)$$

3.6 Query Modification

In our method, the inference function takes an important role to find out characteristic words which are just applicable to the target pages. The possibility that the words whose absolute values in evaluation criterion are large may be the characteristic words for relevance or irrelevance may be positively confirmed. Also, it may be said that when retrieved results are a few the words are effective to select the results with the high ratio of accuracy. Thus, we can choose the suitable words to modify the existing query, using the product between the order of retrieved results and that of absolute values in evaluation criterion vector.

$$SCORE(w_j) = ECRank(w_j) \cdot ResultNumRank(w_j) \qquad (12)$$

$ECRank(w_j)$ represents the order of index word w_j in ranking on the basis of large absolute value in evaluation criterion vector, and $ResultNumRank(w_j)$ does individual order of index words w_j in their descending order with respect to the number of index words included in results, retrieved by Google. We use a few words one by one with low values in Expression (12) as candidates in query modification. In case that the value of evaluation criterion vector is positive, AND retrieval is applied after having added the appropriately-selected word into a new query. On the other hand, in case that the value is negative we retrieve after having applied the minus operator to the head of corresponding word and also having added new words to the query. Table 3 shows an example of modification under the same situation as the previous evaluation criterion vector.

Table 3. Example of query modification (query: Apple, intention: maker)

#	query	product value
1	ｱｯﾌﾟﾙSTORE(AppleSTORE)	3 (=1×3)
2	ｱｯﾌﾟﾙ (Apple lodging)	10 (=10×1)
3	ｱｯﾌﾟﾙﾘﾘｰｽ(Apple release)	15 (=3×5)
4	ｱｯﾌﾟﾙ (Apple purchase)	18 (=2×9)
5	ｱｯﾌﾟﾙAPPLE(Apple APPLE)	18 (=9×2)
6	ｱｯﾌﾟﾙ (Apple product)	24 (=4×6)
7	ｱｯﾌﾟﾙ採 (Apple adoption)	24 (=6×4)
8	ｱｯﾌﾟﾙ相 (Apple consultation)	35 (=5×7)
9	ｱｯﾌﾟﾙ (Apple usage)	56 (=7×8)
10	ｱｯﾌﾟﾙﾌﾟﾗｲﾊﾞｼｰﾎﾟﾘｼｰ (Apple privacy policy)	80 (=8×10)

4 Experiment and Evaluation

Here, we describe the experiments and experimental results on re-ranking method and query modification method. Figure 3 shows the interface in our prototype system. User can specify the indication of "relevance" or "irrelevance" by the buttons in the left sides of individually retrieved pages. The system re-ranks the retrieved results according to user-specified indications.

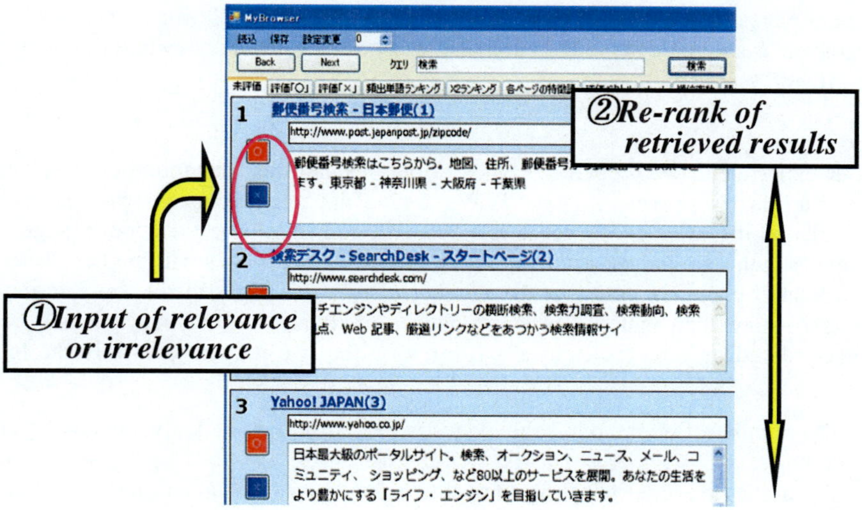

Fig. 3. Interface in prototype system

4.1 Evaluation in Re-ranking Method

Before describing our experiments, we use Okapi BM25 [15] as a resolution method for page vector, and Rocchio method [7] as a re-ranking means. Okapi BM25 is a typical method for assigning the weights to index words, and various researches related to this method have been reported. The weight $w(q)$ for word q appeared in a sentence $\boldsymbol{D}$ is:

$$W(q) = \boldsymbol{log}\ ((N\text{-}n(q)+0.5)/(n(q)+0.5) \\ \cdot ((freq(q, \boldsymbol{D}) \cdot (k+1))/(freq(q, \boldsymbol{D})+k(1\text{-}b+b \cdot |\boldsymbol{D}|/avgdl))) \tag{13}$$

Here, N is the total number of documents, $n(q)$ is the number of documents which contain word q, $freq(q, \boldsymbol{D})$ is the occurrence number of words q in the sentence $\boldsymbol{D}$. Also, $|\boldsymbol{D}|$ is the number of words in $\boldsymbol{D}$, $avgdl$ is the average for word lengths in sentences. k and b are parameters, and are heuristically assigned to k=2.0 and b=0.75. Okapi BM25 is a probability-based excellent method to reduce the reflection of ***tf*** in ***tf-idf*** by the volume of sentences. Each page vector is computed by the weight value.

Rocchio method can infer the common features for relevant pages by using the average of page vectors. However, it is not always possible to extract the common features for irrelevant pages because the retrieved results do not only contain a number of irrelevant pages, but also various kinds of pages. Thus, there is the problem in Rocchio method that the modification accuracy is too strongly dependent on the evaluation number of relevant pages. The expression in Rocchio method as a re-ranking means is:

$$Q_{n+1}= Q_n+ (\alpha/N^+)\cdot \Sigma_{i=1,N^+} (D_i^+)- (\beta/N^-)\cdot \Sigma_{j=1,N^-} (D_j^-) \quad (14)$$

Q_n is n-th query vector, N^+ or N^- is the number of relevant pages or the number of irrelevant pages, and D_i^+ or D_j^- is a page vector evaluated as relevance or irrelevance. Parameters α and β are set to 0.75 and 0.15 respectively, from a viewpoint of heuristic experiment.

Experiment

In our experiment, we prepared 9 types of queries and the corresponding 24 evaluation criterion, as shown in Table 4. We decide the retrieved pages as "relevance" when the words adaptable to these relevant criterion should be included in pages. In our experiment, we judged as "irrelevance" for irrelevant pages with highest order in un-evaluated pages, in case that the number of relevant pages in the top-7 retrieved results is more than that of irrelevant pages; otherwise, we decided the most relevantly-ordered page as "relevance". The reason that we judged with only the top-7 pages is:

- The number of our recognizable pages is 7±2 at time as the known chunking concept;
- The higher the order of retrieved results is the easier the evaluation is;
- It is natural to evaluate irrelevant pages and get rid of them in case that relevant pages almost occupy all of top-ranked pages.

Similarly, the relevant pages also are evaluated. In this experiment, the top-200 retrieved pages are usefully applied to our evaluation. Also, since the fact that almost users refer to several pages on the top-2 retrieved results has been reported in many researches, we think that in our purpose it is sufficient if 10 times of 20 retrieved results should be referred.

Our evaluation measurements are ***recall*** and ***precision***. ***Precision*** is defined as the ratio about how many relevant pages are included in N retrieved pages; and ***recall*** is the ratio about how many relevant pages are acceptable to the total number of relevant pages, included in the top-200 retrieved pages.

Experimental Result

Figure 4 shows average ***recall*** for number of evaluations in our method and existing method, respectively. Also, Figure 5 and Figure 6 show respectively average ***precision***'s of top-N results of our proposed method and existing method in each evaluation. Fig.4 indicates that the more the number of evaluations becomes the larger the difference between ***recall*** in our proposed method and that in the existing method is. Also, the more the number of evaluations increases the more the ***recall***'s in our method and existing method increase similarly. Also, in Fig.5 and Fig.6 the similar features in ***precision*** are observed.

Table 4. Experiment-used query and page of relevant criterion

query	relevance criterion	query	relevance criterion
(east-west line)	(tokyo) (kyoto) 札幌(sappolo) (sendai)	ﾌｨｷﾞｭｱ(figure)	ｽｹｰﾄ(skate) 形(dollar)
		ﾗｲｵﾝ(lion)	学 (chemical product) 物(animal)
ｱｯﾌﾟﾙ (Apple)	ﾒｰｶ(maker) (car dealership)	ｽﾋﾟｰﾄﾞ(speed)	ﾄﾗﾝﾌﾟ(trump) 度(line speed) 学 (learning)
ｱﾏｿﾞﾝ (amazon)	ネット (net sailing) 地 (area name)	条(sanjo)	(kyoto) (niigata)
ｼﾞｬｶﾞｰ (jaguar)	ﾌﾟﾛﾚｽﾗｰ(wrestler) (watch) 自 (automobile) (comic)	中克 (katsumi tanaka)	ﾋﾟｱﾆｽﾄ(piano player) 学教 (professor) (poem singer)

Table 5. Precision of retrieved results by recommended query

#	top-10	top-20	top-50
1	.713	.710	.709
2	.680	.677	.605
3	.421	.396	.386
4	.553	.547	.511
5	.347	.363	.382
average	.551	.547	.527
initial result	.159	.163	
difference	.392	.388	

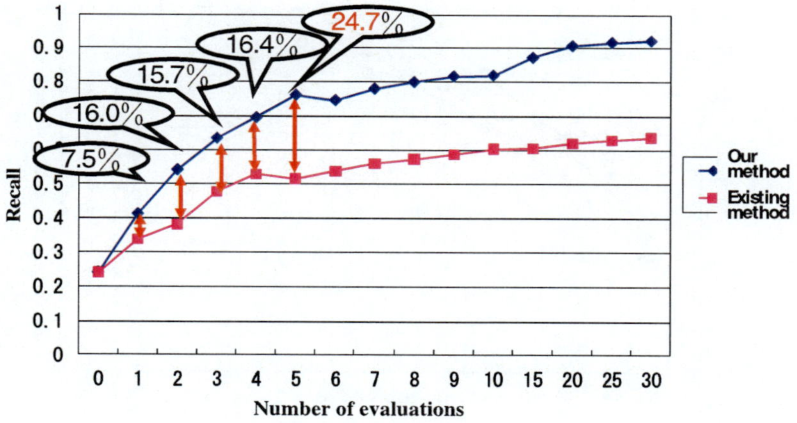

Fig. 4. Recalls in our method and existing method

4.2 Evaluation in Query Modification

In this experiment, 4 evaluations were performed, using 5 words. This retrieval is composed of 15 tasks to be used in Fig.5 after having erased 2 tasks, which were not regarded as "relevance" in 4 evaluations, from 17 tasks whose ***recall*** in initial retrieval results is less than 36% in Fig.4.

The reason that we removed the tasks which are not evaluated as relevance is dependent on what only average features of irrelevant pages are acquired, but the words which represent the features about relevant pages are not yet evaluated. The objective in our experiment is to extract more relevant pages by re-retrieval means. Thus, ***precision***'s for top-10, top-20 and top-50 pages selected from re-retrieved results are respectively computed.

Table 5 is the ***precision***'s of re-retrieved results by using top-5 candidate words. "average" represents average ***precision*** of 5 modified candidate words, "initial result" is an average ***precision*** for initial results before re-retrieval, and "difference" is a gap between average ***precision*** in re-retrieved results and once computed before re-retrieval. Table 6 shows mainly the ratios for query whose ***precision*** is more than 80% and query whose ***precision*** is less than 20%.

Table 6. Ratio of query from precision of retrieved results

#	more than 80%	80%～20%	less than 20%
1	.667	.133	.200
2	.533	.200	.267
3	.267	.133	.600
4	.467	.200	.333
5	.200	.467	.333

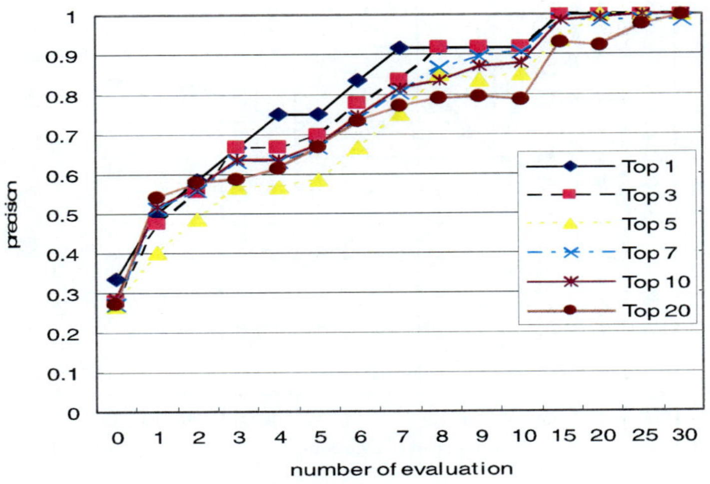

Fig. 5. Precision in our method

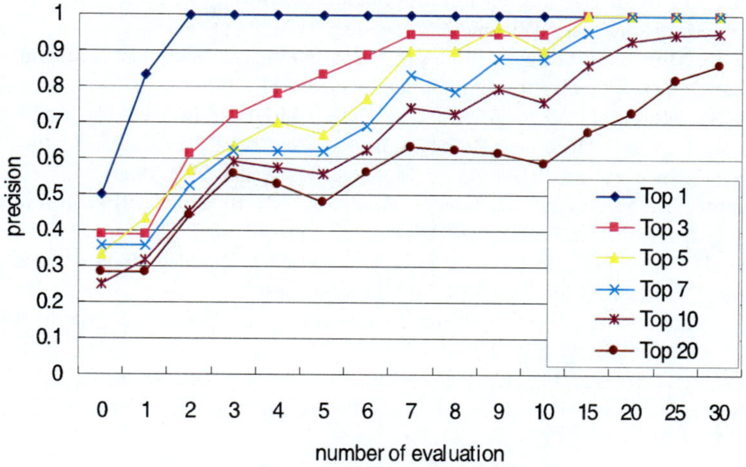

Fig. 6. Precision in existing method

5 Conclusion

In this paper, we proposed a re-ranking method based on the user preference about whether individual pages are relevant or not and also evaluated the execution results. In addition, the query modification procedure was addressed as a powerful function in our re-ranking method. As our future work, the following problem must be resolved: the case that the similar pages are included in the top level as the first retrieved result, and the case that the number of user operations increases when the feature-less pages are included. In this case, the desirable means may be first to cluster retrieved results in advance, and then to look upon typical pages selected from each cluster as the top pages of initially retrieved results.

References

1. Jansen, B.J., Spink, A., Saracevic, T.: Real life, Real users, and Real Needs: A Study and Analysis of User Queries on the Web. Information Processing and Management 36(2), 207–227 (2000)
2. Miller, G.A.: The Magical Number Seven, Plus or Minus Two: Some Limits on our Capacity for Processing Information. The Psychological Review 63, 81–97 (1956)
3. Biittcher, S., Clark, C.L.A., Cormank, G.V.: Information retrieval – Implementing and Evaluating Search Engines, p. 606. The MIT Press, Cambridge (2010)
4. Candam, K.S., Sapino, H.L.: Data Management for Multimedia Retrieval, p. 489. Cambridge Univ. Press, Cambridge (2010)
5. Fang, H., Tao, T., Zhai, C.: A Formal Study of Information Retrieval Heuristics. In: Proc.of 27th Int'l Conf. ACM SIGIR, pp. 49–56 (2004)
6. Krovetz, R., Croft, W.B.: Lexical Ambiguity and Information Retrieval. ACM Trans.on Information Systems (TOIS) 10(2), 115–141 (1992)

7. Rocchio, J.J.: Relevance Feedback in Information Retrieval. The Smart Retrieval System-Experiments in Automatic Document Processing, 313–323 (1971)
8. Onoda, T., Murata, H., Yamada, S.: SVM-based Interactive Document Retrieval with Active Learning. New Generation Computing 26(1), 49–61 (2008)
9. Onoda, T., Murata, H., Yamada, S.: One Class Classification Methods Based on Non-Relevance Feedback Document Retrieval. In: Proc.of 2006 IEEE/WIC/ACM Int'l Conf.on Web Intelligence and Intelligent Agent Technology, pp. 393–396 (2006)
10. Yamamoto, T., Nakamura, S., Tanaka, K.: Rerank-By-Example: Efficient Browsing of Web Search Results. In: Proc. of 18th DEXA 2007, pp. 801–810 (2007)
11. Karube, T., Shizuki, B., Tanaka, J.: A Ranking Interface Based on Interactive Evaluation of Search Results. In: Proc. of WISS (2007) (in Japanese)
12. Jeh, G., Widom, J.: Scaling Personalized Web Search. In: Proc.of 12th World Wide Web Conference (WWW), pp. 271–279 (2003)
13. Matsuo, Y., Ishizuka, M.: Keyword Extraction from a Document Using Word Co-occurrence Statiscal Information. Journal of Japanese Society for Artificial Intelligence 17(3), 213–227 (2002)
14. MeCab: `http://mecab.sourceforge.net/`
15. Robertson, S.E., Walker, S., Jones, S., H-Beaulieu, M., Gatford, M.: "Okapi at TREC-3". In: Proc.of 3rd Text Retrieval Conference, pp. 109–126 (1994)

Chapter 7
Approximately Searching Aggregate k-Nearest Neighbors on Remote Spatial Databases Using Representative Query Points

Hideki Sato

School of Informatics, Daido University, 10-3 Takiharu-cho, Minami-ku, Nagoya, 457-8530 Japan
hsato@daido-it.ac.jp

Abstract. Aggregate k-Nearest Neighbor (k-ANN) queries are required to develop a new promising Location-Based Service (LBS) which supports a group of mobile users in spatial decision making. As a procedure for computing exact results of k-ANN queries over some Web services has to access remote spatial databases through simple and restrictive Web API interfaces, it suffers from a large amount of communication. To overcome the problem, this paper presents another procedure for computing approximate results of k-ANN queries. It relies on a Representative Query Point (RQP) to be used as a key of a k-Nearest Neighbor (k-NN) query for searching spatial data. According to the experiments using synthetic and real data (objects), *Precision* of *sum* k-NN query results using a minimal point as RQP is less than 0.9 in the most case that the number of query points is 10, and over 0.9 in the other most cases. On the other hand, *Precision* of *max* k-NN query results using a minimal point as RQP ranges 0.47 to 0.93 according to the experiments using synthetic data (objects). The experiments using real data (objects) show that *Precision* of *max* k-NN query results is less than 0.8 in case that k is 10, and over 0.8 in the other cases. From these results, it is concluded that accuracy of *sum* k-NN query results is allowable and accuracy of *max* k-NN query results is partially allowable.

1 Introduction

Wireless networks and powerful mobile devices, along with location positioning systems, digital maps, etc., have been continuously developed. Additionally, a large volume of spatial data has been available on the World Wide Web (WWW). Mobile computing has been motivated and made a reality by these technological backgrounds. Location-Based Services (LBSs) are major applications of mobile computing, which provide mobile users with location-dependent information and/or functions. Consider, for example, a single user at a specific location (query point) wants to obtain information of restaurants (data objects), which leads to the top-k minimum distance that he/she has to travel. In order to support him/her, the client program running on his/her mobile terminal captures the data indicating his/her location with a positioning device (e.g., GPS receiver) and sends a k-Nearest Neighbor (k-NN) query[1], [2] using the location as key

T. Watanabe and L.C. Jain (Eds.): Innovations in Intell. Machines – 2, SCI 376, pp. 91–102.
springerlink.com

to some Web services on WWW for gathering restaurant data. Then, the query result is returned to the client and given to him/her for supporting his/her spatial decision making. This kind of LBSs are useful to mobile users and have been already available on many Web sites.

We consider that a new LBS for supporting a group of mobile users is potentially promising, which is an extension of the LBS for supporting a single mobile user, mentioned in the above. Consider, for example, a group of mobile users, each being at a different location (query point), wants to obtain information of restaurants (data objects) to meet together, which leads to the top-k minimum sum of distances that they have to travel. In order to support them, Aggregate k-Nearest Neighbor (k-ANN) queries regarding a set of query points have to be answered, for gathering restaurant information which some Web services disseminate. Of course, the data indicating each user's location can be captured with a positioning device and sent to a location management server. Therefore, all the location data regarding a group of mobile users are available from the location management server. However, there are difficulties in realizing the LBS for two reasons.

First, the Web service receiving k-ANN queries has to access the corresponding spatial databases for answering them. If the spatial databases to be queried are local, and the query processing algorithms have direct access to their spatial indices (i.e., R-trees[3] and its variants), it can answer queries efficiently. However, this assumption does not hold, when k-ANN queries are processed by accessing remote spatial databases that operate autonomously. Although some or all the data from the remote databases can be replicated in a local database and a separate index structure for them can be built, it is infeasible when the database is huge, or large number of remote databases are accessed.

Secondly, accesses to spatial data on WWW are limited by certain types of queries, due to simple and restrictive Web API interfaces. A typical scenario is devoted to retrieving some of the restaurants nearest to the address given as query point through a Web API interface. Unfortunately, Web API interfaces are not supported for answering k-ANN queries on remote spatial databases. In other words, a new strategy for efficiently answering k-ANN queries is required in this setting.

In this paper, we propose a procedure for efficiently answering k-ANN queries, which is useful for developing a new LBS to support a group of mobile users in spatial decision making. Since it relies on a Representative Query Point (RQP) over a set of query points and a k-NN query, it efficiently computes approximate results of k-ANN queries , not exact ones. Accordingly, what point should be chosen as RQP is important for obtaining k-ANN query results of high accuracy. Among several candidates, the minimal point of an aggregate distance function is chosen as RQP based on experimental evaluation. Additionally, experiments using synthetic and real data (objects) evaluate accuracy of k-ANN query results.

The remainder of this paper is organized as follows. Sect.2 describes k-ANN queries and the problem in answering them for the later discussion. Sect.3 presents a procedure for answering k-ANN queries. Sect.4 experimentally evaluates accuracy of k-ANN query results using synthetic and real data, which the proposed procedure computes. Sect.5 presents the related work. Finally, Sect.6 concludes the paper and gives our future work.

2 Preliminaries

In this section, we describe k-ANN queries and the problem in answering k-ANN queries for the later discussion.

2.1 Aggregate k-Nearest Neighbor Queries

Recently, there has been an increasing interest in Nearest Neighbor (NN) queries. Given a set P of data objects (e.g., facilities) and a location q, the NN query returns the nearest object of q in P. Formally, the query retrieves a point $p \in P$, such that $d(p,q) \leq d(p',q), \forall p' \in P$, where $d(\,)$ is a distance function.

Aggregate Nearest Neighbor (ANN) query is a generalized version of NN queries. Let p be a point and Q be a set of query points. Then, aggregate distance function $d_{agg}(p,Q)$ is defined as $agg(\{d(q,p)|q \in Q\})$, where $agg(\,)$ is an aggregate function (e.g., *sum*, *max*, *min*). Given a set P of data objects and a set Q of query points, ANN query retrieves the object p in P, such that $d_{agg}(p,Q)$ is minimized.

k-ANN query is an extension of ANN queries into top-k queries[4,5]. Given a set of data objects P, a set of query points Q, and aggregate distance function $d_{agg}(p,Q)$, k-ANN query $k\text{-}ANN_{agg}(P,Q)$ retrieves $S \subset P$ such that $|S| = k$ and $d_{agg}(p,Q) \leq d_{agg}(p',Q), \forall p \in S, p' \in P - S$ for some $k < |P|$. From the standpoint of spatial decision making support for mobile users, plural number of query results are preferable to a single one, even if it is the best (nearest neighbor) one.

Consider the example of Fig. 1, where $P(= \{p_1, p_2, p_3, p_4\})$ is a set of data objects (e.g., restaurants) and $Q(= \{q_1, q_2\})$ is a set of query points (e.g., locations of mobile users). The number on each edge connecting a data object and a query point represents any distance cost between them. Table 1 presents $d_{agg}(p,Q)$ for each p in P, ANN query result, 3-ANN query results over aggregate function such as *sum*, *max*, and *min*.

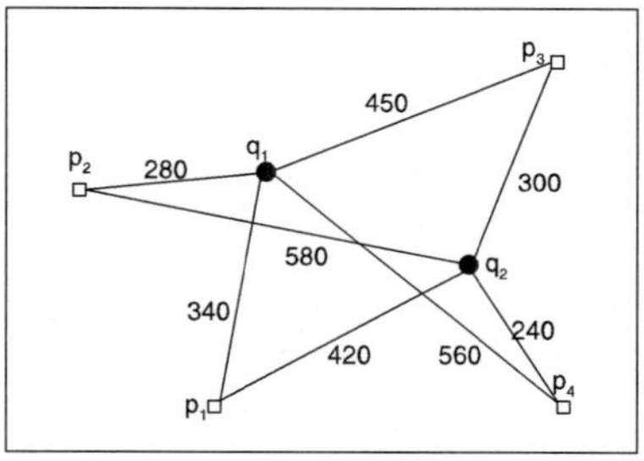

Fig. 1. Example of ANN queries (data object (hollow square), query point (solid circle))

2.2 Problem in Answering k-ANN Queries

Figure 2 presents relations P and Q which contain tuples of data objects and tuples of query points, respectively. Figure 3 shows the SQL statement which expresses a k-ANN

Table 1. Results of ANN query and 3-ANN query shown in Fig. 1

aggregate function	*sum*	*max*	*min*
$d_{agg}(p_1,Q)$	760	420	340
$d_{agg}(p_2,Q)$	860	580	280
$d_{agg}(p_3,Q)$	750	450	300
$d_{agg}(p_4,Q)$	800	560	240
ANN query result	p_3	p_1	p_4
3-ANN query result	$\{p_3,p_1,p_4\}$	$\{p_1,p_3,p_4\}$	$\{p_4,p_2,p_3\}$

Relation *P(identifier, location, ⋯)*

identifier	**location**	⋯
p_1	(x_{p_1},y_{p_1})	⋯
p_2	(x_{p_2},y_{p_2})	⋯
p_3	(x_{p_3},y_{p_3})	⋯
p_4	(x_{p_4},y_{p_4})	⋯

Relation *Q(identifier, location, ⋯)*

identifier	**location**	⋯
q_1	(x_{q_1},y_{q_1})	⋯
q_2	(x_{q_2},y_{q_2})	⋯

Fig. 2. Respective relations for *P* and *Q* of Fig. 1

```
select P.identifier
from P,Q
group by P.identifier
order by agg(d(P.location,Q.location))
limit k
```

agg(): aggregate function such as sum, max,and min
d(): distance functions such as the *Manhattan* one and the *Euclidean* one

Fig. 3. SQL statement describing *k*-ANN query

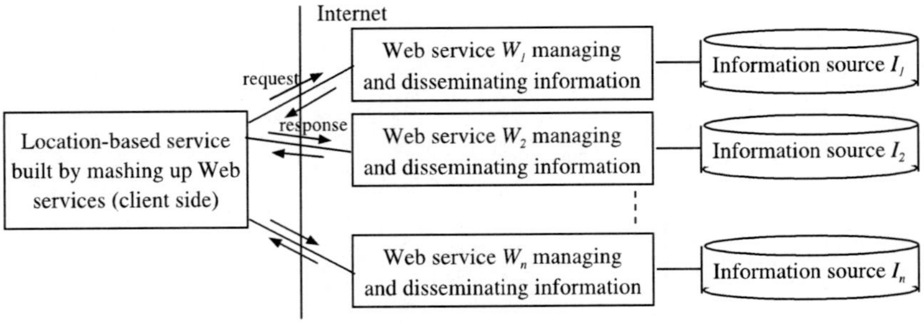

Fig. 4. System model for answering *k*-ANN queries

query over *P* and *Q*. Processing the SQL statement requires Cartesian product over *P* and *Q*.

We take the system model shown in Fig.4 for investigating a procedure to process *k*-ANN queries. Each Web service manages and disseminates its unique information being stored in its own information source. Consider two Web services, which are related to processing the SQL statement shown in Fig.3. One manages the relation of data objects and the other manages the relation of query points. In this setting, Cartesian product has to be computed over data of information sources, each of which is located at a different site. If LBS requires the exact results of k-ANN queries, a large amount of communication is required to obtain them, because the entire data of at least one must be sent to other sites to compute Cartesian product.

3 Procedure for Answering k-ANN Queries

As is mentioned in Sect.2.2, computing exact results of k-ANN queries requires a large amount of communication. To reduce this cost, we present a procedure for computing approximate k-ANN query results, not exact ones. It relies on a RQP over a set of query points and a k-NN query. From the standpoint of spatial decision making support for mobile users, approximate results can be accepted, if their accuracy is practically allowable. Since accuracy highly depends on what point should be chosen as RQP, we first discuss aggregate distance functions and their minimal points before presenting the procedure. In the rest of the paper, we confine distance function to Euclidean distance and aggregate function to *sum* and *max*.

3.1 Aggregate Distance Function

Let p be a point (x, y) and Q be a set of query points. *Sum* distance function $d_{sum,Q}(x,y)$ over Q is defined in Eq. 1 and *maximum* distance function $d_{max,Q}(x,y)$ over Q is defined in Eq. 2. Eq. 2 defines the maximum distance between a point (x, y) and each query point (x_i, y_i) belonging to Q. In other words, it computes the distance between a point (x, y) and a query point (x_i, y_i), if the former point lies in the furthest-point Voronoi region[7] of the latter point. Figure 5 presents contour graphs of two aggregate distance functions over a set of 10 query points whose locations are randomly generated, $d_{sum,Q}(x,y)$ shown in Fig.5(a) and $d_{max,Q}(x,y)$ shown in Fig.5(b). Either of the two functions is convex, which implies that it is single-peak. However, both functions are not differentiable at a point (x_i, y_i) belonging to Q.

$$d_{sum,Q}(x,y) = \sum_{(x_i,y_i)\in Q} \sqrt{(x-x_i)^2+(y-y_i)^2} \tag{1}$$

$$d_{max,Q}(x,y) = max(\{\sqrt{(x-x_i)^2+(y-y_i)^2} | (x_i,y_i) \in Q\}) \tag{2}$$

The minimal point of $d_{sum,Q}(x,y)$ lies inside the convex hull of Q. Although it cannot be computed neither by an analytical method nor by an algorithm based on differentiation, it can be obtained by employing Nelder-Mead method[6] for nonlinear programming problems, which does not rely on gradients of a function. On the other hand, the minimal point of $d_{max,Q}(x,y)$ corresponds exactly with the center of Minimum Covering Circle (MCC) of Q, as is shown in Fig.6. Accordingly, the point can be computed using a Computational Geometric algorithm[7].

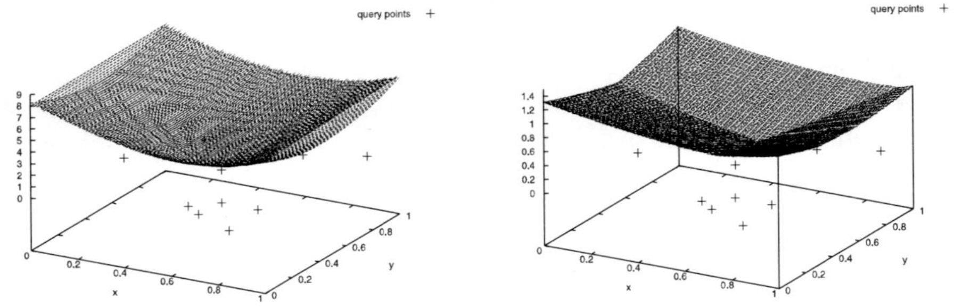

(a) sum distance function $d_{sum,Q}(x,y)$ **(b) maximum distance function** $d_{max,Q}(x,y)$

Fig. 5. Aggregate distance function (Euclidean distance, number of query points=10)

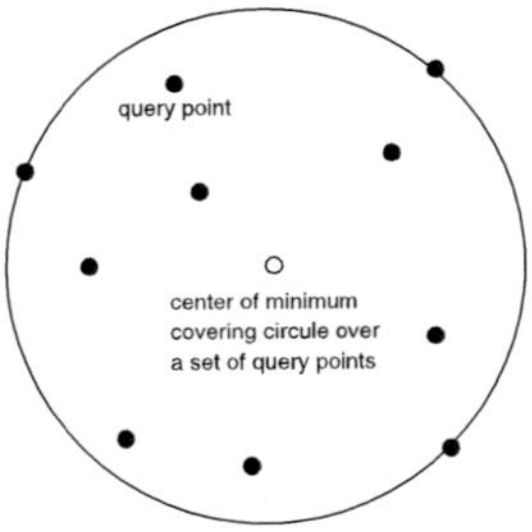

Fig. 6. Minimal point regarding maximum distance of query points

3.2 Processing Scheme Using Representative Query Point and k-NN Query

The followings are assumptions imposed upon to develop the procedure for answering *k*-ANN queries.

assumption 1. A Web service disseminating information of data objects is able to answer *k*-NN queries.

assumption 2. Any a priori knowledge regarding the spatial distribution of data objects is not known.

In order to reduce a large amount of communication being involved in answering a *k*-ANN query over several Web services, RQP is introduced to represent a set of query points and used to be a key of a *k*-NN query which substitutes for the corresponding *k*-ANN query. A *k*-NN query, which is available under **assumption 1**, can be requested to a Web service disseminating information of data objects. Figure 7 shows the procedure using a RQP for answering a *k*-ANN query, which is executed by the client shown in Fig.4. Since the procedure relies on a RQP and a *k*-NN query (See **Step 4** of Fig.7), it can really reduce amounts of communication in answering *k*-ANN queries. However, it is not guaranteed that the procedure computes the exact results of *k*-ANN queries, because it just computes the results of a *k*-NN query from a RQP.

(Step1) A request for gathering a set of query points Q is invoked.
(Step2) *Representative query point* (RQP) over Q is calculated.
(Step3) A *k*-NN query with RQP as its query key is invoked.
(Step4) A *k*-ANN query answer is computed by using the query answer of Step 3.

Fig. 7. Procedure for solving *k*-ANN queries

From the standpoint of spatial decision making support for mobile users, however, approximate results of *k*-ANN queries can be accepted if their accuracy is practically allowable. Therefore, what point should be chosen as RQP is important, which leads to *k*-ANN query results of high accuracy.

4 Experimental Accuracy Evaluation

In this section, accuracy of *k*-ANN query results is experimentally evaluated, which the procedure shown in Fig.7 computes. Since the procedure relies on a RQP and a *k*-NN query, accuracy of *k*-NN query results using a RQP as key is evaluated really. *Precision* is used as criteria to evaluate accuracy. Firstly, several RQPs are compared regarding *Precision*. Secondly, several distributions over the locations of data objects are compared regarding *Precision*. Thirdly, real data objects are used to evaluate *Precision*. Note that the experimental data to be presented in this section are averages of 10 trials conducted for each parameter combination.

4.1 *Precision* Evaluation on Representative Query Points

As is mentioned in Sect. 3.2, what point should be chosen as RQP is important for obtaining *k*-ANN query results of high accuracy. There are several candidates to be considered as RQP under **assumption 2** presented in Sect. 3.2. In order to make comparison, the minimal point of *sum* distance function over Q, the middle point of Q [1], and the centroid of Q are used as RQP of *sum* *k*-NN queries. As for *max* *k*-NN queries, the minimal point of *max* distance function over Q and the center of the Minimum Bounding Rectangle (MBR) of Q are used as RQP.

Precision of *k*-ANN query results is used to evaluate accuracy, which is defined in Eq. 3, where $R_{k-NN,RQP}$ is a *k*-NN query result using RQP, R_{k-ANN} is an original *k*-ANN query result, and k is the size of $R_{k-NN,RQP}$ and R_{k-ANN}. Although the procedure really returns $R_{k-NN,RQP}$ in the ascending order of aggregated distances, Eq. 3 does not take the order into consideration. This is because re-ordering can be exactly done later in Step4 of the procedure (See Fig. 7), if $R_{k-NN,RQP}$ includes the exact results of R_{k-ANN}.

$$Precision(k) = \frac{|R_{k-NN,RQP} \bigcap R_{k-ANN}|}{k} \tag{3}$$

[1] The middle point of Q is defined to be a point $(median(\{x_i|(x_i,y_i) \in Q\}), median(\{y_i|(x_i,y_i) \in Q\}))$, where $median()$ is a function returning the middle value of elements in a set.

Precision of *sum* k-NN queries and *max* k-NN queries is measured for 10000 data objects and several query points whose locations are uniformly distributed. From Tables 2 and 3, the followings are clarified.

1. *Precision* is directly proportional to either the number of query points or k.
2. The minimal point of *sum* distance function excels as RQP the middle point of Q and the centroid of Q. *Precision* of *sum* k-NN query results using minimal points as RQP less than 0.9 in case that the number of query points is 10 and k is 100 or less, and over 0.9 in the other cases. It is allowable for *sum* k-NN query results.
3. The minimal point of *max* distance function excels as RQP the center of MBR of Q. Though *Precision* of *max* k-NN query results using minimal points as RQP ranges from 0.56 to 0.93, it is partially allowable for *max* k-NN query result in some combinations of k and the number of query points.

Table 2. *Precision* of *sum* k-NN query results using each RQP (synthetic data according to uniform distribution. number of data objects=10000)

representative query point (RQP)	*sum* 10-NN query				*sum* 100-NN query				*sum* 1000-NN query			
	number of query points (Q)											
	10	100	1000	10000	10	100	1000	10000	10	100	1000	10000
minimal point of *sum* distance	0.860	0.940	0.990	0.994	0.893	0.962	0.991	0.995	0.910	0.974	0.992	0.996
middle point of Q	0.060	0.350	0.700	0.880	0.431	0.746	0.900	0.963	0.783	0.908	0.969	0.988
centroid of Q	0.000	0.350	0.820	0.920	0.291	0.788	0.937	0.980	0.808	0.943	0.980	0.994

Table 3. *Precision* of *max* k-NN query results using each RQP (synthetic data according to uniform distribution. number of data objects=10000)

representative query point (RQP)	*max* 10-NN query				*max* 100-NN query				*max* 1000-NN query			
	number of query points (Q)											
	10	100	1000	10000	10	100	1000	10000	10	100	1000	10000
minimal point of maximum distance	0.560	0.570	0.770	0.820	0.685	0.717	0.862	0.911	0.813	0.885	0.922	0.930
center of MBR of Q	0.120	0.130	0.540	0.800	0.417	0.560	0.788	0.912	0.746	0.839	0.910	0.931

4.2 *Precision* Evaluation on Skewed Data

In this subsection, *Precision* of k-ANN query results using a minimal point as RQP is evaluated by using 10000 data objects whose locations are generated according to two-dimensional Gaussian distribution. Let a location of a data object be a point (x, y) ($x \in [0,1)$, $y \in [0,1)$). The mean point of Gaussian distribution is randomly generated and the standard deviation (σ) is changed. *Precision* of *sum* k-NN queries and *max* k-NN queries is measured for several query points whose locations are uniformly distributed. Roughly speaking, *Precision* is not affected by distributions over the locations of data objects (See Tables 4 and 5).

Table 4. *Precision* of *sum* k-NN query results using minimal point (number of data objects=10000)

distribution of synthetic data objects	*sum* 10-NN query				*sum* 100-NN query				*sum* 1000-NN query			
	number of query points (Q)											
	10	100	1000	10000	10	100	1000	10000	10	100	1000	10000
Uniform distribution	0.860	0.940	0.990	0.994	0.893	0.962	0.991	0.995	0.910	0.974	0.992	0.996
Gaussian distribution												
$\sigma = 0.06$	0.820	0.950	0.980	0.990	0.824	0.957	0.985	0.991	0.862	0.969	0.987	0.992
$\sigma = 0.09$	0.810	0.960	0.981	0.986	0.840	0.964	0.982	0.987	0.867	0.966	0.984	0.995
$\sigma = 0.12$	0.790	0.950	0.980	0.989	0.848	0.952	0.986	0.990	0.861	0.966	0.989	0.994

Table 5. *Precision* of *max* k-NN query results using minimal point (number of data objects=10000)

distribution of synthetic data objects	*max* 10-NN query				*max* 100-NN query				*max* 1000-NN query			
	number of query points (Q)											
	10	100	1000	10000	10	100	1000	10000	10	100	1000	10000
Uniform distribution	0.560	0.570	0.770	0.820	0.685	0.717	0.862	0.911	0.813	0.885	0.922	0.930
Gaussian distribution												
$\sigma = 0.06$	0.620	0.640	0.670	0.670	0.755	0.757	0.774	0.779	0.826	0.852	0.855	0.858
$\sigma = 0.09$	0.470	0.740	0.770	0.830	0.604	0.743	0.810	0.841	0.745	0.850	0.864	0.875
$\sigma = 0.12$	0.570	0.620	0.780	0.820	0.669	0.702	0.850	0.868	0.779	0.857	0.901	0.908

Table 6. *Precision* of *sum* k-NN query results using minimal point (real data. number of data objects=2003)

sum k-NN queries	number of query points (Q)									
	10	20	30	40	50	60	70	80	90	100
$k = 10$	0.760	0.840	0.950	0.920	0.890	0.940	0.900	0.930	0.920	0.920
$k = 30$	0.857	0.917	0.957	0.927	0.953	0.967	0.950	0.953	0.967	0.973
$k = 50$	0.886	0.942	0.954	0.964	0.954	0.970	0.974	0.976	0.954	0.972

4.3 *Precision* Evaluation Using Real Data

In this subsection, *Precision* of k-ANN query results using a minimal point as RQP is evaluated by using real data, not synthetic data. The data is concerned with restaurants located in Nagoya, which is available at Web site and accessible via Web API[2]. There are 2003 corresponding restaurants, which are concentrated in the downtown of Nagoya. *Precision* of *sum* k-NN query results and *max* k-NN query results is measured for several query points whose locations are uniformly distributed. From Tables 6 and 7, the followings are clarified.

1. *Precision* of *sum* k-NN query results is less than 0.9 in case that the number of query points is 10, and mostly it is over 0.9 in the other cases. It is allowable for *sum* k-NN query results.

[2] *http://webservice.recruit.co.jp/hotpepper/gourmet/v1/*

Table 7. *Precision* of *max* k-NN query results using minimal point (real data. number of data objects=2003)

max k-NN	number of query points (Q)									
queries	10	20	30	40	50	60	70	80	90	100
$k = 10$	0.670	0.710	0.550	0.650	0.720	0.680	0.740	0.700	0.630	0.720
$k = 30$	0.803	0.847	0.760	0.900	0.897	0.870	0.924	0.854	0.837	0.843
$k = 50$	0.812	0.898	0.808	0.956	0.928	0.924	0.937	0.863	0.846	0.870

2. *Precision* of *max* k-NN query results is less than 0.8 in case that k is 10, and mostly it is over 0.8 in the other cases.

5 Related Work

The existing literature in the field of location-dependent queries is extensively surveyed in the article[8]. It presents (1) description of technological contexts and support middleware, (2) definition and classification of location-based services and location-dependent queries, and (3) review and comparison of different query processing approaches.

Among many location-dependent queries, NN queries[1], [2] and their variants such as Reverse NN[9], Constrained NN[10], and Group NN[11], [12] are considered to be important in supporting spatial decision making. A Reverse k-NN query retrieves objects that have a specified object/location among their k nearest neighbors. A Constrained NN query retrieves objects that satisfies a range constraint. For example, a visible k-NN query retrieves k objects with the smallest visible distance to a query object[13].

Since a Group NN query retrieves Aggregate NN objects, the work[11], [12] is much related to ours. First, it has been dedicated to the case of Euclidean distance and *sum* function[11]. Then, it has been generalized to the case of the network distance[12]. Their setting is that the spatial database storing data objects are local to the site where the database storing query objects resides. However, we deal with k-ANN queries, where each database is located at a remote site.

Both of the work[14], [15] are much related to ours, because they provide users with location-dependent query results by using Web API interfaces to remote databases. The former[14] proposes a k-NN query processing algorithm that uses one or more Range queries[3][16], [17], [18] to retrieve the nearest neighbors of a given query point. The latter[15] proposes two *Range* query processing algorithms by using k-NN queries. However, our work differs from theirs in dealing with k-ANN queries, not either k-NN queries or Range queries.

6 Conclusion

In this paper, we have proposed the procedure for efficiently answering k-ANN queries, which is useful for developing a new LBS to support a group of mobile users in spatial

[3] A Range query retrieves the objects located within a certain range/region.

decision making. Since it relies on a RQP over a set of query points and a k-NN query, it efficiently computes approximate results of k-ANN queries, not exact ones. Accordingly, what point should be chosen as RQP is important for obtaining k-ANN query results of high accuracy. Among several candidates, the minimal point of an aggregate distance function has been chosen as RQP based on experimental evaluation.

According to the additional experiments using synthetic and real data (objects), *Precision* of *sum* k-NN query results using a minimal point as RQP is mostly less than 0.9 in case that the number of query points is 10, and over 0.9 in the other cases. On the other hand, *Precision* of *max* k-NN query results using a minimal point as RQP ranges 0.47 to 0.93 according to the experiments using synthetic data (objects). The experiments using real data (objects) show that *Precision* of *max* k-NN query results is less than 0.8 in case that k is 10, and over 0.8 in the other cases. From these results, it is concluded that accuracy of *sum* k-NN query results is allowable and accuracy of *max* k-NN query results is partially allowable.

Our future work is as follows. First of all, it includes (1) development of an efficient k-ANN query processing procedure which answers exact query results, not approximate ones. In order to realize the procedure, a k-NN query using a minimal point as RQP is considered to be the first step. Our future work also includes (2) investigation of the optimal method for computing RQPs for the combination of a distance function and an aggregate function. (3) investigation of *min* k-NN queries whose distance function is multi-modal differently from those of *sum* and *max*, and (4) investigation of an aggregate version of Range queries[16], [17], [18] and their processing procedures.

References

1. Roussopoulos, N., Kelly, S., Vincent, F.: Nearest Neighbor Queries. In: Proc. ACM SIGMOD Int'l Conf. on Management of Data, pp. 71–79 (1995)
2. Hjaltason, G.R., Samet, H.: Distance Browsing in Spatial Databases. ACM Trans. Database Systems 24(2), 265–318 (1999)
3. Guttman, A.: R-trees: A Dynamic Index Structure for Spatial Searching. In: Proc. ACM SIGMOD Int'l Conf. on Management of Data, pp. 47–57 (1984)
4. Fagin, R., Lotem, A., Naor, M.: Optimal Aggregation Algorithms for Middleware. In: Proc. Symp. Principles of Database Systems, pp. 102–113 (2001)
5. Ilyas, H.F., Beskales, G., Soliman, M.A.: A Survey of Top-k Query Processing Techniques in Relational Database Systems. ACM Computing Survey 40(4), Article 11 (2008)
6. Nelder, J.A., Mead, R.: A Simplex Method for Function Minimization. Computational Journal, 308–313 (1965)
7. Berg, M.D., Kreveld, M.V., Overmars, M., Schwarzkopf, O.: Computational Geometry: Algorithms and Applications. Springer, Heidelberg (1997)
8. Ilarri, S., Menna, E., Illarramendi, A.: Location-Dependent Query Processing: Where We Are and Where We Are Heading. ACM Computing Survey 42(3), Article 12 (2010)
9. Korn, F., Muthukrishnan, S.: Influence Sets Based on Reverse Nearest Neighbor Queries. In: Proc. ACM SIGMOD Int'l Conf. on Management of Data, pp. 201–212 (2000)
10. Ferhatosmanoglu, H., Stanoi, I., Agrawal, D., Abbadi, A.E.: Constrained Nearest Neighbor Queries. In: Proc. Seventh Int'l Symp. Advances in Spatial and Temporal Databases, pp. 257–278 (2001)
11. Papadias, D., Shen, Q., Tao, Y., Mouratidis, K.: Group Nearest Neighbor Queries. In: Proc. Int'l Conf. Data Eng., pp. 301–312 (2004)

12. Yiu, M.L., Mamoulis, M., Papadias, D.: Aggregate Nearest Neighbor Queries in Road Networks. IEEE Trans. on Knowledge and Data Engineering 17(6), 820–833 (2005)
13. Nutanong, S., Tanin, E., Zhang, R.: Visible nearest neighbor queries. In: Proc. Int'l Conf. DASFAA, pp. 876–883 (2007)
14. Liu, D., Lim, E., Ng, W.: Efficient k-Nearest Neighbor Queries on Remote Spatial Databases Using Range Estimation. In: Proc. SSDBM, pp. 121–130 (2002)
15. Bae, W.D., Alkobaisi, S., Kim, S.H., Narayanappa, S., Shahabi, C.: Supporting Range Queries on Web Data Using k-Nearest Neighbor Search. In: Proc. W2GIS, pp. 61–75 (2007)
16. Xu, B., Wolfson, O.: Time-Series Prediction with Applications to Traffic and Moving Objetcs Databases. In: Proc. Third ACM Int'l Workshop on MobiDE, pp. 56–60 (2003)
17. Trajcevski, G., Wolfson, O., Xu, B., Nelson, P.: Managing Uncertainty in Moving Objects Databases. ACM Trans. Database Systems 29(3), 463–507 (2004)
18. Yu, P.S., Chen, S.K., Wu, K.L.: Incremental Processing of Continual Range Queries over Moving Objects. IEEE Trans. Knowl. Data Eng. 18(11), 1560–1575 (2006)

Chapter 8
Design and Implementation of a Context-Aware Guide Application "Kagurazaka Explorer"

Yuichi Omori, Jiaqi Wan, and Mikio Hasegawa

Tokyo University of Science,
1-14-6, Kudankita, Chiyoda-ku, Tokyo, Japan
{omori,man}@haselab.ee.kagu.tus.ac.jp,
hasegawa@ee.kagu.tus.ac.jp
http://haselab.ee.kagu.tus.ac.jp/

Abstract. We propose a context-aware guide application, which provides appropriate information selected by a machine learning algorithm according to the preference and the situation of each user. We have designed and implemented the proposed system using the off-the-shelf mobile phones with a built-in GPS module. The machine learning algorithm enables our system to select an appropriate spot based on the user's real-time context such as preference, location, weather, time, etc. As a machine learning algorithm, we use the support vector machine (SVM) to decide the appropriate information for the users. In order to realize high generalization performance, we introduce the principal component analysis (PCA) to generate the input data for the SVM learning. By our experiments in real environments, it is shown that the proposed system works correctly and the correctness of recommendation can be improved by introducing the PCA.

Keywords: Context-Aware Applications, Ubiquitous Computing, Mobile Networks, Machine Learning Algorithms, Support Vector Machines.

1 Introduction

Various wireless network systems have been developed and commercialized, and ubiquitous network access has been available. As new applications which effectively utilize such ubiquitous network access, various ubiquitous computing applications have been studied and developed [1]-[3]. As one of the targets of those applications, context-aware recommendation applications to provide useful information to the mobile users have been proposed [4]-[7]. Although the conventional recommendation applications select information to be provided to the user based only on the static context information such as the user profile and the preference, recently proposed context-aware recommendation applications based on ubiquitous computing technology decide appropriate information based not only on the static information but also on the real-time and real-world context information, such as location, weather and so on. As such a context-aware recommendation application, Blue Mall[5] is a recommendation system that notifies

T. Watanabe and L.C. Jain (Eds.): Innovations in Intell. Machines – 2, SCI 376, pp. 103–115.
springerlink.com

the mobile users of advertisements about nearby shopping stores based on the user's location estimated by the Bluetooth RSSI. A system called Bookmark Handover[6] is a context aware reminding application, which reminds mobile phone users about the events or the visiting spots which they had registered before with the notifying timing context information, such as location, timing, etc. A system called i-concir[7] is one of examples of commercialized services, which provides useful information in daily life based on user's context. In these systems, appropriate information selection from huge amount of candidate data is very important, because too much uninteresting information annoys the users. In order to select appropriate information correctly according to user's context, various algorithms have been developed[8]-[13]. As one of such approaches, the Context-Aware SVM[13] using the machine learning algorithm to decide appropriate information from context data has been proposed, and it has been shown that the support vector machine (SVM)[14] has better performance than other learning algorithms for context-aware information selection[15].

In our research, we apply this context-aware recommendation technique to the real environment. We design and implement a context-aware guide application called Kagurazaka Explorer, which guides mobile users about the Kagurazaka street in Tokyo, Japan according to their context. It provides appropriate information about visiting spots in the Kagurazaka street area selected by a machine learning algorithm. However, the conventional recommendation algorithms using machine learning algorithms for the context modeling require very large amount of training data to achieve high performance, because the feature space of the learning model, which deals with so many types of the context, becomes very high dimensional. This problem becomes so serious for the users who use the system only a few times, or for the recommendation systems for a local area where it is difficult to collect training data in adequate amount. As one of the approaches to solve such an issue, reduction based methods which reduce the dimensions of the feature space, have been proposed. In Ref. [15][16], the reduction based methods removing the non-effective features from feature space by using exhaustive searches to detect features, which do not affect the user's decision, are proposed for context-aware recommendation. However, such approaches using exhaustive searches take much processing time for large feature space because these approaches have to calculate the performance in all combination patterns of the feature parameter. As another approach to reduce the feature space for the SVM, Ref. [17] applied the principal component analysis (PCA) to extract the low-dimensional feature of the training data, and showed that it improves the precision of the estimation and reduces the processing time by decreasing the number of dimensions of feature space. In the proposed guide system, we introduce the PCA to construct an appropriate low-dimensional feature space from the high-dimensional and the small number of the training samples.

The rest of this paper is organized as follows. In section 2, we describe the overall concept of our proposed guide system. In section 3 and 4, we show the detailed technique and the design of our proposed guide system. We evaluate the implemented system in section 5. We conclude the paper in section 6.

2 A Context-Aware Guide System with a Machine Learning Algorithm

We develop a context-aware guide system which provides specific regional information to the mobile users. It adaptively guides the mobile phone users according to their real-time contexts, such as user's preference, location, situation and so on. The most appropriate information, which should be provided to the user, is selected by a learning algorithm with the feedbacks from the users. As a machine learning algorithm, we introduce the Support Vector Machine (SVM) which is shown the effectiveness for context-aware recommendation in Ref. [15]. However, such learning algorithms have serious issues that they require a large number of feedbacks from the users for the data whose feature space is high-dimensional. Furthermore, since our proposed system is supposed to be used to provide specific information of some specialized areas (e.g. commercial avenues, sightseeing spots), it may be difficult to collect sufficient amounts of feedbacks to model user's context in the high-dimensional data space. It will degrade correctness of recommendation, because the feature space dimension is too high compared to the number of training data for extracting characteristics using the SVM. In order to solve such issues we introduce the PCA, which extracts low-dimensional feature from the high dimensional data. In our proposed system, the PCA makes the low-dimensional input data for the learning algorithm, with preserving the features of the user preference. By using such models, the proposed system automatically selects appropriate information for the target users.

Fig. 1 shows the two phases of our proposed guide system: a recommendation phase and a navigation phase. In the recommendation phase, this system recommends information for the mobile users according to their context (e.g. location, weather and so on). In the navigation phase, this system sends the detailed map of the spot selected by the user to navigate them. At the same time, this system regards this spot as the user's favorite in the current context and registers this

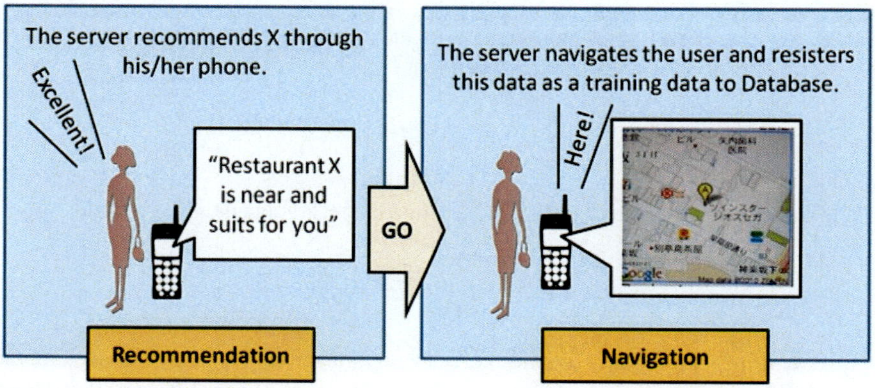

Fig. 1. Two phases of the proposed guide system.

data to the database as a new training data for machine learning algorithms automatically. This system is a kind of real-time learning system, which collects feedbacks from the users during providing selected recommendation information and improves the model using those collected feedbacks.

3 Context-Aware SVM with Principal Component Analysis

We apply a separation technique to classify visiting spots into two classes, appropriate ones to make recommendation and others, according to the context information. We introduce the SVM, which has been shown to have high performance in the context-aware recommendation in Ref. [15], by comparing with several classification algorithms, such as neural networks, decision tree algorithms and so on, for the real data sets. The SVM has better generalization performance by maximizing the margin between training data and separating hyper plane. The training data is the set of the feature vector $\boldsymbol{x}_{i=(1,2,..,m)} = (x_1, x_2, ..., x_n)$ which has n dimensions and its correct output y_i which is defined by

$$y_{i=(1,2,..,m)} = \begin{cases} 1 & \text{— satisfactory data,} \\ -1 & \text{— unsatisfactory data,} \end{cases} \tag{1}$$

where m is the number of training data. The separating hyper plane for the data classification is formulated as follows,

$$g(\boldsymbol{x}) = \sum_{i=1}^{m} a_i y_i K(\boldsymbol{x}_i, \boldsymbol{x}) + b, \tag{2}$$

where $K(\boldsymbol{x}_i, \boldsymbol{x}_j)$ is the kernel function. The optimal user preference vector $\boldsymbol{a} = (a_1, a_2, ..., a_m)$ and the bias parameter b can be obtained by minimizing the following objective function,

$$L(\boldsymbol{a}) = \sum_{i=1}^{m} a_i - \frac{1}{2} \sum_{i=1}^{m} \sum_{j=1}^{m} a_i a_j y_i y_j K(\boldsymbol{x}_i, \boldsymbol{x}_j), \tag{3}$$

$$\text{subjectto} : \sum_{i=1}^{m} a_i y_i = 0, a_i \geq 0 (i = 1, ..., m). \tag{4}$$

As the kernel function for the SVM, we use the RBF kernel[18] defined by

$$K(\boldsymbol{x}_i, \boldsymbol{x}_j) = \exp\left(- \frac{\| \boldsymbol{x}_i - \boldsymbol{x}_j \|^2}{\sigma^2} \right). \tag{5}$$

In our system, we optimize the kernel parameter σ by the cross-validation method. The features of the spots (e.g. cost, location) and the users' situations (e.g. accompany, weather) are used for the feature space generation as shown in Fig. 2. The proposed method models the users' preference by classifying the

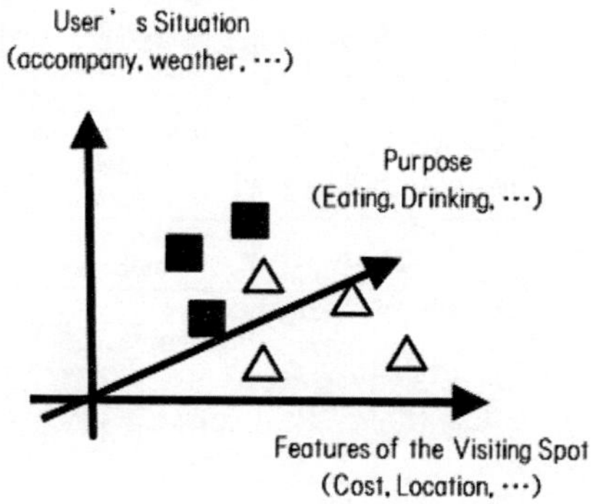

Fig. 2. High-dimensional feature space. The triangles and the square plots indicate training data correspoiding interesting and uninteresting data, respectively.

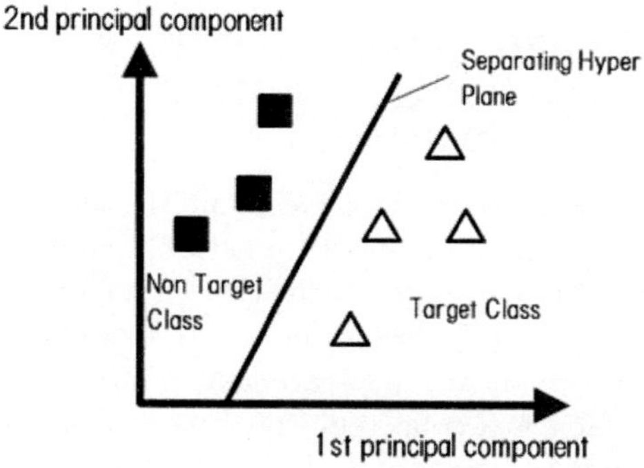

Fig. 3. The SVM in low-dimensional SVM feature space. The triangles and the square plots indicate training data correspoiding interesting and uninteresting data, respectively.

visiting spots into appropriate ones to be recommended and others. The system provides the contents which are regarded as appropriate ones for the user's current contexts.

Since the number of dimensions of feature space is high, a large number of training data is required to extract feature correctly. So, we employ the PCA to decrease the number of dimensions. In the PCA, the principal components, calculated from the high-dimensional feature vectors, are used for extracting low-dimensional feature vectors. The low-dimensional feature vector shown in Fig. 3 is composed of the principal components. We use these low-dimensional feature vectors for the SVM learning.

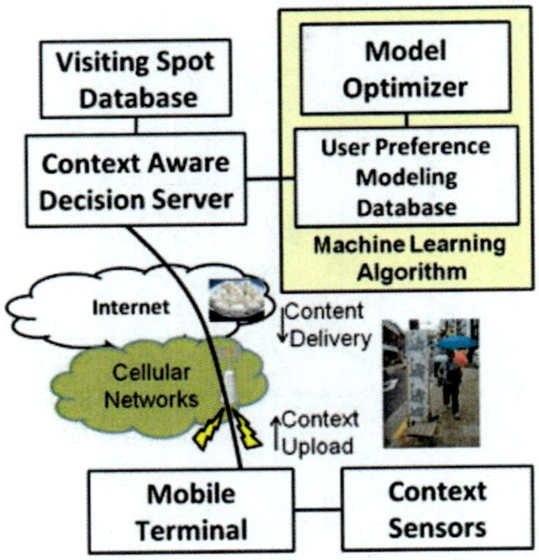

Fig. 4. System structure of the proposed guide system.

4 Design and Implementation

4.1 Design of Proposed System

There are two possible approaches to implement the learning algorithm on our proposed guide system. One is to put all functional components in the mobile terminal, and the other is to put them in the server on the network side[19]. The advantage of the former is the lower computational and traffic loads on the server side. The latter's advantages are that a large amount of the training data are available, which are collected from many other terminals, and that the lower computational load is required on the mobile terminal. In the latter one, we do not have to install any software components to each terminal, by utilizing preinstalled web browser. Therefore, we focus on the second approach, i.e. preserving all training data and learning function on the network side so that a large amount of training data can be used, and the users can use various mobile terminals without installing any additional software components.

As shown in Fig. 4, our proposed guide system consists of four main components: the mobile terminals, a context aware decision server (CADS), a database (DB) for storing training data and learning parameters, and a model optimizer. The mobile terminals are used to receive interesting location information and to upload user's context and feedbacks. The CADS has the functions, to select the contents to be recommended based on the learning algorithm, to deliver the selected contents to the mobile terminals, to formulate new training samples from uploaded feedbacks from the mobile terminals, and to register them to the DB. The model optimizer optimizes the learning model based on the new training data, and updates the learning variables preserved in the DB.

4.2 Implementation of a Context-Aware Guide Application: Kagurazaka Explorer

We implement the learning algorithm and the database functionalities of the proposed guide system on a server machine, which is connected to the Internet and reachable from off-the-shelf mobile phones via cellular networks. The mobile terminal used in this implementation is the 3G mobile phones which are generally available in Japan provided by main three operators. They are equipped with the GPS, and can access to the Internet by web browsers. The mobile terminals communicate with the proposed system by HTTP. We have prepared contents information about restaurants, sightseeing spots, and souvenir stores at Kagurazaka street neighboring to Kagurazaka campus, Tokyo University of Science in Tokyo, Japan, so we call this guide system Kagurazaka Explorer. The high-dimensional feature vector $\boldsymbol{x}_{i=(1,2,..,m)} = (x_1, x_2, ..., x_n)$ is composed by n dimensions. These elements are used to characterize the real spots. By using the PCA, the low-dimensional feature space is extracted from high-dimensional feature space. Values in each dimension are normalized between 0 and 1.

Fig. 5 shows the screen shots of the user interface on the mobile terminal connecting to the Kagurazaka Explorer. Fig. 5 (a) shows the start page of Kagurazaka Explorer. From this page, the user starts search of the recommended spots. After getting the position of the user by built-in GPS module, the user is required to input the number of people in his/her group in the screen shown in Fig. 5 (b). Although the system requires manual input here, it leads much improvement on the rate of correct recommendation as shown in Ref. [20]. After sending such context information, the user gets context-aware recommendation (Fig. 5 (c)). By selecting one of the recommended spots, the user can access to more detailed information (Fig. 5 (d)). When further information including the detailed map shown in Fig. 5 (e), showing the screen of the detailed map to navigate the user to the spot, is requested by the user pressing another one click, the selected spot is regarded as a favorite one for the user and the corresponding training data is updated on the server side.

5 Experiments

We evaluate our proposed system in the real environment. We use the experimental data provided by 9 subjects, who use this system in the Kagurazaka street. Between 29 and 184 experimental data are generated by the subjects in various situations (a rainy day, a fine day, various positions, time and group size). Table 1 shows the number of real data collected from each subject. We experiment the proposed system using these data to find effective feature parameters and to evaluate the effectiveness of the PCA for the proposed learning system.

5.1 Experiments for Selecting Effective Feature Parameters

In this subsection, we investigate the performances of the proposed system with different feature parameters to decide effective feature input setting used for the SVM. We compare three definitions of feature space as follows,

(a) Features of Visiting Spots and User Situation including positional relation between the user and the visiting spot,
(b) Features of Visiting Spots and User Situation including closeness between the user and the visiting spots,
(c) Only Features of visiting spots.

Table 2 shows elements of the feature vector in each definition. Each feature vector is composed of visiting spot feature which is static context information, such as user's purposes shown in table 3, cost and history of the target spot, and user situation feature which is real-time context information. As location context in user situation, while the pattern (a) uses position relation (e.g. large sloping roads) between the user and the visiting spots (i.e. using the GPS position of the user and the visiting spot), the pattern (b) uses the distance between the user and the visiting spots. The pattern (c) uses only static features of visiting spots (i.e. using no situational information). We compare these three definitions of feature space in the simple SVM. We evaluate the effectiveness of each feature pattern by the rate of correct recommendation by the 5-fold cross validation of the real data sets. The real data collected from each user is composed of the feature vector and the user's rating (1: satisfaction or -1: unsatisfaction). In the 5-fold cross validation, such real data sets are separated into the 5 data sets, and as the training data, the 4 data sets of them are used to construct the

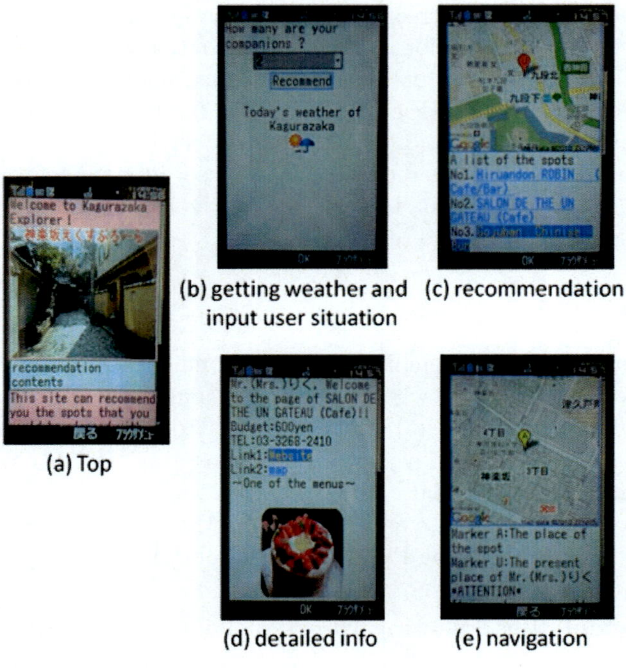

Fig. 5. Screen shots of the implemented system.

Table 1. The number of data collected from each subject.

Subject No.	The number of data
1	118
2	184
3	147
4	56
5	75
6	79
7	66
8	29
9	79

Table 2. Composition of the High-Dimensional Feature Vector in each definition.

	(a) 16 dimensions	(b) 13 dimensions	(c) 8 dimensions
Visiting Spots	User's Purposes(6) Cost(1) History(1)	User's Purposes(6) Cost(1) History(1)	User's Purposes(6) Cost(1) History(1)
User Situation	Time(2) Weather(1) Group Size(1) Positional Relation(4)	Time(2) Weather(1) Group Size(1) Distance(1)	Nothing

user preference model which is tested for the rest data set as the test data. The correct recommendation rate is the number of predicted ratings corresponding to the correct rating of the test data against the number of total test data. Fig. 6 shows the rate of correct recommendation by the 5-fold cross validation of real data set in each feature pattern. The left bars, the central bars, and the right bars for each subject are the pattern of (a), (b) and (c), respectively.

From Fig. 6, we confirm that the patterns of (a) and (b) using situational information have 7% higher performance than that of (c) using only static information, in average. However, in the subjects 1 and 4, the pattern (b) using the distance between the users and the visiting spots has equal or lower performance than the pattern (c). It may because Kagurazaka Street has the distinctive large sloping load, which affects their decisions when they have to go to go up the hill. In order to have the better performance independent on such a relation of locations for every user, we select the pattern (a) using the positional relation parameter as feature parameters in our proposed system.

5.2 Effectiveness of the PCA for the Proposed System

In this subsection, we evaluate the effectiveness of the PCA to improve the performance of the SVM in this system. We use the real data set for the simple SVM and the SVM combined with the PCA. The PCA reduces 16 dimensions of the high-dimensional input data to lower dimensions which are composed of

Table 3. Composition of User's Purposes in Features of Visiting Spots.

User's Purposes (6)	Eating at restaurants(1) Drinking at pubs and bars(1) Relaxing at cafes(1) Taking out foods(1) Buying craft products(1) Enjoying distinctive sceneries(1)

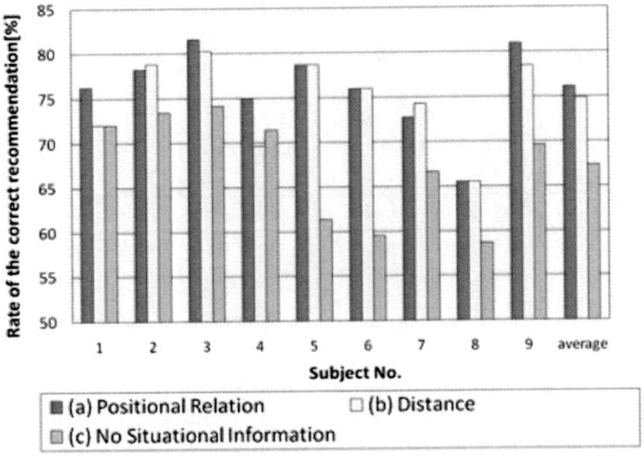

Fig. 6. Comparison of the correct recommendation rate in three feature parameter patterns.

the principal components. By the 5-fold cross validation, we decide the number of the principal components in each subject (table 4). Fig. 7 shows the rate of correct recommendation by the 5-fold cross validation of the real data set, whose feature space is composed of the pattern (a) having the highest performance in the previous experiment. The left bars show the rate of correct recommendation in the simple SVM. The right bars indicate that in the SVM combined with the PCA. In the correct recommendation rate, the SVM combined with the PCA has higher performance than the simple SVM, especially for subjects whose training samples are less than 80 (subject 4, 5, 6, 7, 8 and 9 in Fig. 7). However, in subject 1 and 2, the performance of the SVM combined with the PCA is as same as that of the simple SVM. That is because they have sufficient amount of training data to extract feature from high-dimensional feature space without reducing the dimensions.

From these results, we confirm that in the recommendation using the SVM combined with the PCA is effective for the users have the small number of training data.

Table 4. The number of principal components optimized by 5-fold cross validation.

Subject No.	The number of principal components
1	9
2	11
3	4
4	2
5	4
6	4
7	7
8	2
9	10

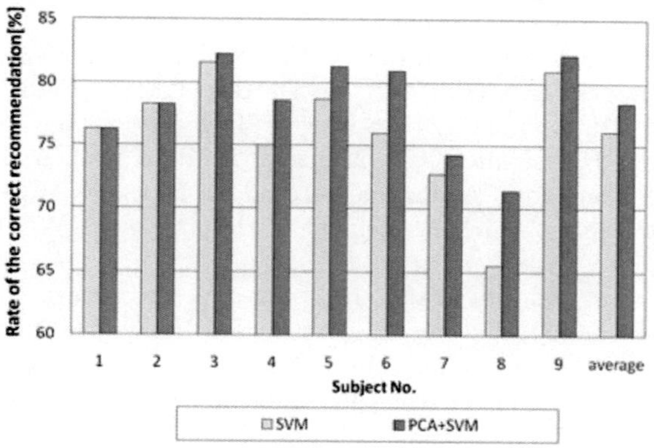

Fig. 7. Comparison of the correct recommendation rate in the simple SVM and the SVM combined with the PCA.

6 Conclusion

In this paper, we proposed a context-aware guide system, which provides appropriate information for each user based on their context. In this system, the SVM is utilized to select the most appropriate information to be notified to each user. We have implemented the real-time context-aware learning system based on the proposed algorithm, using off-the-shelf mobile phone and general 3G cellular networks. Since the conventional context-aware learning algorithms require huge number of training data for the high dimensional feature space, we introduce the PCA to decrease the dimension. By experiment in the real environment, we confirm that our implemented system recommends appropriate information for the mobile users in higher correct recommendation rate, even if users do not have a large amount of training data.

We have evaluated our implemented system in the real shopping street, Kagurazaka street, and its effectiveness could be seen. Therefore, our next target is to perform larger scale experiments with using more contents and more subjects in wider area. Our proposed system may be effective also for larger area information advertisement, because it models user's preference from various context data adaptively and independently on the feature space size.

References

1. Inoue, M., Mahmud, K., Murakami, H., Hasegawa, M., Morikawa, H.: Novel Out-Of-Band Signaling for Seamless Interworking Between Heterogeneous Networks. IEEE Wireless Commun. 11(2), 56–63 (2004)
2. Kawahara, Y., Minami, M., Morikawa, H., Aoyama, T.: Design and Implementation of a Sensor Network Node for Ubiquitous Computing Environment. In: Proc. of Vehicular Technology Conference (2003)
3. Mann, S.: Wearable computing: A first step toward personal imaging. IEEE Computer 30(2) (1997)
4. van Setten, M., Pokraev, S., Koolwaaij, J.: Context-aware recommendations in the mobile tourist application COMPASS. In: De Bra, P.M.E., Nejdl, W. (eds.) AH 2004. LNCS, vol. 3137, pp. 235–244. Springer, Heidelberg (2004)
5. Sanchez, J., Cano, J., Calafate, C.T., Manzoni, P.: Blue Mall: A Bluetooth-based Advertisement System for Commercial Areas. In: Proc. of ACM Workshop on Performance Monitoring and Measurement of Heterogenious Wireless and Wired Networks, pp. 17–22 (2008)
6. Tran, H., Hasegawa, M., Murakami, H., Inoue, M., Morikawa, H., Takayanagi, S., Takahashi, N., Iguchi, K., Ito, H.: Design and Implementation of Bookmark Handover - A Context-aware Reminding Application for Mobile Users. In: Proc. of International Symposium on Wireless Personal Multimedia Communications, pp. 1227–1231 (2006)
7. NTTdocomo, "i-concier" (2010), `http://www.nttdocomo.co.jp/english/service/customize/index.html`
8. Yu, Z., Zhou, X., Zhang, D., Chin, C., Wang, X., Men, J.: Supporting Context-Aware Media Recommendations for Smart Phones. Pervasive Computing 5, 68–75 (2006)
9. Ono, C., Kurokawa, M., Motomura, Y., Asoh, H.: A Context-Aware Movie Preference Model Using a Bayesian Network for Recommendation and Promotion. In: Conati, C., McCoy, K., Paliouras, G. (eds.) UM 2007. LNCS (LNAI), vol. 4511, pp. 247–257. Springer, Heidelberg (2007)
10. Si, H., Kawahara, Y., Morikawa, H., Aoyama, T.: A Stochastic Approach for Creating Context-Aware Services based on Context Histories in Smart Home. In: Proc. of International Conference on Pervasive Computing, pp. 37–41 (2005)
11. Ono, C., Takishima, Y., Motomura, Y., Aso, H., Shinagawa, Y., Imai, M., Anzai, Y.: Context-Aware Users' Preference Models by Integrating Real and Supposed Situation Data. Trans. of IEICE E91-D(11), 2552–2559 (2008)
12. Fukuzawa, Y., Naganuma, T., Onogi, M., Kurakake, S.: Task-ontology Based Preference Estimation for Mobile Recommendation. In: Proc. of CEUR Workshop, vol. 474 (2009)
13. Oku, K., Nakajima, S., Miyazawa, J., Uemura, S.: Context-Aware SVM for Context-Dependent Information Recommendation. In: Proc. of International Conference on Mobile Data Management (2006)
14. Vapnik, V.: The Nature of Statistical Learning Theory. Springer, Heidelberg (1995)

15. Oku, K., Nakajima, S., Miyazaki, J., Uemura, S.: Context-Aware Recommendation System Based on Context-Dependent User Preference Modeling. In: Proc. of Information Processing, vol. 48(SIG11(TOD34)), pp. 162–176 (2007)
16. Adomavicius, G., Sankaranarayanan, R., Sen, S., Tuzhilin, A.: Incorporating Contextual Information in Recommender Systems Using a Multidimensional Approach. ACM Trans. of Information Systems 23(1), 103–145 (2005)
17. Jin, X., Zhang, Y., Yao, D.: Simultaneously prediction of network traffic flow based on PCA-SVR. In: Liu, D., Fei, S., Hou, Z., Zhang, H., Sun, C. (eds.) ISNN 2007. LNCS, vol. 4492, pp. 1022–1031. Springer, Heidelberg (2007)
18. Hsu, C.W., Chang, C.C., Lin, C.J.: A practical guide to support vector classification (2010), http://www.csie.ntu.edu.tw/~cjlin/papers/guide/guide.pdf
19. Iso, T., Kawasaki, N., Kurakake, S.: Personal Context Extractor with Multiple Sensor on a Cell Phone. In: Proc. of International Conference on Mobile and Wireless Communications Networks (2005)
20. Omori, Y., Yamazaki, R., Hasegawa, M.: Design and Implementation of a Context-Aware Recommendation System Based on Machine Learning for Mobile Users. Tech. Rep. of IEICE 109(382), 117–122 (2010)

Chapter 9
Human Motion Retrieval System Based on LMA Features Using Interactive Evolutionary Computation Method

Seiji Okajima, Yuki Wakayama, and Yoshihiro Okada

Graduate School of ISEE, Kyushu University
744, Motooka, Nishi-ku, Fukuoka, 819-0395 Japan
{seiji.okajima,yuki.wakayama,okada}@i.kyushu-u.ac.jp
http://www.isee.kyushu-u.ac.jp/

Abstract. Recently, many motion data have been created because 3D CG animations have become in great demand for movie and video game industries. We need any tools that help us to efficiently retrieve required motions from such a motion data pool. The authors have already proposed a motion retrieval system using Interactive Evolutionary Computation (IEC) based on Genetic Algorithm (GA) and motion features based on Laban Movement Analysis (LMA). In this paper, the authors especially clarify the usefulness of the system by showing experimental results of motion retrievals practically performed by several users. The results indicate that the proposed system is effective for retrieving motion data from a motion database including many motions more than one thousand.

Keywords: Motion Retrieval, Interactive Evolutionary Computation, Genetic Algorithm, Laban Movement Analysis.

1 Introduction

Advances in recent computer hardware technology have made possible 3D rendering images in real time and 3D CG animations have become in great demand for movie and video game industries. Many 3D CG/Animation creation software products have been released so far. However, with the use of such software products, it is still difficult for end-users to create 3D CG animations. For computer animation creation, character design is very important factor but very hard work. Especially, its motion design is very laborious work.

To solve this problem, we have already proposed a motion generation and editing system using Interactive Evolutionary Computation (IEC) [1] based on Genetic Algorithm (GA) [2] that allows us to generate required motions easily and intuitively. However, since the system employs GA for IEC, it needs several existing motion data represented as genes used for the initial generation of GA. The user has to prepare several motion data which are similar to his/her required

T. Watanabe and L.C. Jain (Eds.): Innovations in Intell. Machines – 2, SCI 376, pp. 117–130.
springerlink.com

motions. To prepare such motion data, the easiest way is to retrieve those from a motion database.

Hence, we have been studying motion retrieval systems and already proposed a new motion retrieval system using Interactive Evolutionary Computation [3]. This system allows the user to retrieve motions similar to his/her required motions easily and intuitively only through the evaluation repeatedly performed by scoring satisfaction points to retrieved motions without entering any search queries. The IEC method of the system is based on Genetic Algorithm, so that motion data should be represented as genes practically used as similarity features for the similarity calculation in the system. To extract motion features, we newly defined mathematical expressions of the features using Laban Movement Analysis (LMA) [4]. Because not only the idea of LMA is intuitively understandable for us but also motion features specified in LMA are possible to be represented as mathematical expressions.

In this paper, we describe that the LMA-based motion features are available for the similarity calculation in the system from the results of analyzing them using SOM visualization [3]. Furthermore, we especially clarify the usefulness of the proposed motion retrieval system by showing experimental results of motion retrievals practically performed by several users. The results indicate that the proposed system is effective for retrieving motion data from a motion database including many motions more than one thousand.

The remainder of this paper is organized as follows: First, we introduce the IEC method based on GA and Laban Movement Analysis. Next, we describe related work. And then, a feature extraction method for motion data and gene representation of motions are explained. After that, we explain the detail of our proposed motion retrieval system and present evaluation results to clarify the usefulness of the system. In the last section, we conclude the paper.

2 Interactive Evolutionary Computation and Laban Movement Analysis

In this section, we explain about Interactive Evolutionary Computation (IEC) and Laban Movement Analysis (LMA).

2.1 IEC Method Based on GA

IEC is a general term for methods of evolutionary computation that use human interactive evaluation to obtain optimized solutions [1]. In the IEC method, first of all, a system presents some candidate solutions to the user, and then the user evaluate them by giving a numerical score depending on his/her requirement. After that, the system again presents some solutions more suitable for the user requirement solved by a certain algorithm like GA. After several trials of this operation, the user obtains his/her most desirable solution. In this way, since the IEC method is intuitive and useful to deal with problems depending on human feelings, we decided to employ IEC method based on GA for our motion retrieval system.

2.2 Laban Movement Analysis

LMA is a movement analysis system for the dance which is created by Rudolf Laban. LMA is based on relationships between human body movements and emotions. In LMA, human body movement is explained by features of *Effort* and *Shape* as shown in Table 1. Each feature has two opposite forms which are *Fighting Form* and *Indulging Form. Fighting Form* means a strong, direct(linear) and sudden movement, and *Indulging Form* means a weak, indirect(spiral) and sustained movement.

Effort. *Effort* is a mechanical feature of human movement. *Effort* has three elements which are *Weight*, *Space* and *Time* elements, each of which has two opposite forms. What these elements mean are as follows.

- Weight: Dynamism of body movement, e.g. it can be represented as energy or speed of movement.
- Space: Bias of direction of body movement, e.g. it can be represented as trajectory of movement.
- Time: Temporal Alternation of movement, e.g. it can be represented as the change of acceleration of movement.

Shape. *Shape* is a shape feature of the whole body movement. *Shape* has three elements which are *Table plane*, *Door plane* and *Wheel plane.* Each of them also has two opposite forms. *Shape* feature means spread and movement of body silhouette projected on each of the following three planes.

- Table plane: Spread of body silhouette projected on the transverse plane.
- Door plane: Spread of body silhouette projected on the frontal plane.
- Wheel plane: Movement of body silhouette projected on the sagittal plane.

Table 1. *Effort* and *Shape* elements.

	Weight	Space	Time	Table Plane	Door Plane	Wheel Plane
Fighting Form	Strong	Direct	Sudden	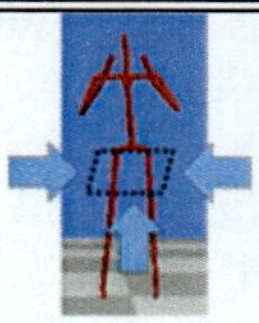 Enclosing	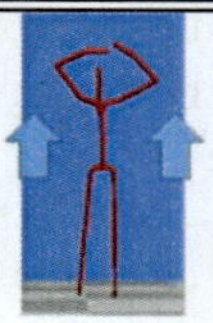 Ascending	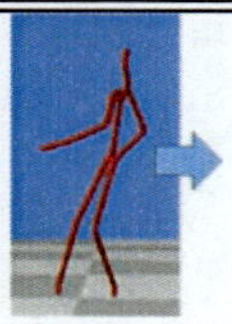 Retreating
Indulging Form	Weak	Indirect	Sustained	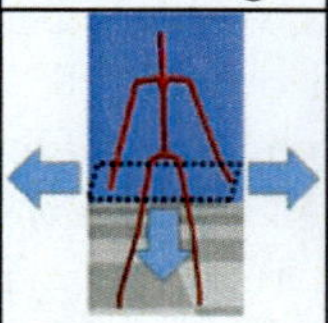 Spreading	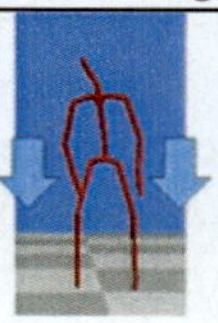 Descending	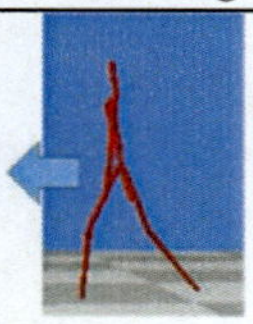 Advancing

3 Related Work

For the motion retrieval, there are some researches. Müller et al. proposed content-based retrieval of motion capture data by using various kinds of qualitative features describing geometric relations [5]. Liu et al. proposed content-based motion retrieval algorithm by partitioning the motion database and constructing a motion index tree based on a hierarchical motion description [6]. These researches are focused on methods of motion indexing or matching. In contrast, our research purpose is to provide a motion retrieval system having an intuitive interface that makes it possible to retrieve motion data interactively and easily.

For the feature extraction method of motions by using LMA, Fangtsou et al. proposed a feature extraction method of motions by using LMA [7]. However, this method does not use *Shape* feature of LMA. Our defined motion features include *Shape* features. Yu et al. proposed a motion retrieval system which allows the user to retrieve motions via Labanotation [8]. This system requests the user to prepare motion data for the queries. Our proposed system does not request any search queries because the system employs IEC method.

IEC is proposed as the interactive calculation method that the user evaluates target data interactively, and finally the system outputs optimized solution based on its evaluated values. The remarkable point where IEC is useful is that the necessitated operation is only the evaluation against data by the user. The data is optimized based on the user's subjective evaluation. So, the system can consider requirements of the user. There are some experimental systems of IEC researches. Ando, et al. proposed a music composition support system for the classical music using IEC [9]. Cho proposed image and music retrieval system using Interactive Genetic Algorithm [10]. Faffi, et al. proposed a design system for Microelectromechanical Systems (MEMS) using IEC [11]. Nishino, et al. proposed an integral 3D-CG contents system based on IEC [12]. By their proposed IEC framework, it is possible to create various attributed 3D-CG contents. Usually, IEC method is based on GA. There is a system [13] that generates some various walk motions using GA. However, there is not any motion data retrieval system using IEC that retrieves and presents motion data according to the user requirement from a motion database. In this paper, we propose such a motion retrieval system using IEC method based on GA.

4 Motion Features Using Laban Movement Analysis

As previously described, we have been developing a motion retrieval system using IEC method based on GA. To use GA, it is necessary to represent motions as their corresponding genes. For that, we newly define motion features as mathematical expressions based on the idea of LMA.

When a human being retrieves a motion, it is thought that the motion is retrieved by focusing on a local part movement such as hands and feet as well as overall movement. Existing LMA-based feature proposed by Fangtsou [7] does not include the information of overall movement. For our LMA feature, *Effort* is focusing on a local part movement of the motion, and *Shape* is focusing on overall movement of the motion.

4.1 LMA-Based Motion Features

To extract body movement features from motion data, we define them as mathematical expressions according to the idea of motion features specified in LMA. In our system, we focus on end-effectors of a human body to extract its features, i.e., its root, left hand, right hand, left foot and right foot.

Feature extraction method for *Effort* is as follows.

1. Weight
   Weight element in LMA represents active emotion derived from the energy and speed of movement. To extract this feature, we focus on speeds of end-effectors in a motion.

   Let F be the number of motion frames and $v_n(f)$ be the speed of an end-effector n in a motion frame f. We calculate Weight feature $Weight_n$ of an end-effector n by the next equation.

$$Weight_n = \sum_{f=1}^{F} |v_n(f)|/F \,. \tag{1}$$

2. Space
   Space element in LMA represents concentrated or unconcentrated emotion derived from the trajectory of movement. To extract this feature, we focus on distributions of speed vectors of end-effectors in a motion and define Space feature value as a norm of a covariance matrix of all speed vectors about each end-effector in a motion.

   Let $V(= [V_1^n V_2^n V_3^n])$ be a speed vector in $\mathbb{R}^3$ and $\mu_i(= E(V_i^n))$ be the mean of V_i^n about an end-effector n. We calculate Space feature $Space_n$ as a norm of a covariance matrix A_n of a speed vector of an end-effector n by the following equations. In the practical calculation, each of V_1^n, V_2^n and V_3^n means a vector about the complete frames in a motion.

$$A_n = \begin{bmatrix} E[(V_1^n - \mu_1^n)(V_1^n - \mu_1^n)] & \cdots & E[(V_1^n - \mu_1^n)(V_3^n - \mu_3^n)] \\ \vdots & \ddots & \vdots \\ E[(V_3^n - \mu_3^n)(V_1^n - \mu_1^n)] & \cdots & E[(V_3^n - \mu_3^n)(V_3^n - \mu_3^n)] \end{bmatrix} . \tag{2}$$

$$Space_n = ||A_n|| = \max_{1 \le j \le 3} \sum_{i=1}^{3} |a_{ij}^n| \,. \tag{3}$$

3. Time
   Time element represents tension emotion derived from sudden or sustained movement. To extract this feature, we calculate the acceleration of a motion. Let F be the number of motion frames and $a_n(f)$ be the acceleration of an end-effector n in a motion frame f. We calculate Time feature $Time_n$ of an end-effector n by the next equation.

$$Time_n = \sum_{f=1}^{F} |\frac{d}{df} a_n(f)|/F \ . \tag{4}$$

As for the feature of *Shape*, we use the mean about all frames of RMS (Root Mean Square) of distances between each end-effector and the root (Center of Mass) of a skeleton in each motion frame.

Let F be the number of motion frames, N be the number of end-effectors and $P(n, f)$ be the coordinate value of an end-effector n in a motion frame F. Then we calculate each Plane feature by the following equations.

$$TablePlane = \frac{1}{F}\sum_{f=1}^{F}\sqrt{\frac{1}{N}\sum_{n=1}^{N}(P_x(n,f) - P_x(root,f))^2} \ , \tag{5}$$

$$DoorPlane = \frac{1}{F}\sum_{f=1}^{F}\sqrt{\frac{1}{N}\sum_{n=1}^{N}(P_y(n,f) - P_y(root,f))^2} \ \ and \tag{6}$$

$$WheelPlane = \frac{1}{F}\sum_{f=1}^{F}\sqrt{\frac{1}{N}\sum_{n=1}^{N}(P_z(n,f) - P_z(root,f))^2} \ . \tag{7}$$

4.2 Gene Representation

We represent motions as their corresponding genes using the LMA-based motion features. As for each of the three types of *Effort* features, we employ the mean of feature values of all end-effectors. Therefore, each chromosome consists of six genes as shown in Fig.1. A chromosome, a gene and an allele are represented as a real vector, a real number and a real value, respectively. For similarity measure of chromosomes, we choose the cosine similarity as a measure of gene similarity. Let x and y be feature vectors and θ is the angle between x and y. Then the cosine similarity sim is defined as

$$sim = \cos\theta = x \cdot y/(|x||y|) \ . \tag{8}$$

In our previous study, as for *Effort* features, we employed the maximum value among corresponding feature values of all end-effectors rather than the mean value of them because in this case we obtain better results of the motion similarity analysis using SOM visualization [3]. However, as described in the next section, we found that users regard the overall movement of a motion rather than its detail as important, so the mean value is better than the maximum value as for *Effort* features.

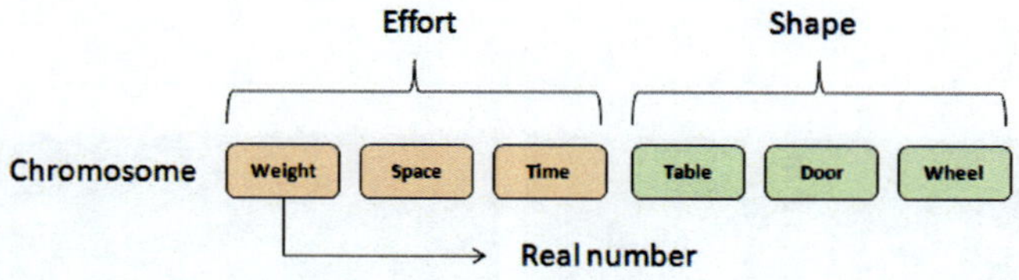

Fig. 1. Gene representation using LMA-based features.

4.3 Visualization and Analysis

To analyze effectiveness of our defined LMA-based motion features for the motion data retrieval, we apply Self-Organizing Maps (SOM) visualization [14] to motion data using their LMA-based features as the feature vectors of SOM. Using SOM layout, similar feature data are located in the same area and it arranges each data in grid, and thus SOM is useful for analyzing similarities among data records of a database. Fig.2 shows SOM layout of our motion database including 296 motions that is a commercial product called "RIKIYA" [15]. Each motion is colored according to its *Effort* and *Shape* features.

The color gradation in Fig.2(a) illustrates that there are positive correlations between *Effort* feature values. Besides, this color gradation indicates emotions expressed in human movements become more active with the color gradient from black at top-right to white at bottom-left. Actually, as shown in Fig.2(c), bottom-left motions become more active compared to top-right motions. By contrast, the color gradation in Fig.2(b) illustrates there are poor correlations between *Shape* feature values. Consequently, motions are divided into similar shape motion groups clearly. For example, motions such as cartwheel, open-arms or something are drawn yellow in Fig.2(d) (upper) which are zoom-in figures of the regions within rectangular lines in Fig.2(b). This means these motions have high TablePlane feature value and DoorPlane feature value. This is intuitively correct. Similarly, motions including mainly walk motions are drawn blue or purple in Fig.2(d) (lower). This means these motions have low TablePlane feature value and WheelPlane feature value. This is also intuitively correct. These observations may clarify that our proposed LMA-based motion features introduced in the previous section are available as similarity features for motion data.

4.4 Genetic Operations

We choose roulette wheel selection algorithm [16] for our system. This selection algorithm calculates probabilities that individuals are selected by GA. We define f_i is a fitness value. The probability p_i of the individual i selected by GA is calculated by

$$p_i = \frac{f_i}{\sum_{k=1}^{N} f_k} . \tag{9}$$

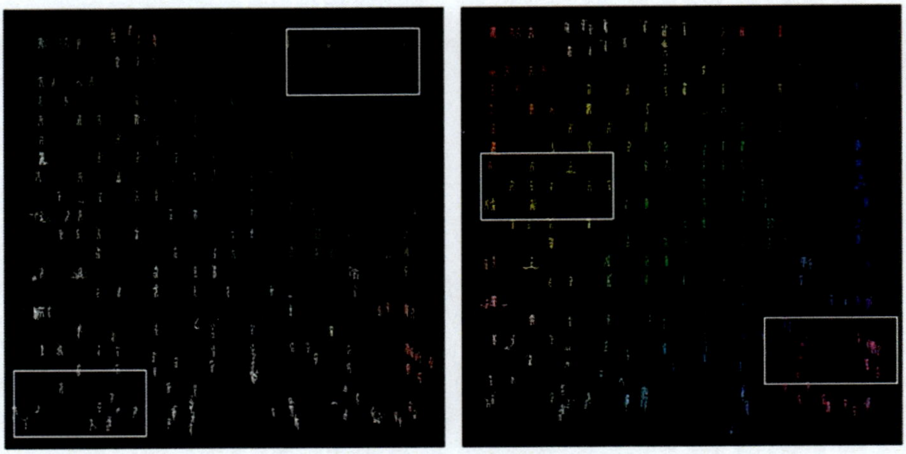

(a) SOM layout of motion database colored by *Effort* feature: Red assigned to the Weight feature value and green assigned to the Space feature value and blue assigned to the Time feature value.

(b) SOM layout of motion database colored by *Shape* feature: Red assigned to the TablePlane feature value and green assigned to the DoorPlane feature value and blue assigned to the WheelPlane feature value.

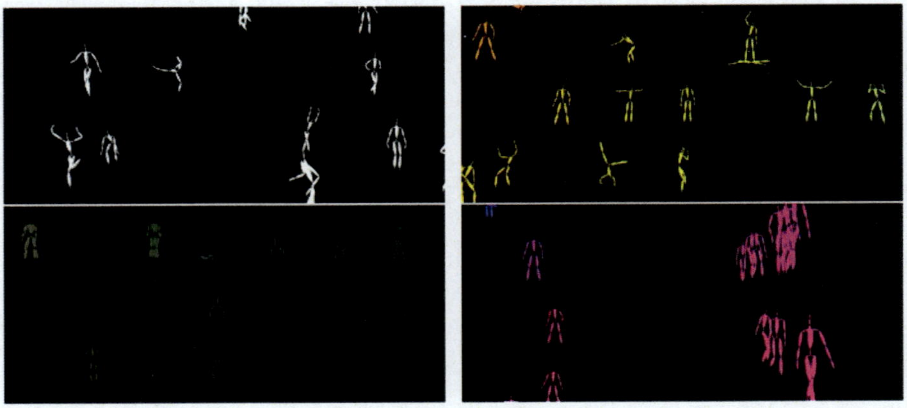

(c) Zoom-in figures of the two regions within rectangular lines in (a).

(d) Zoom-in figures of the two regions within rectangular lines in (b).

Fig. 2. SOM layout of motion database colored by *Effort* (a) and *Shape* (b) feature values. (c) and (d) are zoom-in figures of the regions within rectangular lines in (a) and (b).

In addition, this expression assumes that a fitness value is positive. When the fitness value of an individual is higher, the probability of it becomes higher. If some fitness values are too high rather than others, it causes early convergence which the search settles in the early stages.

There are some crossover operators for real-coded GA such as BLX-α [17] [18], UNDX [19], SPX [20] and so on. In this study, we employ BLX-α because of its simplicity and fast convergence. Let $C_1 = (c_1^1, ..., c_n^1)$ and $C_2 = (c_1^2, ..., c_n^2)$ be parents chromosomes. Then, BLX-α uniformly picks new individuals with a number of the interval $[c_{min} - I \cdot \alpha, c_{max} - I \cdot \alpha]$, where $c_{max} = max(c_i^1, c_i^2)$, $c_{min} = min(c_i^1, c_i^2)$, and $I = c_{max} - c_{min}$.

For a mutation operator, we choose the random mutation operator [18] [21]. Let $C = (c_1, ..., c_i, ..., c_n)$ be a chromosome and $c_i \in [a_i, b_i]$ be a gene to be mutated. Then, c_i' is an uniform number picked from the domain $[a_i, b_i]$.

5 Motion Retrieval System

In this section, we explain our proposed IEC-based motion retrieval system and we also present experimental results of motion retrievals actually performed using the system by several subjects.

5.1 System Overview

There are some typical motion data formats. For example, BVH file format is employed by Biovision Co., Ltd. and ASF-AMC file format is employed by Acclaim Co., Ltd. In our system, we use BVH file format because it is supported by a lot of commercial 3D-CG animation software such as Alias Motion Builder, 3dsMAX Character studio, Poser and so on. This file format consists of two sections: the HIERARCHY section for skeleton information and the MOTION section for motion information. The HIERARCHY section defines an initial pose of a skeleton that includes bone lengths as offset values. The MOTION section defines time series data about sequential poses of a skeleton in a motion.

Fig.3 and Fig.4 show the overview and a screen snapshot of the motion retrieval system, respectively. As the preprocessing, the system creates LMA-based features as a database from the motion database. In this process, index numbers of motions are assigned to each LMA-based feature and the gene is represented as a combination of index numbers. The allele is represented as an index number of a motion. When the user runs the system, it randomly generates genes and retrieves the corresponding twelve motions appeared on a screen. The user evaluates each of these motions by three stage scoring, i.e., good, normal and bad. This evaluation is performed only by mouse clicks on thumbnails of motions. After the evaluation, the system automatically applies GA operations, i.e., selection, crossover and mutation to the genes in order to generate the next generation. And then, the system searches motion data having LMA-based features similar to the features of the newly generated genes to presents them to the user as his/her more desirable motions. After several trials of the evaluation

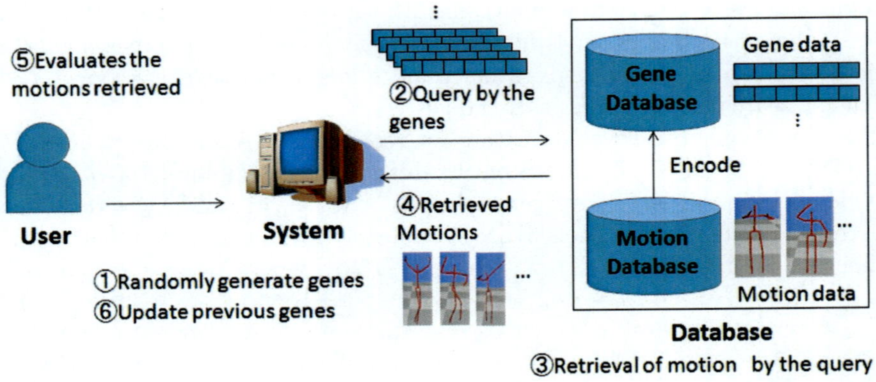

Fig. 3. Overview of motion retrieval system.

process, the user can obtain his/her most desirable motion without any difficult operations.

5.2 Experimental Results

We present experimental results of motion retrievals performed using the proposed system by several subjects. Five students in Graduate School of ISEE, Kyushu University volunteered to participate in the experiment. The experiment is performed on a standard PC with Windows XP Professional, a 2.66 GHz Core 2 Quad processor and 4.0 GB memory.

As a motion database for the experiment, we employed CMU Graphics Lab Motion Capture Database [22]. It contains about 2500 motion data created by recording real human motions using a motion capture system. As for the GA operators, we employed roulette wheel selection operator for the selection, BLX-α crossover operator for the crossover and random mutation operator for the mutation. The value of α is 0.5, crossover rate is 1.0 and mutation rate is 0.01. The fitness values of three stage scoring are 0.8 for good, 0.5 for normal and 0.2 for bad.

For the obtaining the optimum population, we asked the five participants to try to use the system with a different population, i.e., 9, 12 and 16 as shown in Fig.5, and also asked them the question "Which population is preferable for you ?". From the answers to the question, the case of 9 is supported by the two participants, 12 is supported by the three participants and 16 is not supported by any participants. This result means that the case of 16 is obviously too many for the user to scoring them. However, a large number of the population makes it possible to present many motions at once to the user and to reduce a total number of generations. Therefore, we fixed the population is 12. Furthermore, we asked the five participants and found out that users regard the overall movement of a motion as an important factor rather than its detail.

In the experiment for evaluating the usefulness of our proposed system, the participants searched randomly presented target motions using the system. Each

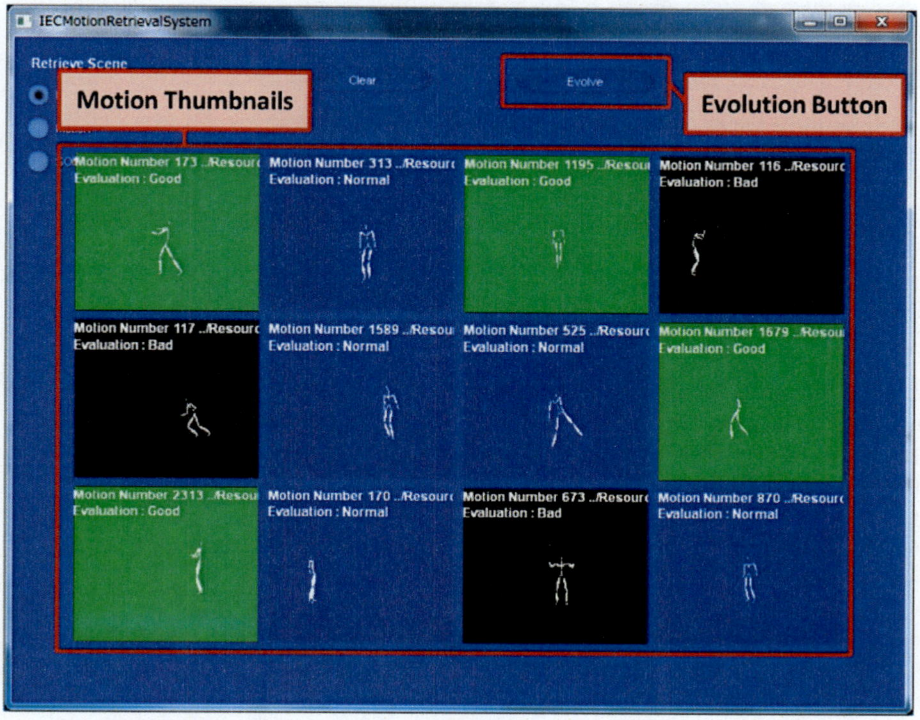

Fig. 4. Screenshot of motion retrieval system.

participant tried to search each of five target motions until 20 generations, and then, we obtained 25 trial results totally. We measured computation and operation times, and we explored retrieved motions. These trials are performed according to the following procedure.

1. Introduction of the motion retrieval system (1 minute).
2. Try to use the system for answering preparation questions (3 minutes).
3. Actual searches for target motions using the system.
4. Answering good points, bad points and comments.

Performance Evaluation. We tried to measure an actual computation time spent for one GA operation and an average user operation time. First, the time spent for one GA operation is less than ten milliseconds and the retrieval time to present next generation is around 1.5 seconds in the case of about 2500 motion data of a database. So, the user manipulates the system without feeling any impatience. Second, the average user operation time until 10, 15 and 20 generations is 6.6 minutes, 9.7 minutes and 12.4 minutes, respectively. As discussed later, it is enough if the user search until around 10 generations or until 15 generations at most. Therefore, it is said that our system allows the user to search his/her desirable motions in a reasonable time.

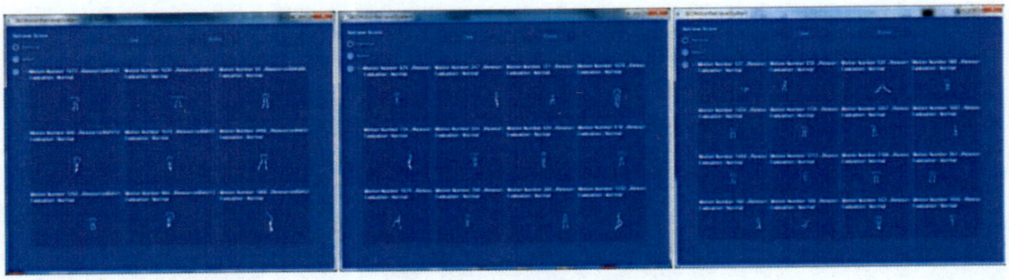

Fig. 5. Screenshots of three motion retrieval systems which have the different population 9 (left), 12 (center) and 16 (right).

Search Results. Next, we explored retrieved motions and classified the results of trials into three types: 1) Retrieval of the same motion as a target motion, 2) Retrieval of the same class motion as a target motion, 3) Retrieval failure. Table. 2 shows the classification of retrieved motion results. Result 1) can be judged from a corresponding file name. Result 2) and 3) are judged from descriptions of CMU Graphics Lab Motion Capture Database and the participants' subjective evaluations.

Table 2. Classification of retrieved motion results.

	Number of Results
1) Retrieval of the same motion as a target motion	4
2) Retrieval of the same class motion as a target motion	17
3) Retrieval failure	4
Sum	25

The motion descriptions of result 3) are *opening a box*, *putting on a skirt*, *story* and *nursery rhyme - "I'm a little teapot..."*. These motions are consisted as the combination of several different motions so the motions are difficult to classify using LMA features and also difficult for users to continue remembering while search operations using the system. These are reasons for the failure of retrieving such target motions.

Fig.6 shows two charts of the average and maximum similarity values to each of the corresponding target motions among motions retrieved as individuals of each generation until 20 generations in the case of result 1). From these charts, it is said that the system appropriately presents various motions according to the user's selection because peaks of the both charts appear before around 10th generation. Therefore, in this case, around 10 generations are enough for users to search his/her desirable motions.

These experimental results indicate that our proposed system is practically useful for retrieving motion data even in the case of a huge database including many motions more than one thousand.

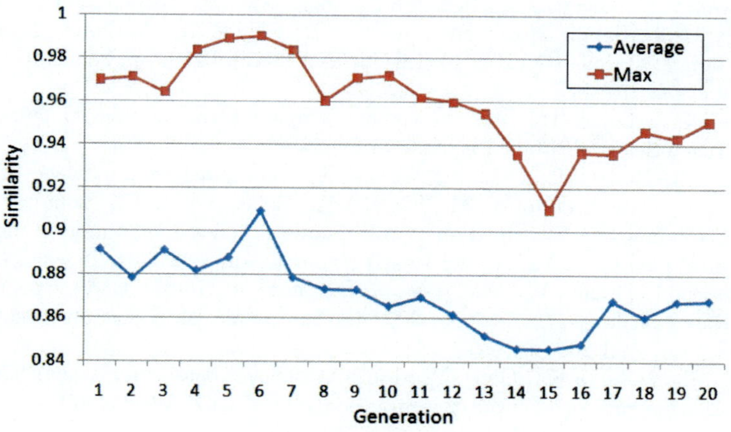

Fig. 6. Average and maximum similarities about result 1).

6 Conclusion and Remarks

In this paper, we introduced the motion retrieval system using IEC based on GA and motion features defined based on LMA which we have already proposed and developed. Our proposed IEC-based motion retrieval system allows the user to retrieve motions similar to his/her required motions easily and intuitively only through the interactive operation to evaluate retrieved motions without any difficult operations. For the motion similarity calculation of the system, we defined LMA-based motion features and clarified that those features are available as similarity features by showing results of analyzing them using SOM visualization. Furthermore, we performed user experiment for evaluating the usefulness of our proposed motion retrieval system. The results indicate that our proposed system is effective for retrieving motion data including many motions more than one thousand.

As future work, there are some improvement points in our system. We will try to find other motion features more available as similarity metrics besides the LMA-based motion features to enhance the motion retrieving accuracy. In addition, we will improve GUI of the system to make it more useful. We also have a plan to provide the proposed system as one of the web services.

References

1. Takagi, H.: Interactive Evolutionary Computation: Fusion of the Capacities of EC Optimization and Human Evaluation. Proc. of the IEEE 89(9), 1275–1296 (2001)
2. Wakayama, Y., Takano, S., Okada, Y., Nishino, H.: Motion Generation System Using Interactive Evolutionary Computation and Signal Processing. In: Proc. of 2009 International Conference on Network-Based Information Systems (NBiS 2009), pp. 492–498. IEEE CS Press, Los Alamitos (2009)

3. Wakayama, Y., Okajima, S., Takano, S., Okada, Y.: IEC-Based Motion Retrieval System Using Laban Movement Analysis. In: Setchi, R., Jordanov, I., Howlett, R.J., Jain, L.C. (eds.) KES 2010. LNCS(LNAI), vol. 6279, pp. 251–260. Springer, Heidelberg (2010)
4. Bartenieff, I., Lewis, D.: Body movement: Coping with the environment. Gordon and Breach Science Publishers, New York (1980)
5. Müller, M., Röder, T., Clausen, M.: Efficient content-based retrieval of motion capture data. In: Proc. of ACM SIGGRAPH 2005, pp. 677–685 (2005)
6. Liu, F., Zhuang, Y., Wu, F., Pan, Y.: 3D motion retrieval with motion index tree. Journal of Computer Vision and Image Understanding 92(2-3), 265–284 (2003)
7. Fangtsou, C., Huang, W.: Analysis and Diagnosis of Human Body Movement Efforts Based on LMA. In: Proc. of 2009 International Conference on Business And Information, BAI 2009 (2009)
8. Yu, T., Shen, X., Li, Q., Geng, W.: Motion retrieval based on movement notation language. Journal of Computer Animation and Virtual Worlds 16(3-4), 273–282 (2005)
9. Ando, D., Dahlstedt, P., Nordahl, M., Iba, H.: Computer Aided Composition for Contemporary Classical Music by means of Interactive GP. Journal of the Society for Art and Science 4(2), 77–87 (2005)
10. Cho, S.B.: Emotional image and musical information retrieval with interactive genetic algorithm. Proc. of the IEEE 92(4), 702–711 (2005)
11. Kamalian, R., Zhang, Y., Takagi, H., Agogino, A.: Reduced human fatigue interactive evolutionary computation for micromachine design. In: Proc. of 2005 International Conference on Machine Learning and Cybernetics, pp. 5666–5671 (2005)
12. Nishino, H., Aoki, K., Takagi, H., Kagawa, T., Utsumiya, K.: A synthesized 3DCG contents generator using IEC framework. In: IEEE International Conference on Systems, Man and Cybernetics, pp. 5719–5724 (2004)
13. Lim, I.S, Thalmann, D.: Pro-actively Interactive Evolution for Computer Animation. In: Proc. of Eurographics Workshop on Animation and Simulation, pp. 45–52 (1999)
14. Kohonen, T.: SELF-ORGANIZING MAPS. Springer, Japan (1996)
15. RIKIYA, `http://www.viewworks.co.jp/rikiya/`
16. Baker, J.E.: Reducing bias and inefficiency in the selection algorithm. In: Proc. of the Second International Conference on Genetic Algorithms on Genetic Algorithms and their Application, pp. 14–21 (1987)
17. Eshelman, L.J., Schaffer, J.D.: Real-Coded Genetic Algorithms and Interval-Schemata, Foundations of Genetic Algorithms 2, pp. 187–202. Morgan Kaufman Publishers, San Mateo (1993)
18. Herrera, F., Lozano, M., Verdegay, J.L.: Tackling Real-Coded Genetic Algorithm: Operators and Tools for Behavioural Analysis. Journal of Artifitial Intelligence Review 12(4), 265–319 (1998)
19. Ono, I., Kobayashi, S.: A real-coded genetic algorithm for function optimization using the unimodal normal distribution crossover. In: Proc. of the Seventh International Conference on Genetic Algorithms, pp. 246–253 (1997)
20. Tsutsui, S., Yamamura, M., Higuchi, T.: Multi-parent Recombination with Simplex Crossover in Real Coded Genetic Algorithm. In: Proc. of the 1999 Genetic and Evolutionary Computation Conference (GECCO 1999), pp. 657–664 (1999)
21. Michalewicz, Z.: Genetic Algorithms + Data Structures = Evolution Program. Springer, Heidelberg (1994)
22. CMU Graphics Lab Motion Capture Database, `http://mocap.cs.cmu.edu/`

Chapter 10
An Exhibit Recommendation System Based on Semantic Networks for Museum

Chihiro Maehara[1], Kotaro Yatsugi[1],
Daewoong Kim[2], and Taketoshi Ushiama[2]

[1] Graduate School of Design, Kyushu University
[2] Faculty of Design, Kyushu University

Abstract. Today, information devices as exhibition guide systems to support visitor's appreciation are introduced in many museums. In this paper, we propose a system for recommending some exhibits suitable for visitor's interests and requirements with an information device. The proposed system supports a user to appreciate an exhibition and arouses interests of the user by recommending exhibits. Exhibits which the system recommends are selected based on the two types of scores: aggregation scores and influence scores. The aggregation score of an exhibit is defined as the sum of the influence scores of exhibits which have influences on it, and the influence score of an exhibit is defined as the sum of the aggregation scores of exhibits influenced by it. Those scores are calculated based on a semantic network on the exhibits in a museum.

Keywords: Personalized recommendation system, Museum, HITS algorithm, Exhibition guide system.

1 Introduction

Recently, there are a variety of digital contents such as web pages, digital photographs, and videos on the Internet, and the number of them is increasing rapidly. To find out suitable contents for a user, some studies on recommendation of digital contents have been reported. Appropriate techniques for recommendation are different depending on the characteristics of target contents. On the other hand, various contents exist not only on the Internet but also in the real world. For example, exhibitions in a museum can be considered as an environment to browse actual contents such as pictures, statues and so on.

Today, many museums use mobile information devices as guide system to support visitors to appreciate an exhibition. For example, the British Museum[1] and the Louvre[2] introduced multimedia guide systems which support some kinds of languages and provide detailed explanations of exhibits, the map of the museum, and themed tours. The Museum of Modern Art[3] also introduced an audio guide system, and we can use the system on our personal devices over Wi-Fi network in museum. "Touch the Museum" in the National Museum of

T. Watanabe and L.C. Jain (Eds.): Innovations in Intell. Machines – 2, SCI 376, pp. 131–141.
springerlink.com

Western Art[4] and "Enosui Navi" in Enoshima Aquarium[5] by Quragelab[6] were developed as the iPod touch application. The systems obtain the location of a visitor over Wi-Fi networks and give the visitor explanation on the exhibits displayed at the location, and show the visitor some images and videos about exhibits which the visitor cannot see usually. Moreover, Louvre - DNP Museum Lab[7] developed a museum guide systems with Augmented Reality (AR) technologies, and The Cite des Sciences et de l'Industrie[8], The National Art Center[9] and Kyoto International Manga Museum[10] introduced social AR application "Sekai Camera"[11] as guide system or communication tool for their visitors.

Such mobile information devices provide visitors detailed explanations of exhibits in which they are interested. Those devices help the visitors to understand exhibitions well.

Conventionally visitors in a museum basically appreciate exhibition along the route composed by curators. Such routes are well organized because they are composed based on domain knowledge. However, there is a problem that it might be inappropriate to show the same route to all visitors because their interest and knowledge on the exhibition are different in each visitor.

In this paper, we propose an approach for recommending some exhibits suitable for visitor's interests and required staying time with a mobile information device. We construct semantic relationship between exhibits to recommend them based on visitor's interests. In addition, to recommend exhibits based on visitor's required staying time, our system change the number of recommended exhibits according to the time. Our goal is to support a visitor to appreciate an exhibition well, and arouse his/her interests by making him/her understand the cultural context surrounding each exhibit.

This paper is composed as followings: Section 2 describes some related works. Section 3 describes the semantic relationship between exhibits. Section 4 explains how to recommend exhibits based on a semantic network. Section 5 shows the development of our prototype system. Section 6 evaluates our system by experiments and discussion on the resuls of the experiments. Section 7 describes conclusion and future works.

2 Related Works

ubiNext[12] is a museum guide system which supports visitor's active learning experience in a museum through Internet services. It recommends some exhibits for the next appreciation to a visitor based on the visitor's interests such as "Gogh" and "the impressionists", the evaluations on the exhibitions by the visitor, and appreciation history of the visitor. Compared with this study, the goal of our study is not to support such visitor's active learning experience.

Koyanagi et al.[13] developed a guide system for visitors of the Hakone Open-Air Museum. It was designed for representing the most recommended path which enables a visitor to appreciate the maximum number of sculptures within his/her given time interval. Our system recommends exhibits based on the structure of semantic relationships between exhibits rather than the layout of them.

Wakiyama et al.[14] proposed a method for information recommendation by implicit construction of user model of preference for paintings based on eye movement. They generate recommendation by detecting the state of being interested with gaze detection; in contrast, we construct semantic relationship between exhibits to recommend.

Moreover, Kadobayashi et al.[15] proposed a method for personalizing the semantic structure of a museum exhibition by mediating curators and visitors. The semantic relations are visualized as a two-dimensional spatial structure based on the viewpoints of the curators and visitors separately.

Abe et al.[16], Abe et al.[17] and Ozaki[18] proposed appreciation supporting systems for digital museums on the Internet using digital archives.

3 Semantic Network on Exhibits

We can find various semantic relationships between exhibits in a museum from viewpoints of such as their shapes, objects drawn in them, their usages and so on. Some conventional systems also use those relationships for recommendations. Typical relationships used in them are about age, region, artist, and technique on the exhibits. These relationships are representative indicators showing characteristic features of the exhibits.

On the other hand, we relate exhibits with cultural influences between them. For example, we suppose that two exhibits A and B are similar in their theme, technique and so on (for example, religious background and production process). If the time when B was produced is after the time when A was produced, we define that B is influenced by A. This cultural influence is represented by the arrow between exhibits (Fig.1). Many of these relationships can be derived from explanation sentences of exhibits containing some characteristic keywords.

Moreover, we utilize relationships between exhibits and conceptual entities. We can find that some exhibits are influenced by the same culture of foreign country, the same religion, and so on (Fig.2). These relationships will also be derived from the explanation sentences about exhibits. Using such relationships, we can relate exhibits in different category such as handicrafts and statues. For a user who is interested in a handicraft, exhibits which are influenced by the same

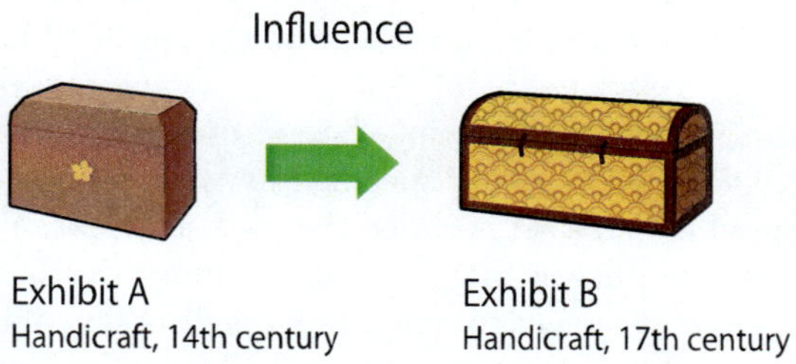

Fig. 1. Cultural influence between exhibits

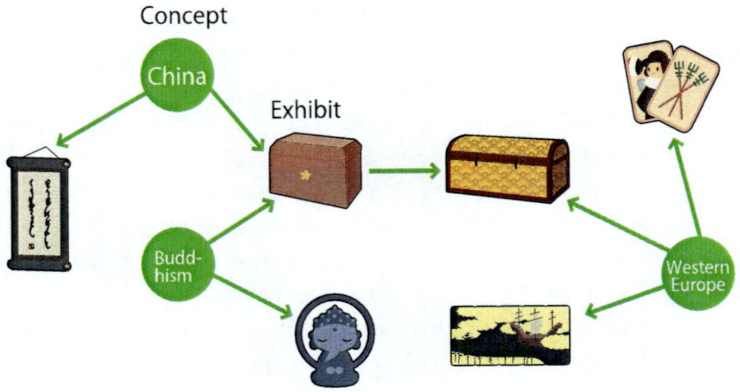

Fig. 2. Influence between exhibits and concepts

culture of the handicraft may arouse the user's interest in another category in which the user has little interest.

By taking account into cultural influences between exhibits, our system can recommend not only the exhibits in categories in which a user is interested, but also other exhibits having similar relationships with them though the user is not much interested in them. Our system enables the user to understand the cultural context surrounding exhibits, and to arouse the user's interest by recommending these exhibits through the interface designed to support the user to understand the relationships between them.

4 Recommendation of Exhibits

This section describes how to recommend exhibits suitable for visitor's interests and required staying time. We use a semantic network described in Section 3 to recommend exhibits based on visitor's interests. We can consider that visitor's situation consists of various factors such as his/her schedule, position, age and so on. In this paper, our system recommends exhibits based on visitor's situation by changing the number of recommended exhibits according to the visitor's required staying time. On the system, the visitor specifies categories and exhibits in which the visitor is interested and the required staying time of the visitor. Recommended exhibits are decided according to the specified information by the visitor.

The recommendation algorithm of our system consists of following steps. First, the set of the exhibits in the category specified by the visitor is constructed from all exhibits in a museum. This set is called the *root set*. Next, a directed graph is constructed from the exhibits in the *root set*. Each node in the graph represents an exhibit or concept, and each edge represents the direction of influence between exhibits or concepts. Additionally, all nodes which have in-edges from any node in the *root set*, and all nodes which have out-edges to any node in the *root set* are added in the set. The directed graph is reconstructed based on these nodes and

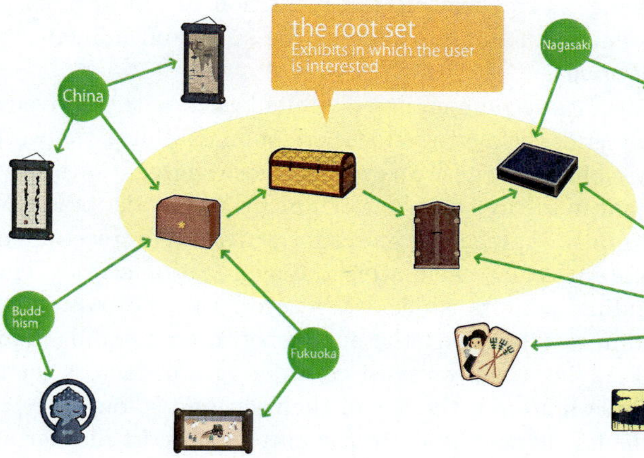

Fig. 3. An example of *base sub-graph*

edges between them. We call this extended directed graph as the *base sub-graph* (Fig.3).

In our method, two types of scores on each node, which represent an exhibit or concept, are used for recommendation: aggregation score and influence score. Both types of scores are derived by the network structure of the *base sub-graph.* An exhibit which is influenced by many exhibits and/or concepts can be considered that its aggregation score is high. An exhibit or concept which has influences on many exhibits can be considered that its influence score is high. This algorithm for deriving the aggregation score and inference score of exhibit was designed based on the HITS algorithm[19]. The HITS is an algorithm for ranking web pages based on the link structure of a target set of web pages[20]. When we regard the relationships between web page and link on the HITS algorithm as the influences between exhibits, we can think that the influence score of an exhibit represents how strongly the exhibit influences other exhibits, and the aggregation score of it represents how strongly the exhibit is influenced by other exhibits whose influence scores are high.

The aggregation score of a node is defined as the sum of the influence scores of the other nodes which influence on the node. The influence score of a node is defined as the sum of the aggregation scores of the other nodes which are influenced by the node. For a node in a *base sub-graph* v, its influence score $Inf(v)$ and its aggregation score $Agg(v)$ are calculated by the following formulas:

$$Inf(v) = \sum_{\omega, v \rightarrow \omega} Agg(\omega) \tag{1}$$

$$Agg(v) = \sum_{\omega, \omega \rightarrow v} Inf(\omega) \tag{2}$$

These scores converge by normalization and repetition of the formula (1) and formula (2). The system calculates the aggregation score and influence score of every node in the *sub-graph*.

Then, the system decides recommending exhibits based on the derived scores. We consider that the higher the aggregation score of an exhibit is the worthier it is to see for the visitor. This is because an exhibit whose aggregation score is high gathers many important influences from the exhibits that the visitor is interested in. Our approach uses only aggregation scores for generating a recommendation and influence scores are used for calculating the aggregation scores. Then, the number of recommending exhibits is decided according to the required staying time of the visitor. In future, to adjust the number of recommending exhibits to the visitor's required staying time, we must consider various factors such as the locations of exhibits in a museum, the age of the user, and so on.

Finally, we recommend these exhibits to the user to support his/her appreciation.

5 Exhibit Recommendation System

5.1 Overview of the System

We assume that our recommendation system would be used in the following steps. First, a visitor borrows a mobile information device at the information desk in a museum, and inputs one or more subjects in which the visitor is interested and their required staying time. Then, the system recommends some exhibits sufficient to the information given by the visitor. Since our system depends on a visitor's interest and his/her required staying time, recommended exhibits might be different for each visitor. For example, if a visitor is interested in a picture, pictures will be mainly recommended to the user. Moreover, to appreciate the whole of exhibition in their required staying time, the number of recommended exhibits will be limited (Fig.4).

The visitor appreciates exhibits referring to the system, and selects some of them that the visitor likes. Then, the system recommends more suitable exhibits for the visitor. The visitor and system repeat this sequence.

5.2 Prototype System

We developed a prototype system to evaluate our method. The system is implemented with the PHP programming language. In the system, the relationships between exhibits and concepts described in Section 3 are registered as a directed graph in advance. The user can select multiple categories and exhibits in which the user is interested, for example pictures and handicrafts. Also the user can specify his/her required staying time. The categories displayed to the user must be properly determined according to the kinds of exhibits in the museum. The result is displayed on the map of the museum (Fig.5).

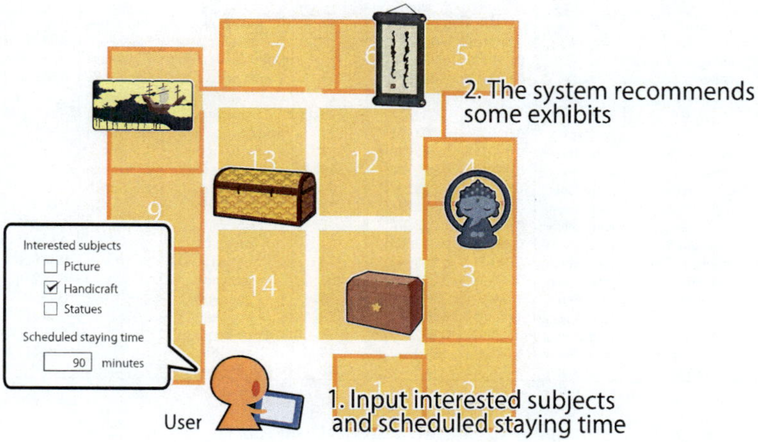

Fig. 4. Snapshot of the recommendation system

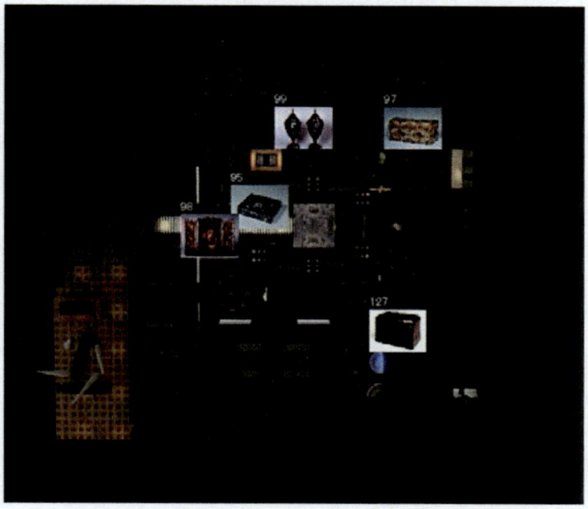

Fig. 5. Recommendation of exhibits

6 Evaluation

6.1 Experiment

We have had an experimental study to evaluate our recommendation method. We used some images and explanation sentences of representative art works in Japanese art history[21] as the exhibits in the museum. In this experiment, we supposed that the user select "Buddhism picture" as the category in which the user is interested. Moreover, we supposed that the appropriate number of the recommended exhibits suitable for the user's situation is 16.

Node num	Category	Age	Inf(v)	Agg(v)
1	craft	Asuka	0.001649	0.001934
2	craft	Asuka	0.001649	0.001934
3	Buddhism picture	Asuka	0.018793	0.004154
4	Buddhism picture	Nara	0.001436	0.006867
5	craft	Nara	0	0.000981
6	craft	Nara	0	0.002259
7	Buddhism picture	Nara	0.010136	0.003621
8	Buddhism picture	early Heian	0.010136	0.025736
9	Buddhism picture	early Heian	0.010136	0.025736
10	Buddhism picture	early Heian	0.010136	0.025736
11	Buddhism picture	early Heian	0.057165	0.025736
12	craft	early Heian	0.015887	0.023674
13	fable picture	late Heian	0.033935	0.043054
14	Buddhism picture	late Heian	0.031113	0.088042
15	picture scroll	late Heian	0.033935	0.039101
16	craft	late Heian	0	0
17	craft	late Heian	0	0.006946
18	Buddhism picture	late Heian	0.033935	0.082733
19	Buddhism picture	late Heian	0.026301	0.090257
20	picture scroll	late Heian	0	0
21	Buddhism picture	Kamakura	0.021211	0.061706
22	Buddhism picture	Kamakura	0.021211	0.054349
23	Buddhism picture	Kamakura	0	0.054349
24	Buddhism picture	Kamakura	0.026301	0.068991
25	Buddhism picture	Kamakura	0	0.068991
26	ink painting	Kamakura	0	0.026355
27	Buddhism picture	Muromachi	0	0.000191
28	ink painting	Muromachi	0	0.000261
29	picture scroll	Muromachi	0	0
30	craft	Muromachi	0	0.000261
31	concept		0.007246	0
32	concept		0.004487	0
33	concept		0.003911	0
34	concept		0.116263	0
35	concept		0.182165	0
36	concept		0.190947	0
37	concept		0.129201	0.001278
38	concept		0.000238	0
39	concept		0.000174	0
40	concept		0	0.039101

Fig. 6. Influence scores and aggregation scores for the test set

Node num	Category	Age	Agg(v)
8	Buddhism picture	early Heian	0.025736
9	Buddhism picture	early Heian	0.025736
10	Buddhism picture	early Heian	0.025736
13	fable picture	late Heian	0.043054
14	Buddhism picture	late Heian	0.088042
15	picture scroll	late Heian	0.039101
16	craft	late Heian	0.042355
18	Buddhism picture	late Heian	0.082733
19	Buddhism picture	late Heian	0.090257
20	picture scroll	late Heian	0.083311
21	Buddhism picture	Kamakura	0.061706
22	Buddhism picture	Kamakura	0.054349
23	Buddhism picture	Kamakura	0.054349
24	Buddhism picture	Kamakura	0.068991
25	Buddhism picture	Kamakura	0.068991
26	ink painting	Kamakura	0.026355

Fig. 7. Recommendation of 16 exhibits

First, 16 Buddhism pictures were selected as the *root set*. Next, the *base sub-graph* was constructed from 14 exhibits that have close relationships between them and 10 concepts taken from explanation sentences of these exhibits. Then, on this *sub-graph*, the influence scores and aggregation scores were calculated by our prototype system. The result is shown in Figure 6. On the basis of this result, 16 exhibits were recommended by extracting exhibits having a high aggregation score (Fig.7).

6.2 Discussion

This section discusses our technique with the results of the experiment. Figure 8 shows a part of the *base sub-graph* constructed in this experiment.

As shown in Figure 8, the exhibits which have high aggregation scores were influenced by the concepts which have high influence scores. Buddhism pictures having high aggregation scores are the nodes 19, 14, and 18. These pictures were influenced by the concepts having high influence scores, the nodes 36 and 35.

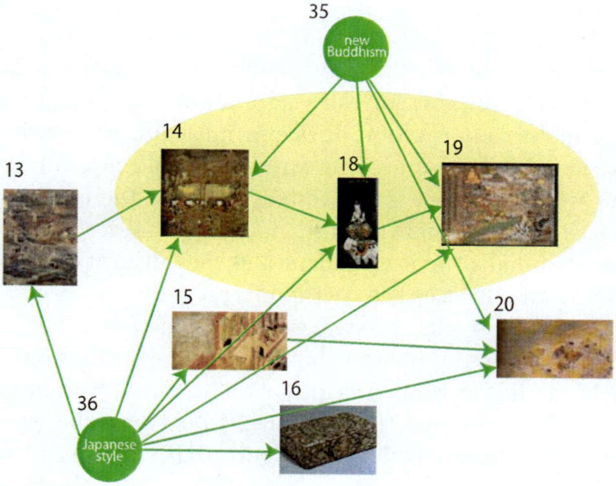

Fig. 8. A part of the *base sub-graph*

The node 36 represents the change of culture from Chinese style to Japanese style in late of the *Heian* era. The node 35 represents the appearance of new kind of Buddhism in late of the *Heian* era. Japanese Buddhism pictures were influenced by these concepts in drawing technique.

In this experiment, Buddhism pictures had high aggregation scores because they were selected as the *root set*. Similarly, some exhibits which were influenced by the same concepts as Buddhism pictures came to have high aggregation scores. The node 20 is a picture which is not drawn as a religious picture, but it was drawn based on a tale of Buddhism. For that reason, it was influenced by the nodes 36 and 35, and its aggregation score is high. Moreover, the nodes 13, 16, and 15 are also influenced by the node 36, and have changed in drawing technique as well as Buddhism pictures. The result of this study showed that by considering cultural influence between exhibits, not only the exhibits in the category that the user is interested in but also the exhibits having close relationships with them though the user has little interest can be recommended.

Secondly, we discuss the selection of exhibits for generating the recommendation. In this experiment, since the *root set* is constructed from one category "Buddhism picture", recommended exhibits are limited to the exhibits produced in late of the *Heian* era to the *Kamakura* era. This is because many Buddhism pictures were produced in those eras. Since Buddhism pictures were evolved in various ways and a lot of them exist today, they have high influence scores and aggregation scores. However, it can be thought that some user want to appreciate various exhibits, not limited to the exhibits produced in particular age. As a solution of such requirement, it is necessary to develop a system which interactively recommends more suitable exhibits for a user according to the selection of exhibits of the user is necessary.

7 Conclusion

In this paper, we proposed an exhibit recommendation system that recommends a visitor some exhibits suitable for the visitor's interests and the required staying time with a mobile information device in museum. By considering cultural influence between exhibits, our system can recommend not only the exhibits in the category in which the user is interested, but also the exhibits having close relationships with them though the user has little interest on them. As a result, our system will make the user understand the cultural context surrounding individual exhibit, and arouses the user's interest.

We have some future works. Firstly, we have a plan of developing a method that automatically derive the relationships between exhibits from their explanation sentences. Secondly, we think it is important to develop an interface designed to support a user to understand the relationships between the recommended exhibits, and a system which calculates an appropriate number of recommending exhibits according to the required staying time of the user. In addition, we have a plan to evaluate the effectiveness of our system by user studies in a museum.

References

1. The British Museum, `http://www.britishmuseum.org/`
2. The Louvre, `http://www.louvre.fr/llv/commun/home.jsp?bmLocale=en`
3. The Museum of Modern Art, `http://www.moma.org/`
4. The National Museum of Western Art, `http://www.nmwa.go.jp/jp/index.html`
5. Enoshima Aquarium, `http://www.enosui.com/`
6. Quragelab, `http://quragelab.jp/`
7. Louvre - DNP Museum Lab, `http://www.museumlab.jp/index.html`
8. The Cite des Sciences et de l'Industrie, `http://www.cite-sciences.fr/en/cite-des-sciences/`
9. The National Art Center, `http://www.nact.jp/`
10. Kyoto International Manga Museum, `http://www.kyotomm.jp/`
11. Sekai Camera, `http://sekaicamera.com/`
12. Masuoka, A., Fukaya, T., Takahashi, T., Takahashi, M., Ito, S.: ubiNEXT: A New Approach to Support Visitor's Learning Experience in Museums. In: HCI International (2005)
13. Koyanagi, F., Kon, T., Higashiyama, A.: The Recommended Path Indication System in Hakone Open-Air Art Museum with Time Designation. The Journal of the Faculty of Science and Technology Seikei University 43(2), 1–8 (2006)
14. Wakiyama, K., Yoshitaka, A., Hirashima, T.: Acquisition of User Interest Model for Painting Recommendation Based on Gaze Detection. Transactions of Information Processing Society of Japan 48(3), 1048–1057 (2007)
15. Kadobayashi, R., Nishimoto, K., Sumi, Y., Mase, K.: Personalizing Semantic Structure of Museum Exhibitions by Mediating between Curators and Visitors. Transactions of Information Processing Society of Japan 40(3), 980–989 (1999)
16. Abe, M., Hada, H., Imai, M., Sunahara, H.: A Proposal of the Automatic Exhibition Scenario Creation System in Digital Museum. ITE Technical Report 26(24), 13–18 (2002)

17. Abe, N., Mitsuishi, T.: Development of Learning System Using Digital Archives from Museums for Spontaneous Learning over Cross-Disciplinary Fields. IPSJ SIG Notes 42, 95–101 (2005)
18. Ozaki, K.: Virtual Art Museum for Art Appreciation Support. Annual bulletin of the Research Institute of Management and Information Science, Shikoku University 9, 23–30 (2003)
19. Baldi, P., Frasconi, P., Padhraic, S.M.: Modeling the Internet and the Web: Probabilistic Methods and Algorithms. Morikita Publishing Co., Ltd (2007)
20. Kleinberg, J.: Authoritative sources in a hyperlinked environment. In: Proc. 9th Ann. ACM-SIAM Symp. on Discrete Algorithms, pp. 668–677. ACM Press, New York (1998); A preliminary version of this paper appeared as IBM Research Report RJ 10076 (May 1997)
21. Tsuji, N.: The Concise History of Japanese Art. Bijutsu Shuppan-Sha Co., Ltd (2003)
22. Bukkyou kaiga - Wikipedia: `http://ja.wikipedia.org/wiki/E4BB8FE69599E7B5B5E794BB`

Chapter 11
Presentation Based Meta-learning Environment by Facilitating Thinking between Lines: A Model Based Approach

Kazuhisa Seta[1] and Mitsuru Ikeda[2]

[1] Faculty of Science, Osaka Prefecture University, Japan
[2] School of Knowledge Science, JAIST, Japan

Abstract. It is difficult to generalize and accumulate experiences of system development as methodologies for building meta-learning support systems because the meaning of "meta-cognition" is vague. Therefore, the importance of a model based system development approach has been recognized. It contributes to systematic refinement of each learning system by iterating a loop that building a model that can clarify design rationale of the system, developing and evaluating each learning system according to the model, and revising the model based on it. Moreover we can accumulate knowledge on meta-learning system development based on it. Thus, we adopt a model-based approach: (i) we extend Kayashima's computational model as a basis to build a meta-learning task model that clarifies factors of difficulties in performing meta-cognitive activities for learning processes, (ii) we specify design concepts for meta-learning scheme as a means to remove/ eliminate the factors of difficulties; then (iii) we embed support functions to facilitate meta-learning processes based on the model. This constitutes a promising approach not only for building learning support systems but also for building human-centric systems in general. In this paper, we firstly describe the philosophy of our research to elucidate our model-oriented approach. Secondly, we present a meta-learning process model as a basis for understanding meta-learning tasks and what factors of difficulty exist in performing meta-learning activities. Thirdly, we explain our conceptualizations as a basis to design sophisticated meta-learning scheme to prompt learners' meta-learning processes. Fourthly, we integrate a meta-learning process model and conceptualizations so that we design our meta-learning scheme based on the deep understanding of meta-learning processes. Fifthly, we present our presentation-based meta-learning scheme designed based on the model and clarify the design rationale of our system based on the model. Then, we present experimental results that suggest that users tightened their criteria to evaluate their own learning processes and understanding states. Furthermore, our support system deepens learners understanding states by prompting their thinking between lines. Finally, we describe the usefulness of the model by characterizing other meta-cognition support schemes.

Keywords: model-oriented system development, meta-learning support system, meta-learning, presentation-based learning.

T. Watanabe and L.C. Jain (Eds.): Innovations in Intell. Machines – 2, SCI 376, pp. 143–166.
springerlink.com

1 Introduction

Our research is designed to produce a meta-learning support system that facilitates learners' learning skill development through reflecting their own learning processes. We designate "learning of learning activities" as meta-learning. Providing meta-cognitively aware instruction is well-known to facilitate meta-learning processes [1]. In learning history, for instance, a student might reflect: "who wrote this document, and how does that affect the interpretation of events?" A physics student might monitor her understanding of the underlying physical principle at work. In learning software development methods, not only memorization of how to depict each diagram in UML and but also considering advantages of the object-oriented system development also prompt internal self-conversation processes. These students strive for reusability and functional extendibility of a designed concrete class structure, which is important to deepen a learner's own understanding. Meta-cognitively aware instruction provides learners domain-specific adequate inquiries from the teacher to deepen their understanding, in other words, it prompts learners to think between lines. It also facilitates their acquisition of domain-specific learning strategies.

Our presentation-based meta-learning support system realized a guidance function that provides meta-cognitively aware instruction to make learners to think between lines [2]. This function is based on knowledge in the educational psychology field. Results of experimental studies suggest that the system can facilitate learners' meta-learning processes: it tightens their criteria to evaluate their learning processes and learning outcomes. It also enhances meta-cognitively aware learning communications among learners in collaborative learning [3, 4]. As a result, participants in experimental group using the system marked higher average score than the ones in control group without our system. Nevertheless, these unique results are insufficient from the viewpoint of accumulating sharable knowledge to develop meta-learning support systems: we should clarify intention to eliminate factors of difficulty related to each embedded function and how each function is generalized.

Sharing and accumulating knowledge is difficult because the meaning of "meta-cognition" [5, 6] is vague. For that reason, the contents of meta-cognitive activities cannot be identified clearly. The contents of "meta-cognition support" implemented in learning systems indicate different kinds of support without explicit analysis or descriptions [7]. This problem also affects the evaluation of meta-cognition support systems: evaluating how each embedded function of the system eliminates obstacles to performing meta-cognitive activities is difficult. Consequently, we cannot evaluate the usefulness of each function in detail, but we can emphasize the system effectiveness by performing exams. Generalizing experiences of system development as methodologies for building meta-learning support systems is also difficult [7, 8]. Some framework to reduce the problem is necessary.

Therefore, the importance of a model oriented system development approach has been recognized. It contributes to systematic refinement of each learning system by iterating a loop of a model that clarifies the system's design rationale, by developing and evaluating each learning system according to the model, and by revising the model based on it. Moreover, the model can accumulate knowledge related to meta-learning system development [9].

Kayashima's model [7] is a sophisticated framework that is useful to clarify factors of difficulties in performing meta-cognitive activities for *problem-solving*. We can refer to it as a basis of system development.

We adopt a model-oriented approach: (i) we extend Kayashima's computational model as a basis to build a meta-learning task model that clarifies factors of difficulties in performing meta-cognitive activities for *learning processes*, (ii) we specify design concepts for meta-learning scheme as a means to remove/ eliminate the factors of difficulties; then (iii) we embed support functions to facilitate meta-learning processes based on the model. This constitutes a promising approach not only for building learning support systems but also for building human-centric systems in general.

As described herein, we present a theoretical foundation to our meta-learning support system for accumulating the knowledge necessary for building meta-cognition support systems. Then, we clarify design rationale of our system based on it. Then, experimental issues and details of concrete functions embedded into the system are explained [2-4, 8, 10].

This paper is organized as follows. Section 2 describes the philosophy of our research to elucidate our model-oriented approach. Section 3 presents a meta-learning process model as a basis for understanding meta-learning tasks and what factors of difficulty exist in performing meta-learning activities. Section 4 explains our conceptualizations as a basis to design sophisticated meta-learning scheme to prompt learners' meta-learning processes. Section 5 integrates a meta-learning process model and conceptualizations so that we design our meta-learning scheme based on the deep understanding of meta-learning processes. Then, we present our presentation-based meta-learning scheme and concrete functions embedded into the system designed based on the model and clarify the design rationale of our system based on the model. Section 6 presents experimental results to show usefulness of our system. In Section 7, we describe the usefulness of the model by characterizing other meta-cognition support schemes.

2 Underlying Philosophy

A learning support system and a learner compose an interaction loop: The system gives stimulations according to the learner's behaviors. Then, prompted by them, learners elicit their own intellectual activities from themselves. Therefore, the learner must be recognized as part of the system to achieve our goal of meta-learning support.

On the other hand, we are unable to consider the learner systematically because human cognitive activities are vague, latent, and context-dependent. A system developer of a human-centric system must design a sophisticated interaction loop between the system and the learner. They tend to design support functions that are apparently *subjectively valid* based on individuals' experiences without making design rationale explicit. Subsequently, they investigate the validity by performing exams. Consequently, the relations among theories clarifying characteristics of cognitive activities in human mind and support functions tend to be weak. For that reason, experiences in developing a system cannot be used or shared well.

A system developer who intends to develop a learning support system must design an adequate interaction loop to encourage learners' intellectual activities according to a theoretical foundation.

In our model-oriented approach, we build a meta-learning process model as a reference model to understand which difficulties we intend to eliminate by adopting and extending Kayashima's computational model, which is specified based on knowledge of the cognitive psychology field. It focuses on factors in a human's head. Furthermore, we specify design concepts for building a meta-learning scheme at the specific system-independent level by referring to experimental knowledge in the cognitive psychology field. They play a guiding role for system developers to embed support functions into meta-learning support systems. Then, we integrate them as a foundation to design our meta-learning scheme. Consequently, system developers can develop meta-learning support systems by realizing support concepts in correspondence with factors of difficulties in performing meta-cognitive activities.

One important difference between ordinary approaches and our approach is that we can clarify design rationale of each support function implemented into the system. This is significant to accumulate and share knowledge to design a meta-cognition support system.

3 Building a Meta-learning Process Model

3.1 Structure of Meta-learning Tasks

Figure 1 presents cognitive activities in performing problem-solving processes (left side) and those in performing learning processes (right side). A problem-solver in the left side performs cognitive activities: the problem-solver reads, understands a given problem, and solves it. At this time, the learner also performs cognitive activities that monitor, re-plan, and control them. These are meta-cognitive activities because they are cognitive activities managing cognitive activities.

Kayashima et al. present a framework by which we can understand factors of difficulties in performing meta-cognitive activities in performing problem-solving processes. They clarify factors of difficulties based on cognitive psychology knowledge, e.g., segmentation of process, invisibility, simultaneous processing with other activities, simultaneous processing with rehearsal, a two-layer working memory, etc. (see [7] for details).

On the other hand, performing meta-cognitive activities in performing *learning processes* (planning and control of learning processes) is more difficult than problem-solving because problem-solving activities and their results in the outside world are visible, whereas those of learning activities and their results (learner's understanding states) are invisible in one's own head. Consequently, the learners do not tend to be aware of the necessity of monitoring and controlling their learning processes. They do not tend to perform meta-cognitive activities spontaneously.

Furthermore, planning learning activities places heavier cognitive loads on learners because they require monitoring activities of their own invisible understanding states and learning processes. It is difficult for ordinary learners to perform even if they try.

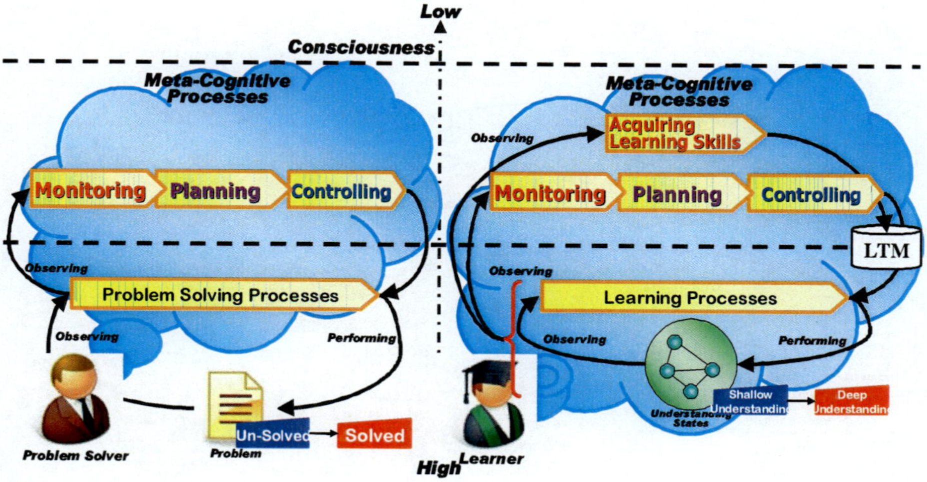

Fig. 1. Structure of Meta-Cognitive Task in Performing Problem-solving Processes (left side) and Learning Processes (right side)

Moreover, acquiring know-how of learning processes (at the upper layer in Fig. 1 (right)) by reflecting learning activities performed is a more latent activity in the meaning that they tend to be unconscious of the necessity of them, whereas ordinary meta-cognitive activities of monitoring and re-planning are prompted if cognitive conflicts occur.

3.2 Meta-learning Process Model

We provide a more detailed model of meta-learning activities in Fig. 2, which depicts a meta-learning process model by extending Kayashima's computational model. It is classified as three layers. At the lowest layer in the figure, i.e., schema level, it represents *"real status"* of a learner's understanding state by performing learning activities. It corresponds to real status at the outside world in case of performing problem-solving activities. We are noncommittal about the boundary between schema and long term memory since there are some opinions and not important for our discussion in this paper.

Upper two layers capture meta-learning processes in a learner's mind (working memory). Changing processes of the learner's understanding state by monitoring own schema are situated at the lower layer in WM. Separate representation of schema level and lower layer of WM makes it possible to represent differences between "leaner's *real* state of his/~ her understanding" and "learner's *belief* on his/~ her own understanding states." Learner's belief on her own understanding states is not always produced/~ modeled by monitoring the real states of them. It is important to characterize meta-cognition in learning because meta-cognitive activity is prompted by the awareness on the gap between the real state and the belief.

Planning of learning processes are represented at the upper layer in WM. Processes of reflecting activities for acquiring learning skills (acquiring domain-specific learning

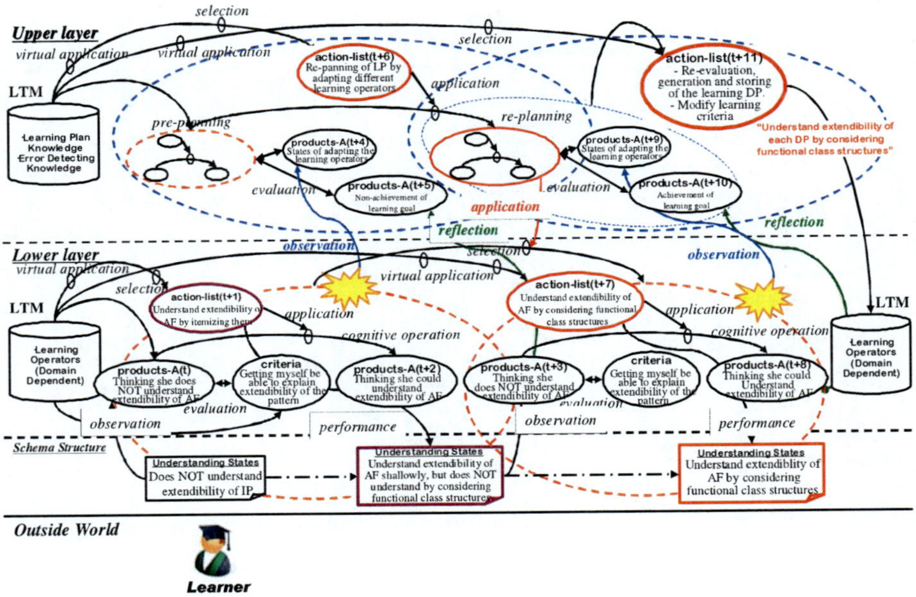

Fig. 2. Meta-Learning Process Model

operators, and modifying criteria to evaluate individual understanding states) are represented at the upper layer. Each ellipse shows a product made at each layer, and "t*" represents time: the "t*" order represents product changes. We separate long term memory (LTM) into the one at the lower layer and the one at the upper layer in WM to characterize knowledge used at each layer.

The model captures the structure of cognitive activities in performing meta-learning task in a domain-independent manner. However, it is presented in learning software design patterns for ease of understanding.

We overview it to elucidate the subjects of this paper. The original model [7] is examined for detailed meaning of operators such as application, selection, and evaluation, which appear in the model. We here summarize them by quoting from [7]. *Observation* is watching something carefully and creating products in WM. *Rehearsal* is a critical task for maintaining contents in WM. *Evaluation* is assessing the state of WM and its subtask is *comparison*. *Virtual application* is applying retrieved operators virtually. *Selection* is choosing appropriate operators among them based on the virtual application results and generating an action-list to WM.

The learner had planned to understand the features of functional extendibility of the Abstract Factory (AF) Pattern in the software design pattern (action-list (t+1)) at the lower layer in WM. Thereafter, she performed learning activities that caused changes of her understanding state and she thought she could understand the extensibility of AF (products-A(t+2)), while she could not understand it actually. She then realized her own lack of understanding (products-A(t+3)) (gap of real states at the schema level and her belief) by performing on-going monitoring or reflective-monitoring (that will be explained in Sect. 4) and created a product (products-A(t+4)) by on-going monitoring or (products-A(t+5)) by reflective-monitoring, respectively, and decided to re-plan her learning processes by adopting different learning operators (action-list (t+6)) at the upper layer. Eventually, she chose other learning operators and generated a new learning plan at the lower layer (action-list (t+7)).

She can understand the topic more deeply if these meta-cognitive activities (learning process planning) are adequately performed, perhaps by making a learning plan to understand functional extendibility of the AF patterns by considering correspondence between the feature of functional extendibility and the concrete class structures.

Learning-skill acquisition processes at the upper layer (action-list(t+11)) require the following cognitive activities: (i) reflecting upon and observing the learning processes at the lower level, (ii) detecting meaningful domain-specific learning operators that had deepened understanding states (e.g. detecting learning operators that show functional extendibility of the AF patterns by considering correspondence between the feature of functional extendibility and concrete class structures in the above case), (iii) re-evaluating, generalizing, and storing them in long-term memory, and (iv) modifying criteria based on them.

It is meaningful that our model captures (i) the difference between "learner's belief and real states of her own understanding" by separating representation of schema level and lower layer of WM, and (ii) learning-skill acquisition activities explicitly by extending Kayashima's model. In contrast, the original model does not capture it because it is built for modeling meta-cognitive activities in problem-solving processes.

3.3 Factors of Difficulty in Performing Meta-learning Activities

Table 1 represents factors of difficulty in performing meta-cognitive activities for learning processes. It is extended based on Kayashima's framework. They clarified primitives to represent factors of difficulty in performing meta-cognitive processes for problem-solving. Based on their framework [7], we add two primitives to represent factors of difficulty for meta-learning. i.e., acquisition of learning operator and acquisition of learning criteria.

By observing objects in the schema, corresponding representation is created in WM: "Observing product of one's understanding state." Because one's own objects in schema are invisible (d2), observing product of one's understanding state becomes incomplete.

Activities from (e)-(g) are performed in WM at the lower layer, while ones from (h)-(k) are performed in WM at the upper layer. We briefly explain factors of difficulty especially in performing meta-cognitive activities in learning ((h)-(k)).

Regarding (h) and (i), they represent factors of difficulty in performing reflective monitoring and on-going monitoring (will be explained in Sect. 4), respectively. In performing reflective monitoring (h), that observes resulting object of one's own learning processes, the difficulty of (d2) invisibility, (d4) inference of cognitive operation, (d5) simultaneous processing with rehearsal and (d7) a two-layer WM exist. In contrast, in performing on-going monitoring (i), that observes one's own learning processes, the difficulty of (d1) segmentation of process, (d2) invisibility, (d3) simultaneous processing with other cognitive activities, (d5) simultaneous processing with rehearsal, (d6) management of resource, (d7) a two-layer WM and (d8) multiple processing exist.

Table 1. Difficulty of performing meta-learning processes

Target	Cognitive Activity	Rehearsal	Observation	Evaluation	Virtual Application	Selection
Schema	Understanding state at own schema (a)		(d2)			
Outside World	Resulting object of others' learning processes (b)		(d2)(d3)			
	Others' learning processes (c)		(d1)(d2)(d3)(d4)			
LTM	Object in LTM (d)	(d5)	(d5)			
Products in WM	Observing product of one's own understanding state (e)	(d5)		(d5)	(d5)	(d5)
	Observing product of resulting object of others' learning processes (f)	(d5)		(d5)	(d5)	(d5)
	Observing product of others' cognitive operation process (g)	(d5)(d6)		(d5)(d6)	(d5)(d6)	(d5)(d6)
	Resulting object of one's own learning processes (h)	(d5)	(d2)(d4)(d5)(d7)			
	One's own learning processes (i)	(d5)(d6)	(d1)(d2)(d3)(d5)(d6)(d7)(d8)			
	Observing product of resulting object (understanding state) of one's own learning (j)	(d5)(d6)		(d5)(d6)		(d5)(d6)
	Observing product of one's own learning processes (k)	(d5)(d6)		(d3)(d5)(d6)(d8)(d9)(d10)(d11)	(d3)(d5)(d6)(d8)(d9)	(d3)(d5)(d6)(d8)

(d1) Segmentation of process
(d2) Invisibility
(d3) Simultaneous processing with other activities
(d4) Inference of cognitive operation
(d5) Simultaneous processing with rehearsal
(d6) Management of resource
(d7) A two-layer WM
(d8) Multiple processing
(d9) planning
(d10) Acquisition of learning operators
(d11) Acquisition of criteria for learning processes

Moreover, regarding (j) and (k), they represent factors of difficulty in performing learning skill acquisition processes: the difficulty of (d10) acquisition of learning operators and (d11) acquisition of criteria for learning processes exist, in addition to the factors of difficulty in performing meta-cognitive activities in problem-solving processes, i.e., (d3) simultaneous processing with their activities, (d5) simultaneous processing with rehearsal, (d6) management of resource, (d8) multiple processing and (d9) planning.

4 Design Concepts for Meta-learning Support Scheme

The meta-learning process model clarifies factors of difficulties in performing learning activities, whereas the conceptualizations described below clarify design concepts from the viewpoint of building a learning scheme to eliminate them.

Table 2 shows five concepts supporting meta-learning: SHIFT, LIFT, REIFICATION, OBJECTIVIZATION, TRANSLATE. They play a guiding role in the design of theory-based meta-learning support systems. We explain to avoid misunderstanding: we don't argue the concepts described in this section are new from the cognitive science viewpoint but we conceptualize from the engineering viewpoint of system development as a basis of functional design for facilitating meta-cognitive learning. By making the concepts as a basis of learning system design explicit and building learning systems based on them, we can accumulate the knowledge for

Table 2. Correspondence Among Concrete functions Based on Support Concepts and Their Targets

Conceptualization	Meaning	Target to eliminate factors of difficulties	Learning Scheme Design
SHIFT	Stagger the time of developing learning skills after performing problem-solving processes	• Simultaneous processing with other activities • Management of resource + Inference of cognitive operation	Task Design (giving a presentation topic the learner had already learned)
LIFT	Make the learner be aware of learning skill acquisition	• Invisibility • Simultaneous processing with rehearsal	Visualization Environment
		• Acquisition of learning operator • Acquisition of criteria for learning	Guidance Function
REIFICATION	Give appropriate language for his/her self-conversation to acquire learning skills	• Segmentation of process	Providing Domain Specific Terms of Learning Activities
TRANSLATE	Transfer the learning skill acquisition task (LSAT) to a problem-solving task that includes same task structure of LSAT.	• A two-layer WM • Multiple Processing	Task Design (giving a presentation task to explain to other learners)
OBJECTIVIZATION	Objectify her/his self-conversation processes by externalizing them for learning communications with other learners	• (triggering cognitive conflicts)	CSCL Environment

building sophisticated learning systems. We explain the meaning of each concept in the following.

SHIFT means to stagger the time of developing learning skills after performing learning processes. By introducing Okamoto's survey of reflection [11], we shall explain SHIFT in detail. He pointed out that reflection of two kinds exist: on-going monitoring and reflective monitoring.

- On-going monitoring means controlling cognitive processes IN problem-solving.
- Reflective monitoring means modifying cognitive processes AFTER solving the problem.

Then, after giving learners math problems expressed in words, he conducted two experiments: he performed interviews stimulating reflective monitoring after solving each problem (Ex. 1); no interview was done after solving each problem (Ex. 2).

An interesting result of this experiment is that the time spent for solving a problem in Ex. 1 increased gradually in comparison with the time in Ex. 2.

He interpreted this result as suggesting that learners who were interviewed tried to read the math problem while integrating information into their schema.

By the following interpretation, we conceptualize his idea as SHIFT as a hypothesis for designing our learning system. The learner in ongoing monitoring performs three different cognitive activities simultaneously. The learner solves a math problem expressed in words, monitors the problem-solving processes, and generalizes the knowledge to transfer it to other problems. These processes are difficult for most learners to perform simultaneously for two reasons: they tend to exhaust their limited cognitive capacity by performing these processes, and they cannot be aware of when and what meta-cognition they must perform or how to perform it.

The SHIFT strategy enhances reflective monitoring by staggering the time of performing the meta-cognitive activities after doing problem-solving/ learning processes.

Furthermore, it is necessary to provide appropriate stimulation to encourage their meta-cognition. Okamoto's monitoring interview corresponds to this stimulation, which can be interpreted as obtaining the meta-cognitive task as easy as a cognitive task by changing an internal self-conversation task to a usual conversation task. Consequently, we conceptualize LIFT as making the learner aware of learning skill acquisition as a principle for system development in this research. For development of meta-cognitive skills, a key issue is realization of SHIFT and LIFT.

REIFICATION gives appropriate language for the subject of meta-cognition: we cannot realize LIFT if we do not give appropriate REIFICATION. Consequently, the concept of REIFICATION is included in the concept of LIFT. However, we cannot always realize appropriate LIFT even with REIFICATION. Therefore, we must give suitable REIFICATION to prompt learners' meta-cognition. For that reason, we separate REIFICATION from LIFT conceptually.

Through OBJECTIVIZATION, we make internal self-conversation processes objective by discussing them with others. It contributes to cognitive conflicts in a learner's mind, which facilitates the learner's meta-cognitive activities, triggered by objective reaction of learning partners to the explanations.

TRANSLATE changes the learning skill acquisition task to a problem-solving task that includes the same task structure of the learning skill acquisition task.

Based on these design concepts, developers can design valid meta-learning support scheme by realizing them.

5 Model Based Development of Presentation Based Meta-learning Support System

Based on the two models, we can build presentation based meta-learning scheme with explicit clarification of its design rationale. It also presents the usefulness of our model using our system as an example.

5.1 Task Design to Facilitate Meta-Learning Activities

The meta-learning process model and conceptualizations are integrated to design our meta-learning scheme and concrete support functions embedded into the system.

Table 2 presents correspondence among support functions based on the conceptualizations and their targets to eliminate factors of difficulties in performing meta-learning processes.

According to SHIFT and TRANSLATE, we design presentation tasks for learners where the learner must make presentation materials based on pre-learned knowledge. Our system presupposes a learner who has already learned a specific topic, UML, and software design patterns [12]. The learner must produce readily comprehensible presentation material for other learners whose academic ability is similar to that of the presenter. This task setting is important for the learner to examine meta-cognitive learning: if the learner must perform both learning and making presentations, the learner cannot allocate sufficient cognitive capacity to perform the meta-cognitive activities. This task setting corresponds to SHIFT. It staggers the time of performing monitoring and generalizing processes after performing learning.

Thereby, SHIFT removes factors of (d3) simultaneous processing with other activities and eliminates (d6) management of resource, although it *increases* (d5) factors of inference of cognitive operation: it does not require on-going monitoring but prompts reflective-monitoring. Furthermore, TRANSLATE reduces factors of two-layer WM and of multiple processing by translating learning process planning and learning skill acquisition tasks to the problem-solving (presentation) task.

5.2 Learning System Design to Facilitate Meta-learning Activities

5.2.1 Scenario of Using the System

Before explaining the design rationale of our learning support system, a learning scenario using our system is outlined.

Learners in our learning scheme perform learning by following three steps.

(i) Learning specific domain contents through self-study or attending lectures until learners think they have understood them
(ii) Producing comprehensive presentation materials to teach other learners who have the same academic level (presentation design phase)
(iii) Collaborative learning using presentation materials (collaborative meta-learning phase)

This system supports learners' activities at phases (ii) and (iii) (Support at (i) phase is beyond our scope.). Outlines of their activities and embedded support are described below.

A learner at (ii) phase produces the following

- Teaching plans (intention structure of the presentation) by referring to terms on domain-specific teaching activities, and
- Presentation material according to the intention structure to solve a given presentation subject.

Then the system provides the following.

- Guidance information to facilitate a learner's reflection on personal learning processes. The information is given by the learner's request to move to a subsequent collaborative meta-learning phase. The learner's request is interpreted as a declaration that the presentation satisfies the presentation subject.

The learner reconsiders:

- whether the presentation satisfies the requirements or not by referring to guidance information that suggests the learning topic might be embedded

A learner in at (iii) phase performs,

- collaborative meta-learning processes. The system analyzes a learner's presentation structures and provides viewpoints to facilitate learners' meta-learning communications.

5.2.2 Design Rationale of Presentation Based Meta-learning Scheme

Figure 3 portrays the system provided at the (ii) phase, which comprises five panels. The system is implemented in Visual Basic (Microsoft Corp.) and Java. It functions cooperatively with PowerPoint (Microsoft Corp.).

In this environment, educational activities are shown in Fig. 3(i):"make the learner consider what functions might be extended," "make the learner understand the functional extendibility of the DP by analyzing the class structure," "make the learner consider which classes need not be modified even in a functional extension," and so forth. This is designed based on REIFICATION. It decreases the difficulty of segmentation of process.

The learner gradually details teaching plans (called the intention structure) in the intention structure, shown in Fig. 3(ii), by referring to such teaching activities, and finally gives concrete shape to each presentation material and makes connections among lowest educational activities and presentation materials. It plays a key role to prompt learners' thinking between lines. This is designed based on the LIFT design principle. It is intended to reduce the difficulty of invisibility and simultaneous processing with rehearsal.

LIFT is also realized as a function that provides the learner with guidance information for checking the validity of designed learning processes. Guidance information of facilitating learner's reflection on personal learning processes is shown in the guidance panel, Fig. 3(v), at the time of moving to following the collaborative learning phase if educational activities that the teacher requires are not embedded into the learner's intention structure (teaching plan) of the presentation. This is designed based on LIFT. It is intended to reduce the difficulty of acquisition of learning operators and acquisition of criteria for learning processes.

5.2.3 Design Rationale of the System at the Collaborative Learning Phase

Figure 4 portrays a screen image at the collaborative learning phase. The window includes six panels: The system is also functioning cooperatively with PowerPoint.

Each learner uses windows of two kinds: one shows the personal intention structure and presentation (Fig. 4); the other shows the learning partner's. Thus, learning partners can view their presentations with an intention structure.

Our system provides viewpoints to discuss teaching and learning methods. This function, which will be explained in Section 5.3.2, is also designed based on LIFT. It is also intended to reduce the difficulty of acquisition of learning operators and acquisition of criteria for learning processes.

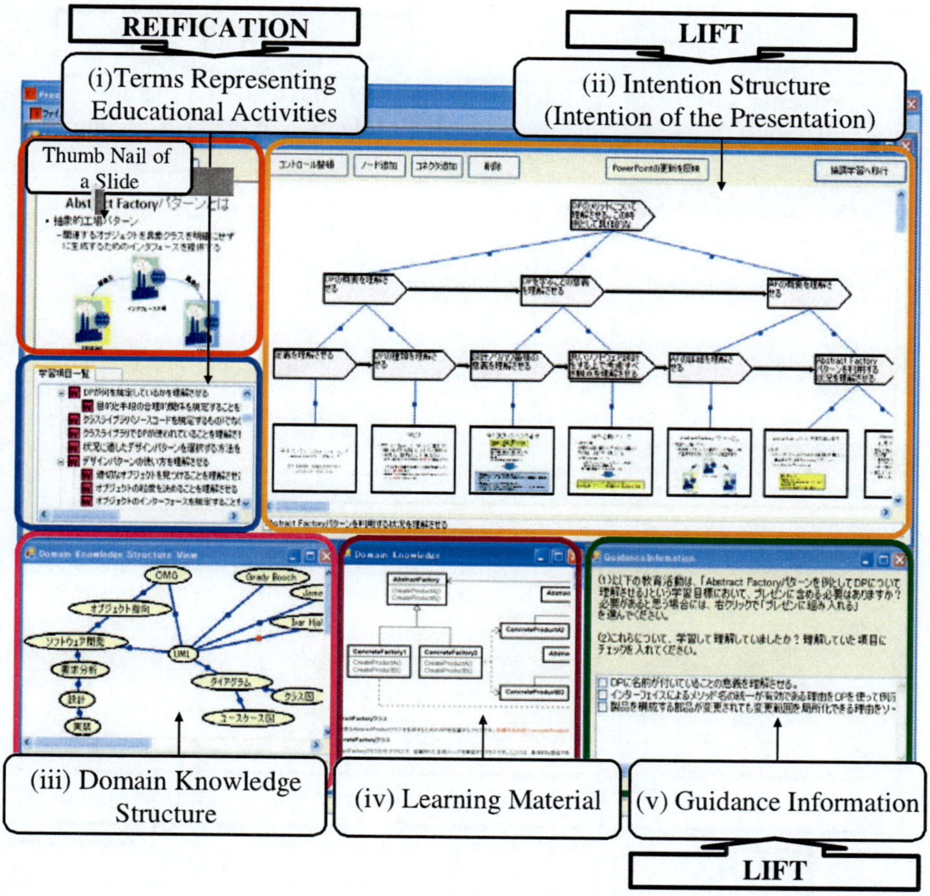

Fig. 3. Model based Design of Meta-Learning Support System: Presentation Design Phase

OBJECTIVIZATION is realized as a CSCL environment. It is intended to trigger cognitive conflicts in the learner's mind through communication with learning partners' reactions, and to reduce the factors of difficulty of acquisition of learning operators and criteria for learning.

5.3 Embedding Support Functions to Facilitate Meta-learning

In this section, we discuss issues on intention structure and support functions embedded into the system at (ii) and (iii) phases to facilitate meta-learning, although phase (i) is beyond our support.

5.3.1 Intention Structure Reflecting Learning Contexts

To encourage meaningful meta-learning communication among learning partners, each learner must (A) become aware of performing meta-learning and (B) share individual

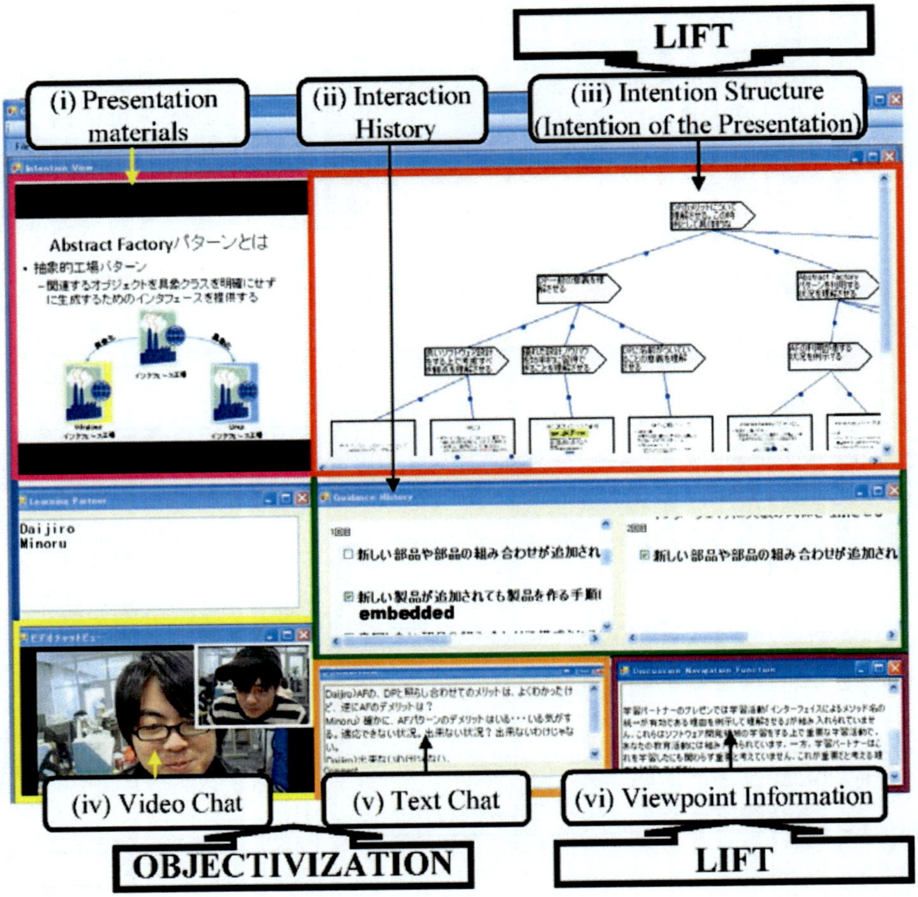

Fig. 4. Model based Design of Meta-Learning Support System: Collaborative Learning Phase

learning contexts. In our learning system, providing a representation to describe their intention of the presentation (intention structure), intention structures and guidance function according to them play roles of enhancing their awareness at the presentation design phase.

At the presentation design phase, we make learners construct intention structures to be aware of learning skill acquisition. Giving appropriate instructions according to learners' learning contexts is significant to facilitate their learning skill acquisition processes. In our task setting of making truly comprehensive presentation materials for use by those who have the same academic level with the presenter, we adopt an assumption that intention structures of presentation reflect learners' learning contexts in their learning. In the intention structure (Fig. 3. (ii)), each node represents an educational goal. Educational goals connected vertically to each other represent that the learner intends to achieve upper goals by performing lower ones, e.g., the learning goal of "Make the learner understand the significance of building DP" is detailed as its sub-learning goals that "Make the learner understand considerable viewpoint of software design" and "Make the learner understand the meaningfulness of that each DP has its own name." These terms are provided from the system to represent the learners' educational goals. Giving this description framework is based on LIFT concept.

5.3.2 Guidance Function to Prompt Meta-cognitive Awareness

Guidance information to facilitate the learner's reflection on personal learning processes is provided when the learner intends to move to the subsequent collaborative learning phase. It represents queries on domain-specific learning activity based on the learner's intention structure. The teacher giving a presentation subjects also constructs an intention structures and indicates required learning (teaching) activities on them that should be embedded into learners' intention structures. The system cannot understand the contents of learners' presentation written in natural language. However, it can process intention structures described by specified terms. Therefore, if learners did not embed them, then the system provides queries by referring the teacher's instruction of required terms:

(1) "Do the following teaching activities need to be included in your presentation to achieve the learning goal 'make the learners understand DP using Abstract Factory pattern as an example?' Choose 'embed into presentation' by right-mouse clicking if you think you need to do so."

(2) "Do you have sufficient understanding of these teaching activities? Check the items you had already understood."

- ☐ Make the learner understand the meaningfulness of the fact that each DP has its own name.
- ☐ Make the learner understand the advantages of object-oriented programming by combining its general theories with concrete examples in the Design Patterns.
- ☐ ... (Required teaching activities that the teacher identified are listed)

The learner is required to examine the importance of each learning activity for constructing comprehensive presentation materials: the learner judges whether the learner's presentation is valid or not and whether each learning activity should be included in the learner's presentation. This guidance is a stimulation to facilitate the learner's reflection on personal learning processes. The fact that the learner did not embed listed teaching activities is interpreted as follows: (a) the learner has no learning activities as domain-specific learning operators in his own consciousness, (therefore the learner cannot perform them) or (b) the learner does not understand the importance of the learning activities even if they have and they had performed their learning processes. The learner's checking activity in query (2) is interpreted as a declaration of whether the learner has them as learning operators.

For (a), the learner must perform the learning activities spontaneously or must be taught from the learning partners at the collaborative learning phase. For (b), the learner must encourage internal self-conversation to consider the importance of each learning activity. The guidance function is embedded based on the LIFT concept. It plays a role of building a foundation to encourage meaningful meta-learning communications among learning partners by stimulating their awareness in meta-learning before starting collaborative learning.

5.3.3 Viewpoint Providing Function to Stimulate Meaningful Learning Communications

The system in the collaborative learning phase provides support of two kinds to facilitate learners' learning skill acquisitions (acquiring learning operators and tightening evaluation criteria) as follows.

1. Support to share learning (teaching) contexts of learning partners by referring to presentation materials with intention structures.
2. Facilitate meaningful discussions to encourage their reflections on their own learning processes by providing discussion viewpoints.

Thinking processes related to one's own learning processes are quite tacit. Therefore it is not easy to externalize and to discuss learners' thinking processes (while teaching processes are externalized as intention structure). Ordinary learners with no support tend to discuss the appearance of illustrations, animations, and so on. To eliminate the problem, our system provides viewpoints to discuss their teaching and learning methods based on the interaction history between the learner and the system at the presentation design phase. As shown in Fig. 4 (vi), the system provides each learner with respective viewpoints to discuss as follows: "You judged the learning activity "Make the learner understand the significance of the fact that an interface specifies the name of each method by taking an example." as important. It is an important learning activity in learning software development domain and you embedded it into your presentation. On the other hand, your learning partner judged it as not important even they performed. Explain why you think this learning activity is important."

Collaborative learners can discuss their domain-specific teaching methods by referring to the viewpoints for meta-learning communication.

6 Experiments

6.1 Objectives and Methods

We conducted an experiment in a course to verify the meaningfulness of our learning scheme and usefulness of support functions embedded into the system. We specifically examine the issues of whether the system can encourage meta-learning communications. The outline of the experimentation is described below.

- **Subjects:** 16 graduate students participated. They had completed software engineering (UML) and object-oriented (Java) programming courses when they were undergraduate students. They were divided into two groups at random: eight students were in the experimental group (ExpG) using the system; eight were in the control group (CtlG).
- **Presentation topic:** Make presentation materials explaining the merits of building design patterns by taking the abstract factory pattern as an example.
- **Terms provided:** We specified 109 terms representing domain-specific teaching activities (including 16 required ones) to describe their intention structures.
- **Flow of the experiments:** Continuous 7 days lecture (90 min lecture each day) without weekend (Fig. 5):

- ✓ **1st – 2nd day:** Self-study of software design patterns until they think they have understood them. We provide same learning materials to all students. (Questionnaire administered at the end.)
- ✓ **3rd – 5th day:** Making presentation materials. Participants in ExpG used our system; those of CtlG used only Power Point (Microsoft Corp.). Thus, SHIFT and TRANSLATE principles are realized as the task setting not only for participants in ExpG but also for ones in CtlG. (Questionnaire administered at the end.) Participants who had not finish making presentation materials in the lecture completed them as homework. In this experiment, we did not use domain knowledge structure pane and hyper text pane to provide same learning materials for both CtlG and ExpG.
- ✓ **6th day:** Collaborative learning for meta-learning. Each participant in CtlG had provided guidance information before coming to collaborative learning. Four pairs in each group are constructed for collaborative learning. Participants in ExpG referred discussion viewpoint if they thought it is meaningful. In this experiment, we did not use video and text chatting function but did adopt face-to-face communication specifically to examine the evaluation of usefulness of viewpoint provides function. They performed CSCL by sitting next to each other. (Questionnaire administered at the end.)
- ✓ **7th day:** Teacher had made a presentation after examination to take a credit of the half-semester course. (Questionnaire administered at the end.)

- **Evaluation methods:** Administered four questionnaires (5-scale: 5. Strongly Agree - 4. Agree - 3. Undecided - 2. Disagree - 1. Strongly Disagree, 52 items for ExpG in total, 30 items for CtlG in total) and analyzed protocol data. We also interviewed if needed.

One of the authors conducted the experiment in his course: he explained the meaningfulness of meta-learning—what it is and the intentions of performing the presentation-based learning for all students—at the beginning of the first day's lecture. He also explained that the learning goal of the lecture is to acquire software development domain specific learning methods. He also instructed all learners to discuss learning methods just before collaborative learning.

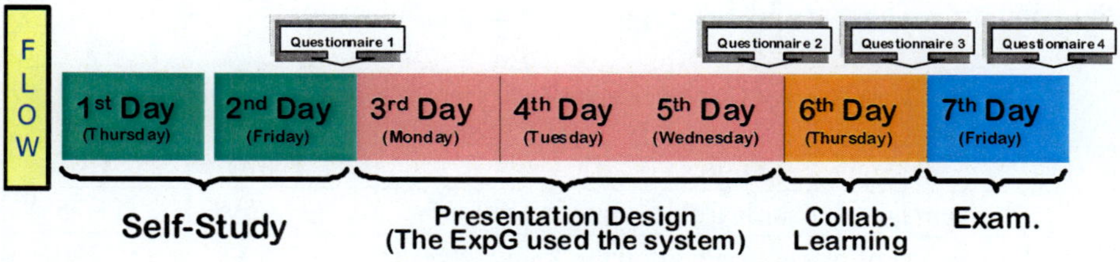

Fig. 5. Flow of the Experiments

6.2 Experimental Results and Analysis

6.2.1 Time Ratio Analysis of Learners' Communication Topics

Communication Topics Table 3 presents a time-based ratio of their communications.

We calculated the time ratio (Tr) by the following formula:

$$Tr = Tmc / Twc$$
$$Tmc = \Sigma \textit{Total Time of Meta-Learning Communication in Each Pair}$$
$$Twc = \Sigma \textit{Total Time of Collaborative Learning Communication in Each Pair}$$

Procedure to analyze protocol data is as follows:

(1) We made text data for all the pairs in each group using recorded spoken dialogue.
(2) Two of authors marked candidates of meta-learning communications independently according to the criteria (a) to (d) described below.
(3) We adopt intersections as meta-learning communications by matching above two data, then measured and added each time of meta-learning communications by listening recorded spoken dialogue.

Categories of meta-learning communication are as follows:

(a) Discussion on whether learning activities should be embedded in the presentation or not
(b) Expression of self-reflection on one's own learning processes
(c) Expression of one's awareness on one's insufficient/ mis-understanding state, or explanation of its reason
(d) Explanation of domain knowledge after expressing one's intention of checking one's own understanding states

Regarding domain knowledge explanation, we didn't include the time without expressing one's intention described in (d), because we cannot judge whether monitoring of one's understanding state were occurred. The average time ratio of meta-learning communication of four pairs in ExpG is drastically more than the ones in CtlG although the teacher had instructed to all participants to perform meta-learning communications for getting them be aware of meta-learning.

Therefore, it suggests that the system was able to encourage learner's meta-cognitively aware learning communications.

The average time ratios of communication for confirming their understanding of fundamental domain-concepts and for trivial things (how to depict the class diagram, illustration and animation of the slides, and so on) in CtlG are significantly higher than those in ExpG. These results also support the meaningfulness of the system.

Table 3. Time Ratio of Communication Topics in Collaborative Learning Phase

Topics	ExpG	CtlG
Percentage of meta-learning communication	31.75%	11.75%
Percentage of discussion on domain knowledge	1.5%	12.5%
Percentage of discussion on appearance of slides	0.5%	20.25%

6.2.2 Questionnaire Analysis

Table 4 presents results of questionnaires after their collaborative learning to consider whether the system facilitates their meta-learning processes. As we described in 6.1, we administered 4 questionnaires in the experiment. We show all the questionnaire items

Table 4. Results of Questionnaire after the Collaborative Learning Phase

Questionnaire Items		ExpG		CtlG	
		Mean	SD	Mean	SD
1	Do you think the collaborative learning after making your presentation materials enhanced your reflection on your own learning processes?	4.375	0.267	4.375	1.982
2	Do you think the intention structures facilitated your analysis on your learning partner's presentation structures (his teaching methods to construct audience's understanding)?	3.375	0.553		
3	Do you think the viewpoint providing function enhanced your consciousness of your learning methods?	3.625	0.839		
4	Do you think the viewpoint providing function facilitates your analysis of your learning processes?	3.625	0.839		
5	Do you think the viewpoint providing function facilitated your discussion?	4	1.142		
6	Do you think collaborative learning changed your criteria to evaluate your understanding of DP?	2.875	1.553	3.375	1.982
7	Do you think you could acquire learning methods using collaborative learning?	3.375	0.839	3.625	1.41
8	Do you think your learning processes for other DPs will change after performing this presentation-based learning?	3.75	1.071	3.5	1.428
9	Do you think you could acquire learning methods by performing this presentation-based learning?	3.625	0.553	4.125	0.982
10	Do you think your consciousness of learnin will change by performing this presentation-based learning?	4.1	0.238	3.875	0.982

administered after collaborative learning (10 items for ExpG and 6 for CtlG) to focus on the subjects of this paper to verify the usefulness of our learning scheme: whether can it facilitate meta-cognitively aware learning communication and whether can learners tighten their criteria to evaluate their own learning processes.

Questionnaire items 1 and 6–10 are for participants in both ExpG and CtlG: item 1 is related to the usefulness of the presentation-based learning scheme and 6–10 are related to learning effects from the viewpoint of meta-learning. Items 2–5 only for participants in ExpG are on usefulness of support functions embedded into the system.

Regarding item 1, participants in both ExpG and CtlG gave quite high marks, which suggests the presentation based meta-learning stimulates learners' reflection on their learning processes. Regarding item 2, participants in ExpG gave high marks, which means that descriptions of intention structures are useful to share their learning contexts. Furthermore, some students mentioned that describing intention structure strongly inspired them to reflect on their own learning processes and to consider teaching processes.

Regarding items 3–5, participants in ExpG almost all gave high marks, suggesting that embedded support according to the LIFT concept is useful to encourage learners' reflections on their learning processes and their meta-learning. Especially, we were able to verify the viewpoint providing function can trigger their meta-learning communications. It is expected that learners will execute better learning processes using the acquired domain-specific learning activities and tightened evaluation criteria if the learners' meta-learning processes are performed successfully.

Items 6–10 inquired the about learners' consciousness of them. Both groups gave high marks to each item. However, CtlG gave higher marks than ExpG for the acquisition of domain-specific learning activities (items 7 and 9), whereas ExpG gave higher marks than CtlG for items related to the consciousness of changes of their own future learning processes (items 8 and 10).

Those responses seem to be mutually contradictory. However, they are not so by the following interpretation: learners in ExpG had tightened their learning criteria to evaluate their learning processes and understanding states; thereby, they also strictly evaluated their meta-learning processes. The results of the average time ratio of meta-learning communication support this. However, the fact that participants in ExpG gave low marks related to item 6 suggests that they noticed that they were unable to perform all meta-learning processes by themselves even though they were able to understand the importance of meta-learning. Actually, we do not embed the functions that support performance of learning activities acquired by meta-learning processes even when the system triggers learning activities. A student mentioned "I feel I could not finish meta-learning yet, thus I need to continue to be aware of acquiring learning methods. Thus, I thought my consciousness of learning will change." On the other hand, participants in CtlG spent less time for meta-learning communications, suggesting that the learners' evaluation criteria had not been tightened through their communications. Consequently, their evaluation results for these items were more tolerant.

The experimental results tend to support our hypothesis that learners using our system tighten their learning criteria according to above interpretation. However, we could not detect statistically significant differences among evaluation results. Therefore, further investigations are required by conducting other experiments.

Furthermore, Table 4 shows the values of SD of CtlG are higher than the ones of ExpG. It suggests that the system has an effect of raising the learner's low standard. We have to carefully address this issue by conducting further experiments.

6.2.3 Results of Examination

We had given tests to check their understanding states after collaborative learning. Table 5 and Fig. 6 show 3 problems given to the participants and results of average scores of each group, respectively. To answer the problem 1, learners can score high points even by only memorizing class diagram without deep understanding. On the other hand, to answer the problem 2 and 3, they have to not only memorize benefits of the AF pattern by printing surface but also understand them with thinking the intentions of a designed class structure (for Q2) and with thinking the roles of software design patterns from more global viewpoint of software development cycle (for Q3). In other words, they require to read not only texts but also to think between the lines by themselves.

In Q1, ExpG and CtlG scores are nearly even score. On the other hand, in problem 2 and Q3, ExpG's score is significantly higher than CtlG's. The differences between ExpG and CtlG get larger according to the difficulty level of the problems does higher.

Table 5. Questions Given to the Learners after Collaborative Learning

	Problems
1	Depict the class diagram of the Abstract Factory pattern.
2	What kinds of advantages are there for client classes by abstracting the points of generating part instances?
3	Summarize the roles of design patterns from the viewpoint of improving software development cycle and describe their reasons.

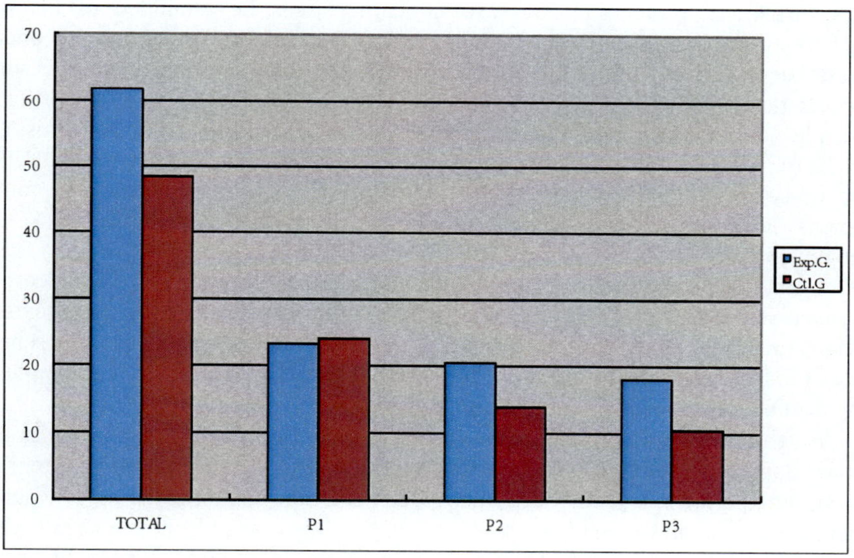

Fig. 6. Average Scores of Each Group

These results show our support system deepens learners understanding states by prompting meta-cognition of their thinking between lines [1].

7 Related Works

By introducing the framework, we characterize the following learning schemes from the viewpoint of learning scheme design according to design concepts: error based simulation (EBS), self-directed exploratory learning scheme, and problem-posing learning scheme.

EBS is a support system to prompt learners' meta-cognitive activities in problem-solving processes [13]. It realizes sophisticated simulation environment called error-based simulation. It visualizes the behavior of mechanics based on the equation of motion that a learner made, thereby it prompts cognitive conflicts with awareness of the gap between his belief on problem-solving and actual behavior. This simulation function is considered as realization of OBJECTIVIZATION concept, whereas we build CSCL environment based on OBJECTIVIZATION concept.

Learning support systems for self-directed exploratory learning embed a function to lighten a learner's cognitive loads of performing meta-cognition [14]. This embedding of functions does not include the SHIFT principle. Furthermore, self-setting of learning goals and acquisition of experience of self-exploratory are important points in this approach. That emphasis differs from our goal because we fix the target domain knowledge for domain-specific meta-learning and we seek to encourage learning skill acquisition through production of presentation materials.

In a problem-posing learning scheme [15, 16], learning skill acquisition tasks are translated to a problem-posing task: that includes SHIFT and TRANSLATE principles. In performing a problem-posing task, a learner must be reminded of his own problem-solving processes, which includes the LIFT principle. Furthermore, secondary effects occur because the posed problem must be solved by other learners, which includes the OBJECTIVIZATION principle. It does not include a REIFICATION principle. In this learning scheme, learners might be unable to follow this task translation because the problem-posing task requires heavier cognitive loads than problem-solving tasks do. Our learning scheme makes it easier to monitor and control learner's learning activities by translating learning skill acquisition task into problem-solving task based on the TRANSLATION principle.

Our learning scheme is a kind of explanation-based learning [17]. This learning scheme essentially embeds OBJECTIVIZATION principle through observing learning partners' reaction. It is characteristic that our presentation based learning scheme embeds SHIFT principle as a task setting, and LIFT and REIFICATION principles as support functions to prompt meta-learning processes.

Through interaction with computer agents, Betty's Brain supports learners as they acquire domain knowledge and self-regulated skills [18]. Learners in their system and our system perform teaching activities. Betty's Brain supports learners' teaching processes on domain knowledge by externalizing the changes of Betty's understanding. It realizes SHIFT, LIFT, and TRANSLATE concepts. In contrast, we embed support functions to stimulate their judgment of the importance of domain-specific learning activities and facilitate their communication on them.

An interaction analysis system for collaborative learning was proposed by Inaba et al. [19]. They systemized concepts that can characterize learning interactions among learners, such as "showing common knowledge" and "showing the way to solve problem." The teacher can characterize interaction logs using these concepts; then the system can understand the situation of the learners' interaction. Therefore, the system can show information related to each learner's state as well as the situation of the group discussion. Consequently, the teacher can instruct the group discussion based on that information. We also would like to develop an interaction analysis system for our presentation-based meta-learning scheme. It is also helpful for learners to analyze their own interactions.

8 Concluding Remarks

As described in this paper, we presented a philosophy of our research to elucidate our model-oriented approach. Then, we proposed a meta-learning process model by extending Kayashima's computational model for characterizing meta-learning activities. Then, we presented our conceptualizations as a basis of learning scheme design.

Furthermore, a meta-learning process model and the conceptualizations were integrated to support design of meta-learning systems based on a deep understanding of meta-learning processes. They play a guiding role in the design of meta-learning support systems. Moreover, the model plays an important role in accumulating and sharing experiences of individual learning system development, because we could understand the design rationale of each support functions embedded in our meta-learning support system based on the model. It also presents the usefulness of the framework by taking our presentation-based meta-learning system.

We implemented and conducted an experimental study, and evaluated the usefulness of each function according to the design rationale. They worked well due to our design rationale, however, it suggested refinement of terms for describing intention structure is needed due to the questionnaire based and protocol analysis. The system could stimulate learners' reflection on their learning processes and enhance meta-cognitively aware learning communications among learners in collaborative learning. It also could tighten their criteria to evaluate their learning processes and learning outcomes through all the processes in our learning scheme. This knowledge can be found by analyzing the data with referring to the model.

Therefore, we could build a foundation of systematic refinement of our meta-learning systems. Further refinement of the models must continue through theoretical and experimental aspects for the basis of meta-learning. Individual support functions embedded into each meta-learning support functions can be characterized by the support concepts, thus it makes it easier to compare usefulness of them from the viewpoint of same design rationale. We would like to address systematic generation of evaluation items for each function according to the model as a future work.

References

[1] Bransford, J., Brown, A., Cocking, R. (eds.): Brain, Mind, Experience, and School, in How People Learn. National Academy Press, Washington (2000)

[2] Maeno, H., Seta, K., Ikeda, M.: "Development of Meta-Learning Support System based on Model based Approach. In: Proc. of the 10th IASTED International Conference on Artificial Intelligence and Applications (AIA 2010), pp. 442–449 (2010)

[3] Noguchi, D., Seta, K., Ikeda, M.: Presentation Based Learning Support System to Facilitate Meta-Learning Communications. In: Proc. of International Conference on Computers in Education, pp. 137–144 (2010)

[4] Seta, K., Noguchi, D., Ikeda, M.: Presentation-Based Collaborative Learning Support System to Facilitate Meta-Cognitively Aware Learning Communication. The Journal of Information and Systems in Education (in Press, 2011)

[5] Brown, A.L., Bransford, J.D., Ferrara, R.A., Campione, J.C.: Learning, Remembering, and Understanding. In: Markman, E.M., Flavell, J.H. (eds.) Handbook of child psychology. Cognitive Development, 4th edn., vol. 3, pp. 515–529. Wiley, New York (1983)

[6] Flavell, J.H.: Metacognitive aspects of problem solving. In: Resnick, L. (ed.) The Nature of Intelligence, pp. 231–235. Lawrence Erlbaum Associates, Mahwah (1976)

[7] Kayashima, M., Inaba, A., Mizoguchi, R.: What Do You Mean by to Help Learning of Metacognition? In: Proc. of the 12th Artificial Intelligence in Education (AIED 2005), Amsterdam, The Netherlands, pp. 346–353 (2005)
[8] Seta, K., Fujiwara, M., Noguchi, D., Maeno, H., Ikeda, M.: Building a Framework to Design and Evaluate Meta-Learning Support Systems. LNCS (LNAI), pp. 163–172. Springer, Heidelberg (2010)
[9] Hayashi, Y.: Strategy-centered Modeling for Better Understanding of Learning/ Instructional Theories. International Journal of Knowledge and web Intelligence 1(3&4) (in press, 2010)
[10] Seta, K., Ikeda, M.: Conceptualizations for Designing a Learning System to Facilitate Metacognitive Learning. In: Proc. of World Conference on Educational Multimedia, Hypermedia &Telecommunication (ED-MEDIA), Vienna, Austria, pp. 2134–2143 (2008)
[11] Okamoto, M.: Review of Metacognitive Research – from educational implications to teaching methods. Journal of Japanese Society for Information and Systems in Education 19(3), 178–187 (2002) (in Japanese)
[12] Gamma, E., Helm, R., Johnson, R.: Design Patterns: Elements of Reusable Object-Oriented Software, illustrated edition. Addison-Wesley Professional, Reading (1994)
[13] Horiguchi, T., Imai, I., Toumoto, T., Hirashima, T.: A Classroom Practice of Error-based Simulation as Counterexample to Students' Misunderstanding of Mechanics. In: Proc. of International Conference on Computers in Education (ICCE 2007), pp. 519–525 (2007)
[14] Kashihara, A., Taira, K., Shinya, M., Sawazaki, K.: Cognitive Apprenticeship Approach to Developing Meta-Cognitive Skill with Cognitive Tool for Web-based Navigational Learning. In: Proc. of the IASTED International Conference on Web-Based Education (WBE 2008), Innsbruck, Austria, pp. 351–356 (2008)
[15] Nakano, A., Hirashima, T., Takeuchi, A.: Developing and evaluation of a computer-based problem posing in the case of arithmetical word problems. In: The Fourth International Conference on Computer Applications, ICCA 2006 (2006)
[16] Kojima, K., Miwa, K.: Case Retrieval System for Mathematical Learning from Analogical Instances. In: Proc. of the International Conference on Computers in Education (ICCE), pp. 1124–1128 (2003)
[17] Chi, M.T.H., Vassok, M., Lewis, P.J., Glaser, R.: Self-explanations: How students study and use examples in learning to solve problems. Cognitive Science 13, 145–182 (1989)
[18] Schwartz, D.L., et al.: Interactive Metacognition: Monitoring and Regulating aTeachable Agent. In: Hacker, D.J., Dunlosky, J., Graesser, A.C. (eds.) Handbook of Metacognition in Education, pp. 340–358. Routledge, New York (2009)
[19] Inaba, A., Ohkubo, R., Ikeda, M., Mizoguchi, R.: An Interaction Analysis Support System for CSCL. Proc. Transactions of Information Processing Society of Japan 44(11), 2617–2627 (2004) (in Japanese)

Chapter 12
Case-Based Reasoning Approach to Adaptive Modelling in Exploratory Learning

Mihaela Cocea[1,2], Sergio Gutierrez-Santos[2], and George D. Magoulas[2]

[1] School of Computing, University of Portsmouth,
Buckingham Building, Lion Terrace, Portsmouth, Hampshire, PO1 3HE, UK
mihaela.cocea@port.ac.uk
[2] London Knowledge Lab, Birkbeck College, University of London,
23-29 Emerald Street, London, WC1N 3QS, UK
{sergut,gmagoulas}@dcs.bbk.ac.uk

Abstract. Exploratory Learning Environments allow learners to use different strategies for solving the same problem. However, not all possible strategies are known in advance to the designer or teacher and, even if they were, considerable time and effort would be required to introduce them in the knowledge base. We have previously proposed a learner modelling mechanism inspired from Case-based Reasoning to diagnose the learners when constructing or exploring models. This mechanism models the learners' behaviour through simple and composite cases, where a composite case is a sequence of simple cases and is referred to as a strategy. This chapter presents research that enhances the modelling approach with an adaptive mechanism that enriches the knowledge base as new relevant information is encountered. The adaptive mechanism identifies and stores two types of cases: (a) inefficient simple cases, i.e. cases that make the process of generalisation more difficult for the learners, and (b) new valid composite cases or strategies.

Keywords: user modelling, knowledge base adaptation, exploratory learning environments, case-based reasoning.

1 Introduction

Exploratory learning environments (ELEs) are built upon a constructionist pedagogical approach [1], which is characterised by two core ideas: (a) learning is seen as a reconstruction of knowledge rather than as a transmission of knowledge and (b) learning is most effective when it is part of an activity in which learners feel they are constructing a meaningful product [1]. The constructionist approach is inspired by Piaget's constructivist theory [2] which states that learners construct mental models to understand the world around them. Consequently, based on these principles, exploratory learning environments allow learners a high degree of freedom and encourage learners to explore and experiment with different models within the particular learning system. Therefore,

T. Watanabe and L.C. Jain (Eds.): Innovations in Intell. Machines – 2, SCI 376, pp. 167–184.
springerlink.com

these environments are radically different from Intelligent Tutoring Systems in which the learning activities are highly structured and the learner is guided in a stepwise manner.

Exploratory learning environments provide activities that involve constructing [2] and/or exploring models, varying their parameters and observing the effects of these variations on the models' behaviour. When provided with guidance and support ELEs have a positive impact on learning compared with other more structured environments [3]; however, the lack of support may actually hinder learning [4]. Therefore, to make ELEs more effective, intelligent support is needed, despite the difficulties arising from their open nature.

To provide intelligent support, a mechanism for diagnosing the learner is needed, which in Intelligent Learning Environments is done through user/learner modelling. The typical approach is based on concepts of the domain: learners are required to study materials about a concept and then their knowledge level is assessed through testing. In ELEs the emphasis is on the process of learning by means of constructionist activities rather than on the knowledge. Therefore, the focus is on the actions the learners perform in the educational system than on answers to tests, and, consequently, the learner modelling process should focus on analysing the learners' interactions with the system.

To address this, we have proposed a learner modelling mechanism for monitoring learners' actions when constructing/exploring models by modelling sequences of actions that reflect different strategies in solving a task [5]. An important problem, however, remains: only a limited number of strategies are known in advance and can be introduced by the designer/teacher. In addition, even if all strategies would be known, introducing them in the knowledge base of a system would take considerable time and effort. Moreover, the knowledge about a task evolves over time - students may discover different ways of approaching the same task, rendering the knowledge base suboptimal for generating proper feedback, even if initially it had a good coverage. To address this issue, we employ a mechanism for adapting the knowledge base in the context of *eXpresser* [6], an exploratory learning environment for mathematical generalisation.

The knowledge base adaptation involves a mechanism for acquiring inefficient simple cases, i.e. cases which include actions that make it difficult for students to create a generalisable model, and a mechanism for acquiring new strategies. The former could be potentially useful to enable targeted feedback about the inefficiency of certain parts of a construction, or certain actions of the student; this approach could also lead gradually to creating a library of inefficient constructions produced by students that could be analysed further by a researcher/teacher. Without the latter a new valid strategy will not be recognised as such, and, consequently, the learner modelling module will diagnose the learner to be still far from a valid solution and any potential feedback will be confusing as it will guide the learner towards the most similar strategy stored in the knowledge base.

The rest of the chapter is structured as follows. The next section briefly introduces *eXpresser* and the problem of mathematical generalisation. Section 3 describes the case-based reasoning cycle for *eXpresser* and gives a brief overview of the knowledge representation and the identification mechanism employed. Section 4 presents our proposed approach for adapting the knowledge base. Section 5 describes the validation of this approach and, finally, Section 6 concludes the chapter and presents some directions for future work.

2 Mathematical Generalisation with *eXpresser*

Mathematical generalisation has been defined or described in several ways, varying from philosophical views that could be applied to any type of generalisation to views very specific to mathematics. Examples from the first category are: (a) "an object and a means of thinking and communicating" [7](p. 63), and (b) "applying an argument in a broader context" [8](p. 38). An example from the second category is: "Generalizing problems, also known as numeric sequences or geometric growing sequences, present patterns of growth in different contexts. Students are asked to find the underlying structure and express it as an explicit function or 'rule'." [9](p. 442).

Mathematical generalisation is at the centre of algebraic expressions, as "algebra is, in one sense, the language of generalisation of quantity. It provides experience of, and a language for, expressing generality, manipulating generality, and reasoning about generality" [10](p. 105). This relation, however, together with the idea of recognising and analysing patterns and articulating structure, seems to be elusive to students who fail to understand algebra and its purpose [11]. Students are unable to express a general pattern or relationship in natural language or in algebraic form [12].

Students, however, are able to identify and predict patterns [10] and there are claims that it is not the generalisation problems that are causing difficulties to students, but the way these are presented and the limitations of the teaching approaches used [9]. Typically, "generalising problems are usually presented as numeric or geometric sequences, and typically ask students to predict the number of elements in any position in the sequence and to articulate that as a rule" [9](p. 443). A common strategy is "the construction of a table of values from which a closed-form formula is extracted and checked with one or two examples" [13](p. 7), introducing a tendency towards pattern spotting and emphasising its numerical aspect [14], [15]. This approach obscures the variables involved, "which severely limits students ability to conceptualise the functional relationship between variables, explain and justify the rules that they find, and use the rules in a meaningful way for problem solving" [9](p. 444).

Another approach that affects students' understanding of generalisation is the focus on mathematical products rather than mathematical processes [16], [17]. Malara and Navarra[17] argue that students should be taught to distance themselves from the result and the operations needed to obtain that result, and to reach a higher level of thinking by focusing on the structure of a problem.

Another difficulty encountered in teaching mathematical generalisation is the students' difficulty to use letters that stand for the unknown [18] and to realise that letters represent values [19]. Secondary school students also tend to lack a mathematical vocabulary for expressing generality [11] and their written responses lack precision [16].

Taking these aspects into account, a system called *eXpresser* [6] was developed using an iterative process that involved designing with students and teachers. The main aim was to develop an environment that provides the students with the means for expressing generality rather than considering special cases or spotting patterns.

eXpresser enables constructing patterns, creating dependencies between them, naming properties of patterns and creating algebraic-like rules with either names or numbers. It is designed for classroom use and targets 11 to 14 years old pupils. Each task involves two main phases: building a construction and deriving an algebraic-like rule from it.

Fig. 1 illustrates the system, the *properties list* of a pattern (linked to another one) and an example of a rule. The screenshot on the left includes two windows: (a) the students' world, where the students build their constructions and (b) the general world that displays the same construction with a different value for the variable(s) involved in the task, and where students can check the generality of their construction by animating their pattern (using the Play button).

We illustrate here a task called 'stepping stones' (see Fig. 1) displayed in the students' world with a number of 3 red (lighter colour) tiles and in the general world with a number of 8 red tiles; the task requires to build such a construction and to find a general rule for the number of blue (darker colour) tiles needed to surround the red ones. The construction for this task can be built in several ways

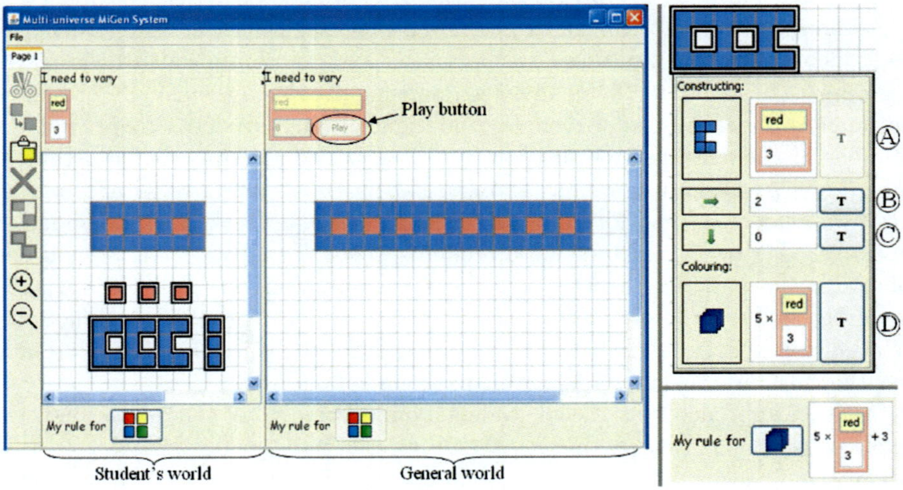

Fig. 1. *eXpresser* screenshots. The screenshot on the left includes a toolbar, the students' world and the general world. The screenshot on the top right shows the property list of a pattern. The bottom right screenshot illustrates a rule.

that we call *strategies.* Here we illustrate the 'C strategy', named after the shape of the building-block, i.e. the basic unit of a pattern. The components of this strategy are displayed separately in the students' world for ease of visualisation: a red pattern, having 3 tiles, a blue one made of a C-shape pattern repeated 3 times, and 3 blue tiles.

The property list of the C-shape pattern is displayed in the screenshot on the top right. The first property (Ⓐ) specifies the number of iterations of the building-block; the value for this attribute is set to the value of the iterations of the red pattern by using a T-box (that includes a name and a value); by using a T-box, the two (or more) properties are made dependent, i.e. when the value in the T-box changes in one property, it also changes in the other one(s). The next properties are *move-right* (Ⓑ), which is set to 2, and *move-down* (Ⓒ), which is set to 0. The last property (Ⓓ) establishes the number needed to colour all the tiles in the pattern - in our case 5 times the iterations of the red pattern. The bottom right screenshot displays the rule for the number of blue tiles: $5 \times red + 3$, where red stands for the T-box displayed in Ⓐ (a T-box can be displayed with name only, value only or both).

The construction in Fig. 1 and the rule in the bottom-right corner constitute one possible solution for the 'stepping stones' task. Although in its simplest form the rule is unique, there are several ways to build the construction and infer a rule from its components. Thus, there is no unique solution and students follow various kinds of strategies to construct their models (i.e. construction and rule). Two examples of such different constructions and rules are illustrated in Fig. 2. The following section presents our approach for modelling and identification of strategies.

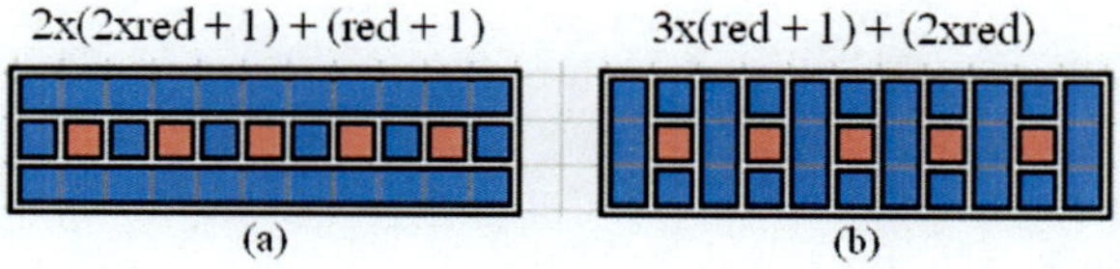

Fig. 2. (a) 'HParallel' Strategy; (b) 'VParallel' Strategy.

3 Modelling Learners' Strategies Using Case-Based Reasoning

In case-based reasoning (CBR) [20] knowledge is stored as cases, typically including the description of a problem and its solution. When a new problem is encountered, similar cases are retrieved and the solution is used or adapted from one or more of the most similar cases. The CBR cycle typically includes four processes [20]: (a) *Retrieve* cases that are similar to the current problem; (b) *Reuse* the cases (and adapt) them in order to solve the current problem; (c) *Revise* the proposed solution if necessary; (d) *Retain* the new solution as part of a new case (see Fig. 3).

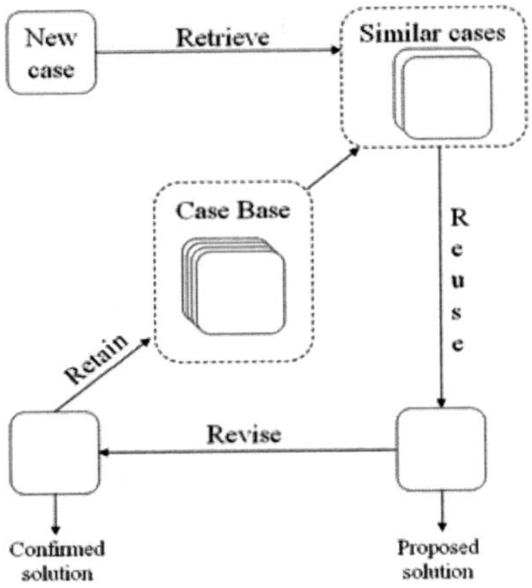

Fig. 3. CBR cycle.

In exploratory learning the same problem has multiple solutions and it is important to identify which one is used by the learner or whether the learner has produced a new valid solution. To address this for *eXpresser* each task has a case-base (or knowledge base) of solutions (i.e. strategies). When a learner is building a construction, it is transformed into a sequence of simple cases (i.e. strategy) and compared with all the strategies in the case-base for the particular task that the learner is working on; the case-base consists of strategies, i.e. composite cases, rather than simple cases. To *retrieve* the strategies that are most similar to the one used by the learner, appropriate similarity metrics are employed (see below). Once the most similar strategies are identified, they are used in a scaffolding mechanism that implements a form of *reuse* by taking this information into account along with other information, such as the characteristics of the learner (e.g. knowledge level, spatial ability), completeness of solution and state within a task. The *reuse*, *revise* and *retain* steps are part of the knowledge base adaptation described in Section 4: simple cases are modified and then stored in a set of inefficient cases; new strategies are stored without modifications.

We use the term knowledge base adaptation in the sense that the knowledge base changes over time to adapt to new ways in which learners approach tasks - ways that could be either efficient or inefficient. This is referred to as 'adaptation to a changing environment' [21]. It is not, however, the same as adaptation in the CBR sense, although this is present to a certain degree in the acquisition of inefficient cases, as it involves the processes of *reuse* and *revise* which are generally referred to as case adaptation [22]. The acquisition of new strategies corresponds to case-base maintenance in CBR terminology [20], as it involves adding a new case for which no similar case has been found.

The following paragraphs briefly present the knowledge representation and the similarity metrics used for strategy identification.

3.1 Knowledge Representation

In our approach, strategies for building a construction are represented as a series of simple cases with certain relations between them. A simple case is defined as $C_i = \{F_i, RA_i, RC_i\}$, where C_i represents the case and F_i is a set of attributes. RA_i is a set of relations between attributes and RC_i is a set of relations between C_i and other cases respectively.

The set of attributes of a given case C_i is defined as $F_i = \{\alpha_{i_1}, \alpha_{i_2}, \ldots, \alpha_{i_N}\}$, where N represents the number of attributes. It includes three types of attributes: (a) variables (the first v attributes), (b) numeric (attributes from $v+1$ to w) and (c) binary (attributes from $w+1$ to N). The numeric attributes correspond to the values in the property list and the variables correspond to the type of those properties: number, T-box, expression with number(s) or expression with T-box(es). The binary attributes refer to the membership of a case to a strategy and is defined as a $PartOfS$ function which returns 1 if the case belongs to the strategy and 0 if it does not. There are S binary attributes, where S is the number of strategies in the knowledge base.

The set of relations between attributes of a given case C_i and attributes of other cases (as well as attributes of C_i) is represented as $RA_i = \{RA_{i_1}, RA_{i_2}, \ldots, RA_{i_M}\}$, where M represents the number of relations between attributes and at least one of the attributes in each relation $RA_{i_m}, \forall m = \overline{1, M}$, is from F_i, the set of attributes of C_i. Two types of binary relations are used: (a) *dependency relations* such as the one illustrated in Fig. 1 where the number of the iterations of the blue pattern depends on the iterations of the red pattern through the use of a T-box; these relations are formally represented as $\alpha_{i_k} = DEP(\alpha_{j_l})$, where $\alpha_{i_k}$and α_{j_l} are variables of cases i and j and means that α_{i_k} depends on α_{j_l}; (b) *value relations* such as the fact that the value of the colouring property of the blue pattern in Fig. 1 is 5 times the value of the iterations of the red pattern. A case is considered *specific* when it does not have dependency relations and is considered *general* when it has all the dependency relations required by the task.

The set of relations between cases is represented as $RC_i = \{RC_{i_1}, RC_{i_2}, \ldots, RC_{i_P}\}$, where P represents the number of relations between cases and one of the cases in each relation $RC_{i_j}, \forall j = \overline{1, P}$ is the current case (C_i). Two time-relations are used: (a) $Prev$ relation indicates the previous case and (b) $Next$ relation indicates the next case, with respect to the current case. Each case includes at most one of each of these two relations.

A strategy is defined as $S_u = \{N_u(C), N_u(RA), N_u(RC)\}$, $u = \overline{1, S}$, where S represents the number of strategies in the knowledge base, $N_u(C)$ is a set of cases, $N_u(RA)$ is a set of relation between attributes of cases and $N_u(RC)$ is a set of relations between cases.

To illustrate how a learner's construction is transformed into the knowledge representation detailed above, we use the 'stepping stones' task introduced in Section 2, which requires to find the number of tiles that surround a pattern like the red one displayed in Fig. 1. There are several strategies for constructing the

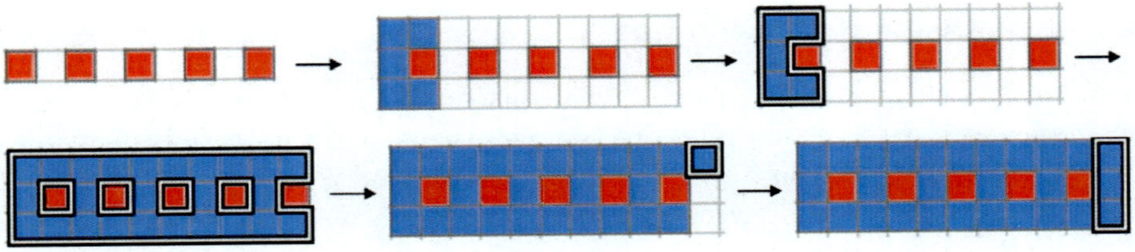

Fig. 4. Possible steps for 'C strategy'.

surrounding for that pattern as illustrated in Fig. 1 (the 'C strategy') and Fig. 2 ('HParallel' and 'VParallel' strategies).

Besides multiple possible constructions, there are several ways of reaching the same construction. A possible trajectory for the 'C strategy' is illustrated in Fig. 4. The learner may start with the footpath (the red tiles) and then build a group of five blue tiles around the leftmost red tile having the form of a 'C'. Next, the group is iterated five times (the number of red tiles) and, finally, a vertical pattern of three tiles is added at the right of the footpath. The details for most steps of this particular strategy are displayed in Table 1. This table includes a list a patterns, the relations between attributes and the relations between cases.

The first step includes only one case: the red tiles pattern. After some intermediate steps, not illustrated here, the second step includes 6 cases, i.e. the red pattern and five single blue tiles, which are in a given order as expressed by the set of *Prev* and *Next* relations. In the third step, the 5 blue tiles are grouped in one pattern which now becomes C_2; consequently, at this point there are 2 successive cases. In the fourth step, the second case, i.e the group of 5 blue tiles, is repeated 5 times (the number of red tiles), so now there is also a value and a dependency relation. In the fifth step a new blue tile is added, becoming C_3 and in the sixth step this tile is iterated 3 times; in the last two steps, the relations between attributes and between cases are the same as in step 4.

Table 1. S_u definition for each step of the 'C strategy'.

S_u	$N_u\,(C)$	$N_u\,(RA)$	$N_u\,(RC)$
Step 1	C_1	-	-
Step 2	$C_1, C_2, C_3,$ C_4, C_5, C_6	-	$Prev(C_{i+1}) = C_i$ for $i = \overline{1,5}$ $Next(C_i) = C_{i+1}$ for $i = \overline{1,5}$
Step 3	C_1, C_2	-	$Next(C_1) = C_2$ $Prev(C_2) = C_1$
Step 4	C_1, C_2	$\alpha_{2_3} = \alpha_{1_3}$ $\alpha_{2_3} = DEP(\alpha_{1_3})$	$Next(C_1) = C_2$ $Prev(C_2) = C_1$
Step 5	C_1, C_2, C_3	$\alpha_{2_3} = \alpha_{1_3}$ $\alpha_{2_3} = DEP(\alpha_{1_3})$	$Next(C_i) = C_{i+1}$ for $i = \overline{1,2}$ $Prev(C_{i+1}) = C_i$ for $i = \overline{1,2}$
Step 6	C_1, C_2, C_3	$\alpha_{2_3} = \alpha_{1_3}$ $\alpha_{2_3} = DEP(\alpha_{1_3})$	$Next(C_i) = C_{i+1}$ for $i = \overline{1,2}$ $Prev(C_{i+1}) = C_i$ for $i = \overline{1,2}$

The attributes for each pattern were not included in Table 1 as the focus is on the representation of the strategy and a list of attributes for each pattern in every step would hinder the understanding of the high level representation. The difference between Step 5 and Step 6, however, is not clear without knowing that the difference lies in the attribute values, i.e. for C_3, the iterations attribute has changed from 1 to 3 and the move down attribute has changed from 0 to 1, which is not shown in Table 1.

3.2 Similarity Metrics

Strategy identification is based on scoring elements of the strategy followed by the learner according to the similarity of their attributes and their relations to strategies previously stored. Thus, to identify components of a strategy, four similarity measures are defined:

(a) *Numeric attributes* - Euclidean distance: $D_{IR} = \sqrt{\sum_{j=v+1}^{N}(\alpha_{I_j} - \alpha_{R_j})^2}$ (I and R stand for input and retrieved cases, respectively); attributes from $v+1$ to N are used, i.e. the numeric and binary attributes described in the previous section.
(b) *Variables*: $V_{IR} = \sum_{j=1}^{v} g(\alpha_{I_j}, \alpha_{R_j})/v$ (attributes from 1 to v are variables), where g is defined as: $g(\alpha_{I_j}, \alpha_{R_j}) = 1$ if $\alpha_{I_j} = \alpha_{R_j}$ and $g(\alpha_{I_j}, \alpha_{R_j}) = 0$ if $\alpha_{I_j} \neq \alpha_{R_j}$.
(c) *Relations between attributes* - Jaccard's coefficient: $A_{IR} = \frac{|RA_I \cap RA_R|}{|RA_I \cup RA_R|}$. A_{IR} is the number of relations between attributes that the input and retrieved case have in common divided by the total number of relations between attributes of the two cases;
(d) *Relations between cases* - Jaccard's coefficient: $B_{IR} = \frac{|RC_I \cap RC_R|}{|RC_I \cup RC_R|}$, where B_{IR} is the number of relations between cases that the input and retrieved case have in common divided by the the total number of relations between cases of I and R.

To identify the closest strategy to the one followed by a learner during construction, cumulative similarity measures are used for each of the four similarity types:

(a) Numeric attributes - as this metric has a reversed meaning compared to the other ones, i.e. a smaller number means a greater similarity, the following function is used to bring it to the same meaning as the other three similarity measures, i.e. a greater number means greater similarity:

$$F_1 = \begin{cases} \frac{z}{\sum_{i=1}^{z} D_{I_i R_i}} & \text{if } \sum_{i=1}^{z} D_{I_i R_i} \neq 0 \\ z & \text{if } \sum_{i=1}^{z} D_{I_i R_i} = 0, \end{cases}$$

(b) Variables: $F_2 = (\sum_{i=1}^{z} V_{I_i R_i})/z$;
(c) Relations between attributes: $F_3 = (\sum_{i=1}^{z} A_{I_i R_i})/y$;
(d) Relations between cases: $F_4 = (\sum_{i=1}^{z} B_{I_i R_i})/z$,

where z represents the minimum number of cases among the two compared strategies and y represents the number of pairs of cases in the retrieved strategy that have relations between attributes; for example, the 'C strategy' has three cases and only one relation between an attribute of case C_1 and an attribute of C_2 (see Table 1); therefore there is only one pair of cases that have a relation between attribute, i.e. $y = 1$.

As the similarity metric for numeric attributes has a different range from the other metrics, normalisation is applied to have a common measurement scale, i.e. $[0, 1]$. This is done using linear scaling to unit range [23] by applying the following function: $\overline{x} = \frac{x-l}{u-l}$, where x is the value to be normalised, l is the lower bound and u is the upper bound for that particular value. The range of the values that can be taken by the similarity metric for the numeric attributes, i.e. F_1, is $[0, z]$. Consequently, to transform the values so that they are within the $[0, 1]$ range, the following normalisation function is applied: $\overline{F_1} = F_1/z$.

Weights are applied to the four similarity metrics to express the central aspect of the construction, the structure. This is mostly reflected by the $\overline{F_1}$ metric and, to a lesser extent, by the F_3 metric. Therefore, we agreed on the following weights: $w_1 = 6, w_2 = 1, w_3 = 2, w_4 = 1$. Consequently, the similarity metric for strategies is: $Sim = 6 * \overline{F_1} + F_2 + 2 * F_3 + F_4$, which can take values in the range of $[0, 10]$.

The metrics have been tested for several situations of pedagogical importance: identifying complete strategies, partial strategies, mixed strategies and non-symmetrical strategies. The similarity metrics were successful in identifying all these situations (details can be found in [5]).

4 Adaptation of the Knowledge Base

Adaptive systems refer to systems that change over time to respond to new situations. There are three levels of adaptation depending on the complexity and difficulty of the adaptation process, with the first level being the least difficult and the third being the most complex and difficult [21]: (a) adaptation to a changing environment; (b) adaptation to a similar setting without explicitly being ported to it; (c) adaptation to a new/unknown application. Our adaptive modelling mechanism involves adaptivity at the first level, meaning that the system adapts itself to a drift in the environment by recognising the changes and reacting accordingly [21].

Before going into the details of our approach, we would like to point out the structure of the knowledge base. As mentioned in Section 3, for each task, there is a corresponding knowledge base which consists of strategies. The strategies are represented as a list of simple cases; each case is represented as a list of attributes, a list of relations between attributes and a list of relations between cases. We are not using indexing as for our purpose the similarity matching is not computationally expensive; moreover, because there is a separate knowledge base for each task, the size of the knowledge bases is relatively small.

Our proposed approach for adapting the knowledge-base of *eXpresser* includes acquiring inefficient simple cases and acquiring new strategies. Fig. 5 shows some examples from the 'stepping stones' task introduced previously; the constructions in Fig. 5a and 5c have been broken down into the individual components used by the students for ease of visualisation. These examples, with the adaptation rationale and mechanism are discussed below.

4.1 Acquiring Inefficient Simple Cases

The goal of this mechanism is to identify parts of strategies constructed in inefficient ways and store them in a set or library of 'inefficient constructions', i.e. constructions that pose difficulties for the learners in their process of generalisation. The library could be further used for automatic generation of feedback or could be analysed by a researcher or teacher. The results of such an analysis could be then used to design better interventions or make other design decisions for the current system, could be presented as a lesson learned to the scientific community of mathematics teachers and researchers, or even discussed further in class (e.g in the case of an inefficient construction that is frequently chosen by the pupils of that class).

The construction in Fig. 5a illustrates an inefficient pattern within the "HParallel" strategy of the 'stepping stones' task: the middle bar of blue tiles is constructed as a group of two tiles repeated twice - this can be seen in the property list of this pattern displayed in Fig. 5b. The efficient way to construct this component is one tile repeated four times or, to make it general, one tile repeated the number of red tiles plus one. The efficient and the inefficient way of constructing the middle row of blue tiles lead to the same visual output, i.e. there is no difference in the appearance of the construction, making the situation even more confusing. The difficulty lies in relating the values used in the construction of the middle row of blue tiles (C_i) to the ones used in the middle row of red tiles (C_j). If the learner would relate the value 2 of iterations of C_i to the value 3 of iterations of C_j, i.e. the value 2 is obtained by using the number of red tiles (3) minus 1, this would work only for a 'stepping stones' task defined for 3 red tiles. In other words, this will not lead to a general model.

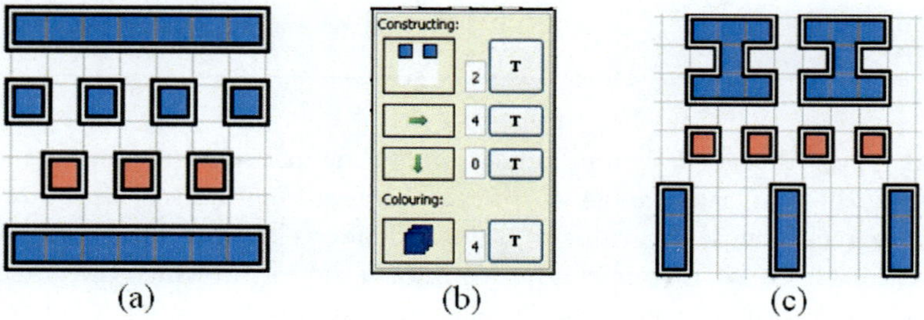

Fig. 5. (a) HParallel strategy with one inefficient component (blue middle row) ; (b) property list of the inefficient component; (c) a new strategy

Algorithms 1, 2 and 3 illustrate how inefficient simple cases are identified and stored. First, the most similar strategy is found. If there is no exact match, but the similarity is above a certain threshold θ, the process continues with the identification of the inefficient cases; for each of these cases, several checks are performed (Alg. 2). Upon satisfactory results and if the cases are not already in the set of inefficient cases, they are then stored (Alg. 3).

Algorithm 1. Verification($StrategiesCaseBase$, $InputStrategy$)

```
Find most similar strategy to InputStrategy from StrategiesCaseBase
StoredStrategy ← most similar strategy;
if similarity > θ then
  Find cases of InputStrategy that are not an exact match to any case of
  StoredStrategy
  for each case that is not an exact match do
    InputCase ← the case that is not an exact match
    Compare InputCase to all cases of the set of inefficient cases;
    if no exact match then
      Find the most similar case to InputCase from the cases of StoredStrategy
      StoredCase ← the most similar case
      if Conditions(StoredCase, InputCase) returns true then // see Alg. 2
        InefficientCaseAcquisition(StoredCase, InputCase) // see Alg. 3
      end if
    end if
  end for
end if
```

Algorithm 2. Conditions($C1$, $C2$)

```
if (MoveRight[C1] ≠ 0 and Iterations[C1] * MoveRight[C1] = Iterations[C2] *
MoveRight[C2]) or
(MoveDown[C1] ≠ 0 and Iterations[C1] * MoveDown[C1] = Iterations[C2] *
MoveDown[C2]) then
  return true
else
  return false
end if
```

What is stored is actually a modification of the most similar (efficient) case, in which only the numerical values of *iterations*, *move-right* and/or *move-down* are updated together with the value and dependency relations. These are the only modifications because, on one hand, they inform the way in which the pattern has been built and its non-generalisable relations, and, on the other hand, it is important to preserve the values of $PartOfS$ attributes, so the researcher/teacher knows in which strategies these can occur. The colouring attributes and the relation between cases are not important for this purpose and, therefore, they are not modified. This has also the advantage of being computationally cheaper.

Algorithm 3. InefficientCaseAcquisition($StoredCase$, $InputCase$)

```
NewCase ← StoredCase
for i = 4 to v − 1 do // attributes from iterations to move-down
  if value of attribute i of NewCase different from that of InputCase then
    replace value of attribute i of NewCase with the one of InputCase
  end if
end for
for all relations between attributes do // value and dependency relations
  replace relations of NewCase with the ones of InputCase
end for
add NewCase to the set of inefficient cases
```

4.2 New Strategy Acquisition

The goal of this mechanism is to identify new strategies and store them for future use. New strategies could be added by the teacher or could be recognised as new from the learners' constructions. In the later case, after the verification checks described below, the decision of storing a new strategy is left with the teacher. This serves as another validation step for the detected new strategy.

Fig. 5c illustrates the so-called "I strategy", as some of its building blocks resemble the letter I. When compared to all stored strategies, this strategy is rightly most similar to the 'VParallel' one (see Fig. 2b), as some parts correspond to it. However, the similarity is low, suggesting it may be a new strategy. Without the adaptation mechanism, the learner modelling module will infer that the learner is using the 'VParallel' strategy, but is still far from having completed it. This imprecise information could be potentially damaging as it could, for example, lead to inappropriate system actions, e.g. providing confusing feedback that would guide the learner towards the 'VParallel' strategy. Conversely, identifying the new construction as a new valid strategy will prevent generating potentially confusing feedback, and storing the new strategy will enable producing appropriate feedback in the future - automatically or with input from the teacher/researcher.

Algorithms 4, 5 and 6 illustrate the process by which an input strategy could be identified and stored as a new strategy (composite case). If the similarity between the input strategy and the most similar strategy from the case-base is below a certain threshold θ_1 (Alg. 4), some validation checks are performed (Alg. 5) and upon satisfaction, the new strategy is stored in the case-base (Alg. 6). If the input strategy has been introduced by a teacher and the similarity is below θ_1, the teacher can still decide to go ahead with storing the new strategy, even if it is very similar to an existing one in the database.

In Algorithm 5 the SolutionCheck($InputStrategy$) function verifies whether $InputStrategy$ 'looks like' a solution by examining if the mask of $InputStrategy$ corresponds to the mask of the task. The following check takes into consideration the number of simple cases in the $InputStrategy$. Good solutions are characterized by a relatively small number of simple cases; therefore, we propose for the value of θ_2 the maximum number of cases among all stored strategies for the

Algorithm 4. NewStrategyVerification($StrategiesCaseBase$, $InputStrategy$)

Find most similar strategy to $InputStrategy$ from the $StrategiesCaseBase$
if $similarity < \theta_1$ **then**
  if ValidSolution($InputStrategy$) returns **true then** // see Alg. 5
    NewStrategyAcquisition($InputStrategy$) // see Alg. 6
  end if
end if

Algorithm 5. ValidSolution($InputStrategy$)

if SolutionCheck($InputStrategy$) returns **true then** // checks if $InputStrategy$ 'looks like' a solution
  if the number of cases of $InputStrategy < \theta_2$ **then**
    if $InputStrategy$ has relations between attributes **then**
      RelationVerification($InputStrategy$) // verifies that the numeric relation corresponds to the task rule solution
      if successful verification **then**
        return true
      end if
    end if
  end if
end if

Algorithm 6. NewStrategyAcquisition($NewStrategy$)

add $NewStrategy$ to the strategies case-base
adjust values of $PartOfS$

corresponding task, plus a margin error (such as 3). If this check is satisfied, the RelationVerification($InputStrategy$) function derives a rule from the value relations of the cases and checks its correspondence to the rule solution of the task. For example, in the construction of Fig. 5c, the rule derived is $3*(\frac{red}{2}+1)+7*\frac{red}{2}$ which corresponds to the solution $5*red+3$. If all checks are satisfied, the new strategy is stored in the case-base and the $PartOfS$ values are adjusted.

5 Validation

The validation of our proposed adaptive modelling mechanisms includes: (a) identifying the boundaries of how far a pattern can be (inefficiently) modified and still be recognised as similar to its original (efficient form); (b) correct identification of inefficient cases within these boundaries and (c) correct identification of new strategies. This low-level testing of the system shows how the adaptation of the knowledge-base and the learner modelling module function together to improve the performance of the system.

To this end, experiments have been conducted using real data produced from classroom use of *eXpresser* as well as artificial data that simulated situations

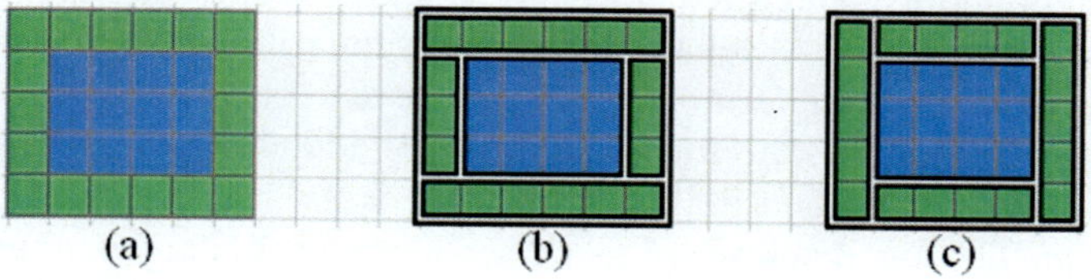

Fig. 6. (a) the construction for the 'pond tiling' task ; (b) 'I Strategy'; (c) 'H strategy'.

observed in the classroom sessions. Simulated situations were based on varying parameters of models produced by learners in order to provide more data.

First, a preliminary experiment using classroom data was conducted to identify possible values for the threshold θ in Algorithm 1 and threshold θ_1 in Algorithm 4. Since our main aim was to test the adaptive modelling mechanism we decided not to seek optimal values for these thresholds, but only to find a good enough value for each one. Two possibilities were quickly identified - for θ: the minimum *overall similarity* (4.50) minus an error margin (0.50) or value 1.00 for the *numerical similarity*; for θ_1: the maximum *overall similarity* (3.20) plus an error margin (0.30) or value 1 for the *numeric similarity.*

Experiment 1: identifying the boundaries of how far a pattern can be inefficiently modified and still be recognised as similar to its original efficient form. As mentioned previously, we consider changes in a pattern that can lead to the same visual output as the original one but use different building-blocks. More specifically, these building-blocks are groups of two or more of the original efficient building-block. This experiment looks for the limits of changes that a pattern can undergo without losing its structure so that it can be still considered to be the same pattern.

For this experiment we used 34 artificial inefficient cases from two tasks: (a) 'stepping stones' that was defined earlier and (b) 'pond tiling' which requires to find the number of tiles needed to surround any rectangular pond. Fig. 6 illustrates the construction for the 'pond tiling' problem and two strategies frequently used by students to solve this task. Our adaptive mechanism was build to work for any task in *eXpresser* rather than for particular tasks. For the two tasks we used in our experiments, the tests were conducted using their corresponding user data and their (separate) knowledge bases; the results were collated.

From the 34 cases, 47% were from the 'stepping stones' task and 53% were from the 'pond tiling' task. Using these cases, the following boundaries were identified: (i) groups of less than 4 building-blocks; (ii) groups of 2 building-blocks repeated less than 6 times and (iii) groups of 3 building-blocks iterated less than 4 times.

Experiment 2: correct identification of inefficient cases within the previously identified boundaries. From the total of 34 inefficient cases used in *Experiment 1*, 13 were outside the identified boundaries and 21 were within. From the 21 cases within the boundaries, 62% were from the 'stepping stones' task and 38% were from the 'pond tiling' task.

Using the previously identified values for θ, we obtained the following results: out of these 21 cases, 52.48% had the overall similarity greater than 4.00 and 100% had the numeric similarity above 1.00. These results indicate that a small modification of a pattern can drastically affect the identification of the strategy the learner is following; hence almost half the cases have an overall similarity less than 4.00. The results obtained using the numeric similarity are much better and consistent with the fact the modifications are just numerical.

Experiment 3: correct identification of strategies. The data for this experiment included 10 new strategies: 7 observed in trials with pupils and 3 artificial. Out of the 10 new strategies, 4 were from the 'pond tiling' task; all of them were observed in trials with pupils. The remaining 6 new strategies were from the 'stepping stones' task, with 3 of them observed and 3 artificial. The knowledge base for the two tasks included originally 4 strategies for the 'stepping stones' task and 2 strategies for the 'pond tiling' task.

Using the previously identified values for θ_1, we obtained the following results: out of the 10 new strategies, 100% had the overall similarity below 3.50 and 70% had the numeric similarity below 1.00. As opposed to Experiment 2, the overall similarity performs better, being consistent with the fact that the overall similarity reflects better the resemblance with the stored strategies than the numeric similarity alone. Given the range that the overall similarity has, i.e. 0 to 10, values below 3.50 indicate a very low similarity and therefore, rightly suggest that the learner's construction is considerably different from the ones in the knowledge base.

6 Conclusions

In this chapter we presented research on modelling users' behaviour in *eXpresser*, an exploratory learning environment for mathematical generalisation. We adopted a case-based formulation of strategies learners follow when building constructions in *eXpresser* and employed similarity metrics to identify which strategy is used by each learner. Due to the open nature of the environment, however, not all strategies are known in advance. Moreover, learners use the system in inefficient ways that lead to difficulties in solving the given tasks. To overcome these problems, we developed an adaptive modelling mechanism to expand an initially small knowledge base, by identifying inefficient cases (i.e. cases that pose additional difficulty to the user's learning process) and new strategies.

For both inefficient patterns and new strategies, the principle is the same: they are compared with data from the knowledge base and if they are not already stored, some task-related checks are performed and upon successful verification, they are added to the knowledge base. With this mechanism, new data can be added to the knowledge base without affecting the recognition of existing data.

To evaluate our proposed adaptive modelling mechanism three experiments were conducted: (a) identifying the boundaries of how far a pattern can be inefficiently modified and still be recognised as similar to its original efficient form; (b) correct identification of inefficient cases within these boundaries and (c) correct identification of new strategies. The evaluation of the proposed approach showed that it is capable of recognising new inefficient patterns within certain boundaries and of recognising new strategies. The boundaries for recognising inefficient patterns are related to the similarity metrics' ability to identify how much they have been modified from their original initial form. When looking at the modifications that learners tend to make, we notice that they take the form of using repetitions of the basic building-block, which modify the structure of the pattern. The similarity metrics, however, were defined to recognise *structural* similarity. Therefore, to improve the metrics' ability to recognise modifications of efficient patterns, they should be enhanced with the capacity to recognise sub-patterns.

Our adaptive modelling mechanism ensures that the learner diagnosis will be accurate even when the researcher or teacher authors only one or two strategies for a new task. Also, it ensures that the learner diagnosis will be accurate when learners' behaviour changes over time even if initially there is a large knowledge base.

The adaptive mechanism that we developed was tailored for *eXpresser* and the domain of mathematical generalisation. We believe, however, that the high level idea can be used in other exploratory learning environments and for domains where several approaches are possible for the same problem.

Future work includes using the information on inefficient cases and new strategies in an automatic way to either incorporate this information in the feedback and/or inform the teachers and allow them to author feedback.

Acknowledgements. This work is partially funded by the ESRC/EPSRC Teaching and Learning Research Programme (Technology Enhanced Learning; Award no: RES-139-25-0381).

References

1. Papert, S.: Mindstorms: children, computers and powerful ideas. BasicBooks, New York (1993)
2. Piaget, J.: The Psychology of Intelligence. Routledge, New York (1950)
3. de Jong, T., van Joolingen, W.: Scientific discovery learning with computer simulations of conceptual domains. Review of Educational Research 68, 179–202 (1998)
4. Kirschner, P., Sweller, J., Clark, R.: Why minimal guidance during instruction does not work: An analysis of the failure of constructivist, discovery, problem-based, experiential and inquiry-based teaching. Educational Psychologist 41(2), 75–86 (2006)
5. Cocea, M., Magoulas, G.D.: Task-oriented modeling of learner behaviour in exploratory learning for mathematical generalisation. In: Proceedings of the 2nd ISEE workshop, in conjunction with AIED 2009, pp. 16–24 (2009)
6. Pearce, D., Mavrikis, M., Geraniou, E., Gutiérrez, S.: Issues in the Design of an Environment to Support the Learning of Mathematical Generalisation. In: Dillenbourg, P., Specht, M. (eds.) EC-TEL 2008. LNCS, vol. 5192, pp. 326–337. Springer, Heidelberg (2008)

7. Dorer, W.: Forms and means of generalisation in mathematics. In: Bishop, A., Mellin-Olsen, S., van Dormolen, J. (eds.) Mathematical Knowledge: Its Growth through Teaching, pp. 63–85. Kluwer Academic Publishers, Dordrecht (1991)
8. Harel, G., Tall, D.: The general, the abstract and the generic in advanced mathematics. For the Learning of Mathematics 11(1), 38–42 (1991)
9. Moss, J., Beatty, R.: Knowledge building in mathematics: Supporting collaborative learning in pattern problems. International Journal of Computer-Supported Collaborative Learning 1(4), 441–465 (2006)
10. Mason, J., Haggarty, L.: Aspects of Teaching Secondary Mathematics: Perspectives on Practice. In: Generalisation and algebra: Exploiting children's powers, Routledge Falmer and the Open University, pp. 105–120 (2002)
11. Geraniou, E., Mavrikis, M., Hoyles, C., Noss, R.: A constructionist approach to mathematical generalisation. In: Joubert, M. (ed.) Proceedings of the British Society for Research into Learning Mathematics. BSRLM Proceedings, vol. 28(2) (2008)
12. Hoyles, C., Küchemann, D.: Students understanding of logical implication. Educational Studies in Mathematics 51(3), 193–223 (2002)
13. Bednarz, N., Kieran, C., Lee, L.: Approaches to algebra: Perspectives for research and teaching. In: Bednarz, N., Kieran, C., Lee, L. (eds.) Approaches to algebra: Perspectives for research and teaching, pp. 3–12. Kluwer Academic Publishers, Dordrecht (1991)
14. Noss, R., Healy, L., Hoyles, C.: The construction of mathematical meanings: Connecting the visual with the symbolic. Educational Studies in Mathematics 33(2), 203–233 (1997)
15. Noss, R., Hoyles, C.: Windows on Mathematical Meanings: Learning cultures and computers. Kluwer Academic Publishers, Dordrecht (1996)
16. Warren, E., Cooper, T.J.: Generalising the pattern rule for visual growth patterns: actions that support 8 year olds' thinking. Educational Studies in Mathematics 67(2), 171–185 (2008)
17. Malara, N., Navarra, G.: ArAl Project: Arithmetic pathways towards favouring pre-algebraic thinking. Pitagora Editrice, Bologna (2003)
18. Küuchemann, D.: Childrens Understanding of Mathematics, pp. 11–16. John Murray, London (1991)
19. Duke, R., Graham, A.: Inside the letter. Mathematics Teaching Incorporating Micromath 200, 42–45 (2007)
20. Kolodner, J.L.: Case-Based Reasoning, 2nd edn. Kaufmann Publishers, Inc., San Francisco (1993)
21. Anguita, D.: Smart adaptive systems: State of the art and future directions of research. In: Proceedings of the 1st European Symposium on Intelligent Technologies, Hybrid Systems and Smart Adaptive Systems, EUNITE 2001, pp. 1–4 (2001)
22. Mitra, R., Basak, J.: Methods of case adaptation: A survey: Research articles. International Journal of Intelligent Systems 20, 627–645 (2005)
23. Aksoy, S., Haralick, R.: Feature normalisation and likelihood-based similarity measures for image retrieval. Pattern Recognition Letters 22(5), 563–582 (2001)

Chapter 13
Discussion Support System for Understanding Research Papers Based on Topic Visualization

Masato Aoki, Yuki Hayashi, Tomoko Kojiri, and Toyohide Watanabe

Graduate School of Information Science, Nagoya University
Furo-cho, Chikusa-ku, Nagoya, 464-8603, Japan
{maoki,yhayashi,kojiri,watanabe}@watanabe.ss.is.nagoya-u.ac.jp

Abstract. When reading a research paper, not only to understand its contents but also to obtain related knowledge is essential. Since knowledge of each student is different, they can acquire related knowledge through discussion with others. However, discussion sometimes falls into the specific topics and students are unable to acquire various knowledge. Our objective is to construct a collaborative discussion support system for promoting effective discussion by visualizing the diversity of discussed topics. If they can notice the discussion situation timely, they may be able to derive different topics. To effectively evaluate a paper, participants should discuss each research aspects. In our research, topics are extracted and discriminated according to the stages by their targets in the paper. In addition, the topics are evaluated from the viewpoints of the similarities between a topic and the paper, and among topics. For expressing the discussion situation, our system visualizes topics (topic nodes) around the core of the circle (section node) which represents stages in the paper. The similarity between a topic and its target section is represented by the distance between topic and section nodes. The similarity among topics is represented by the distance among topic nodes. By organizing topics around the section node, participants can intuitively understand the discussion situation and are encouraged to voluntarily discuss diverse topics. Based on an experimental result, our system can allocate topics appropriately. In addition, participants were able to grasp the discussion situation by observing the discussion visualization.

Keywords: understanding research paper, collaborative discussion, discussion visualization, discussion environment.

1 Introduction

When we clarify the originality of our own research, we must investigate related research. It is useful to read many research papers at an early stage of research for understanding not only the various applicable techniques but also the viewpoints of solving the target problems. It is important for the students, especially who are not used to read research papers, to read research papers according to

T. Watanabe and L.C. Jain (Eds.): Innovations in Intell. Machines – 2, SCI 376, pp. 185–201.
springerlink.com

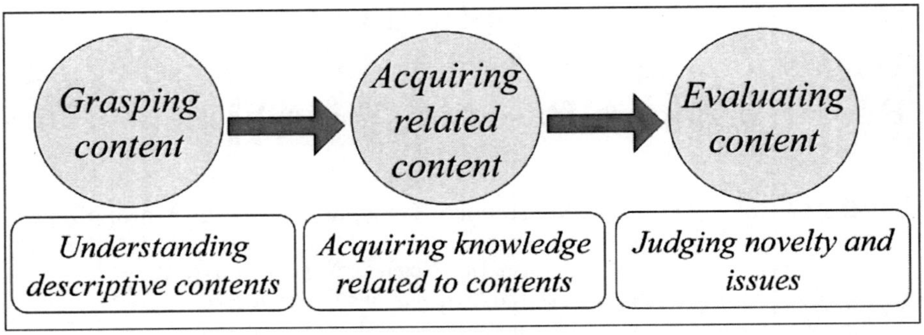

Fig. 1. Process of understanding research papers

the following process. Firstly,they understand its contents by perusing it (*grasping content stage*). Secondly, they consider its related knowledge such as other methods and various viewpoints (*acquiring related content stage*). Finally, they evaluate its novelty and issues (*evaluating content stage*). The process of understanding a research paper is shown in Figure 1. To appropriately evaluate the paper, they must consider related knowledge from various perspectives for every research stage. In many cases, we evaluate the paper by our own knowledge without aggressively acquiring related contents. In such cases, if we do not have enough knowledge of the paper, we cannot evaluate it correctly. One solution to this problem is to discuss the paper in a group. Through a discussion with others, we can acquire related knowledge that others have. Moreover, knowledge from different perspectives may be derived through the discussion.

With the recent development of information and communication technologies, much research supports discussion in distributed environment [1,2]. This research focuses on collaborative discussion in a research group for obtaining related knowledge of research papers in such a distributed environment. To appropriately evaluate a paper, participants should discuss its various research stages, such as background, objective, and method. However, participants cannot always effectively discuss a paper because they sometimes discuss the paper from a limited perspective. If they could notice the discussion situation timely, they may derive new topics from different perspectives. Our objective is to construct a collaborative discussion support system for promoting effective discussion by visualizing the diversity of discussed topics. This research proposes an environment for deriving various knowledge of the paper.

Many researches for discussion visualization have been reported. These researches focus on activeness of participants or temporal relations among topics. However, discussion quality is not often represented by only activeness of participants and active time. In discussing research papers, to derive various knowledge related to the paper is desirable. In our research, topics are evaluated and visualized from the perspective of the related knowledge which they contain. To obtain related knowledge, diverse topics must be discussed from various perspectives. Moreover, topics for acquiring related contents contain not only knowledge written in the paper but also knowledge that is not written in the paper, whereas topics of grasping contents tend to have only knowledge in the paper. Thus, every topic in the discussion is evaluated from the viewpoint of the similarities between the topic and the paper, and among topics.

Currently, we focus on text-based discussion using chat. Discussion participants are members in a laboratory. They read the paper in advance and gather freely. There is no teacher in the discussion. In this research, topics are extracted from messages in the chat which are posted by participants. Relations between topics and the paper and those among topics are estimated. For expressing the discussion situation, the paper is placed in a center of a circle and topics are distributed in a circle. By organizing topics around the contents of the paper, participants can intuitively understand the discussion situation and are encouraged to discuss voluntarily diverse topics.

2 Related Work

Many researches proposed visualization methods of discussion structure after a discussion has finished. Conklin et al. constructed a discussion support system for problem-solving by representing the relationship between messages [3]. In their system, participants' messages are divided into four types (Issue, Position, Argument, and Other) and represented by nodes with different attributes. The relations between messages are represented by labeled links, such as *generalizes*, *specializes*, and *responds-to*. By observing the nodes and the links, participants can detect inconsistencies and neglected topics. However, judging the whole discussion situation is not easy because relations among topics are not provided.

Janssen et al. developed a system for encouraging participants to deepen discussions [4]. Topics are represented as sequences of messages enclosed in squares. Each square is placed on the left or right sides which correspond to the *discussion* and *agreement* states. Each message is classified as *discussion* or *agreement* based on its role in communication. The positions of the squares are determined by the classification of messages. By observing the distributions of squares, participants can understand well discussed and well agreed topics. However, they cannot grasp the variety of discussed topics because they cannot recognize the similarities among topics.

Zhu et al. developed a system for detecting current and ongoing topics by focusing on the similar contents and participations of users [5]. By applying the extended TF-IDF model to a threaded chat, the system detects semantically similar threads. From the common participated user, the topics of identical contents are detected. Kojiri et al. also proposed a system that visualizes the structure of an ongoing discussion [6]. To smoothly integrate the participation of latecomers in discussions, the system extracts important messages from topics based on the number of messages of the topics and the posted time. These systems can indicate current important messages, but they cannot promote active discussions.

Some research has activated discussions by showing the activenesses of participants. Leshed et al. constructed a system for showing the degrees of participant contributions to discussions using a school of fish metaphor [7]. Each participant is displayed as a colored fish. The vertical height of each fish represents the degree of agreement from other participants. The system calculates the positions of each fish every minute, and participant trajectories are represented as bubbles. By observing the fish positions and the bubbles, participants can grasp the degrees of participant contributions for discussions each time. Viegas et al. developed a system called Chat Circles that expresses the messages of each participant in resizable circles [8]. Active participants are displayed as large circles, since each circle becomes smaller according to the duration time of not posting messages. Erickson et al. proposed a visualization method of representing the activenesses of discussions by the positions of circles that correspond to participants [9]. Each circle is placed on a common circle that represents workspaces. The common circle's center corresponds to a high activity level. Xiong et al. constructed a system which represents the activeness of each participant by a flower metaphor [10]. Participants are represented as flowers whose stem lengths express the lengths of their login times. Moreover, the petals of each participant represent the proposed topics, and responses from other participants are displayed as small circles at the distal end of the target topic. Tat et al. constructed a system that expresses discussion activeness and atmosphere [11]. The participants' messages are arranged in the directions of each participant as circles. The emotions of each participant are estimated by emoticons and represented by the colors of translucent planes in directions corresponding to the participants. In addition, this system changes the color strengths of the circles based on the number of message characters. By selecting a certain time, participants can intuitively understand the activeness and atmosphere of the discussion. Lam et al. proposed a method for expressing the activeness of groups based on a metaphor of a continuous movement [12]. Each thread in the discussion is represented as a square. Squares of the threads, which include the newest messages, vigorously move like an ocean wave or volcanic lava. All the above researches represent the activeness of ongoing discussions. However, in these researches, participants cannot always grasp the fruitfulness of the discussion.

Our research is different from these researches because our research focuses on leading effective understanding research papers. We introduce criteria which are specialized in understanding research papers. Then, discussed topics are visualized from a viewpoint of their contributions for understanding research paper.

3 Approach

The target of our research is to read engineering research papers that propose newly developed technologies or systems in limited pages. The discussion participants are researchers who are interested in the target areas of the research papers. Researchers read the papers based on their own research perspectives. Research papers consist of several sections for each research stage, so it is important to discuss all sections.

Followings are factors that are grasped by participants in an ideal discussion situation.

($f1$) Current discussion situation
($f2$) Discussion situation for each section
($f3$) Knowledge related to the paper
($f4$) Various perspectives

A discussion is generally classified as *creative* or *problem-solving*. In *creative discussions*, participants do not always have a common clear goal, but seek various perspectives regarding the discussion theme. In *problem-solving discussions*, participants try to reach a specific goal. A discussion for obtaining related knowledge of the paper is *creative* since a clear goal does not exist. Moreover, *creative discussions* are classified as *focused* or *global* depending on the target of the paper from which participants need to acquire knowledge. The purpose of a *focused discussion* is deep understanding of specific parts of the paper such as its technology and assumed environment. In a *global discussion*, participants gather opinions about any part of the paper from various perspectives. Since each section of the paper corresponds to a particular research stage, obtaining comprehensive knowledge of each section leads participants to consider research aspects such as background, objective, solution, and evaluation. Therefore, in this research, we support *creative* and *global discussions*.

In a *global discussion*, all research aspects should be observed, so discussing all sections is important. In addition, if there are many topics from the same perspective, various topics should be discussed to avoid participant's evaluating the paper based on only limited viewpoints.

In *creative discussions*, developed topics that are associated with the section are required. Topics that are not directly related to the contents of papers do not contain information to evaluate the papers. Thus, in this research, every discussion topic is analyzed from the viewpoint of the similarity among its target section and topics.

Figure 2 shows the concept of visualizing topics in discussions. Visualization should increase participant aware of the discussion situation. In our system, the topics, which are collections of messages, are extracted and placed as "a topic node" around "a section node". A section node indicates the contents of the target section. Two similarities, such as similarities among topics and that among topic and section, are calculated based on keywords in the section that are included in the topic messages.

Similarity between a topic and its target section is represented by the distance between topic and section nodes. Similarity among topics is represented by the distance among topic nodes. Participants become aware of their discussion situation by the distribution of the topic nodes around the section nodes ($f1$). Insufficiently discussed sections can be grasped by the number of the topic nodes around each section node ($f2$). If many topic nodes exist near the section node, participants are urged to derive developed topics ($f3$). In addition, if many topic nodes exist in a certain direction from a topic node, participants are urged to derive topics from other perspectives ($f4$).

Figure 3 shows the processing steps for visualizing topics. Currently, we focus on text-based discussion using chat. Our system extracts the keywords of sections

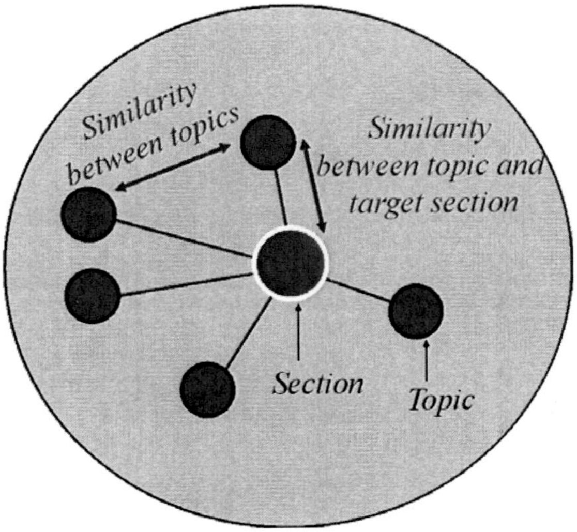

Fig. 2. Concept of visualization

in advance from their texts. The similarity between a topic and its target section is calculated based on the keywords contained in the topic. The similarity among topics is regarded as the difference between the target locations in the section. The target location of a topic means the position in the section that is focused in the discussion. In each section of the paper, related sentences are written near each other. Therefore, if the target sentences of the topics are near each other, their contents may be similar. The similarity among topics is calculated by the distances between target locations. The system determines calculated degrees of similarity between a topic and its target section and the target location of the topics, and displays the topic nodes around the section nodes.

4 Topic Visualization Method

4.1 Extraction of Keywords in Section

Some research has been reported for grasping the discussion. Inaba et al. proposed a method for detecting each participant's degree of participation based on Negotiation Process Model [13]. However, because they focused on just relations among message types, the contents of discussed topics cannot be grasped. Therefore, in this research, we proposed a method for detecting contents of discussed topics based on keywords.

Section contents can be expressed as a set of keywords. When we discuss the paper, we often indicate the target part by pointing out the words written in the part; especially when we do not share the same physical space. To detect the target part in the paper, we define characteristic words (single or successive nouns) as keywords that indicate each section. Of course, it is clear that section

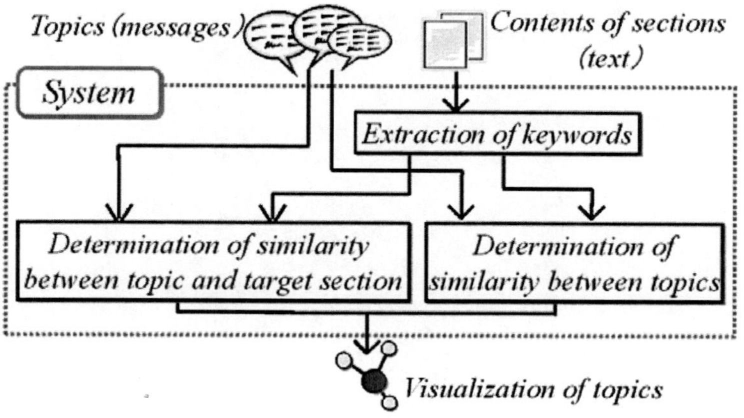

Fig. 3. Processing steps

is not always represented by words in the paper. In addition, the words can be transformed by different words of the same meaning. However, some of the keywords must be included in the messages and it is reasonable to detect the target section by keywords.

To extract keywords, our system detects nouns from the texts of each section in advance by using a morphological analyzer. Such words appear frequently in the section but do not appear in the whole paper. To acquire the keywords for each section, the degrees of importance of each word for the sections are calculated by Equation 1. $value(s, a)$ represents the degrees of importance of word a in section s. $count(s, a)$ is the number of words a that appears in section s, and $N(s)$ is the total number of words in section s. The degree of importance of the word increases if it is used frequently in the section and decreases if it appears throughout the whole paper. Our system extracts words whose degrees of importance are larger than a certain threshold as keywords.

$$value(s, a) = \frac{count(s, a)}{N(s)} \times \log\left(\frac{\sum_i N(i)}{\sum_i count(i, a)}\right) \qquad (1)$$

4.2 Expression of Similarity between Topic and Section

The topics, which have strong connections to the contents of the target section, are placed close to the section node. The distance between topic and section nodes is defined as Equation 2. $distance(s, t)$ represents the distance of topic t from target section s. $\sum_i value(s, i)$ is the sum of the degrees of importance of all keywords in section s. $relation(s, t)$ is the degree of the similarity between topic t and its target section s. Based on the equation, the distance of the topic with a large degree of similarity becomes small, as shown in Figure 4.

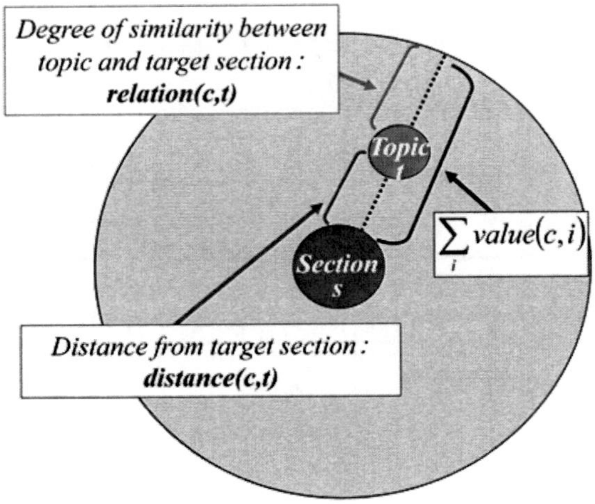

Fig. 4. Expression of similarity between topic and section

The similarity between a topic and its target section is expressed by the ratio of keywords included in a topic. Thus, $relation(s,t)$ is defined as Equation 3. $W(t)$ is the total number of words in topic t. $secIn(s,t)$ is the number of keywords of target section s contained in topic t. α is a constant number that coordinates the effect of $\sum_{i \in t \cap s} value(s,i)$ and takes from 0 to $\sum_i value(s,i)$. As α increases, the effect of $secIn(s,t)$ becomes larger. Based on the equation, the degree of the similarity of the topic that includes a large number of important keywords of the section increases.

$$distance(s,t) = \sum_i value(s,i) - relation(s,t) \tag{2}$$

$$relation(s,t) = \frac{secIn(s,t)}{W(t)} \times (\alpha + \sum_{i \in t \cap s} value(s,i)) \tag{3}$$

4.3 Expression of Similarity among Topics

The similarity among topics can be determined by the target locations in the section. Since research paper is logically structured, the topics are not strongly related if their target locations are not near within the section. We define the degree of the similarity among topics as the distance between their target locations. The target location of a topic can be grasped by keywords of the section and is calculated by Equation 4. $location(s,t)$ represents the target location of topic t in target section s and it takes value from 0 (beginning of section) to 1 (end of section). $position(s,i)$ indicates the appearance position of keyword i in section s. If keyword i appears in multiple locations, $position(s,i)$ is set as the middle point of the appearance position of keyword i. The target location of the

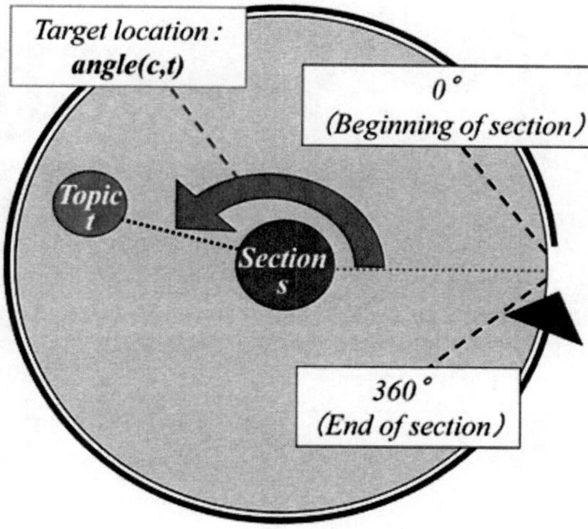

Fig. 5. Expression of target location

topic is represented as the average of the appearance positions of all emerging keywords.

The angle of the topic node is determined based on $location(s, t)$. The beginning of each section corresponds to 0°, and its end is 360° around the section node. The angle of the topic node is calculated by Equation 5. $angle(s, t)$ is an angle of topic t around section s. The angle is determined by multiplying 360° by $location(s, t)$. The topic node is arranged in corresponding $angle(s, t)$, as shown in Figure 5. Based on this expression, the beginning and the end of a section are placed proximally. A main theme of a section is often explained in the first sentence and summarized in its final sentence. Thus, this expression is valid to some extent.

$$location(s,t) = \frac{\frac{1}{secIn(s,t)} \times \sum_{i \in t \cap s} position(s,i)}{N(s)} \quad (4)$$

$$angle(s,t) = 360° \times position(s,t). \quad (5)$$

5 Prototype System

We have constructed a prototype system by embedding a discussion visualization mechanism in a collaborative learning support system developed in our laboratory [14]. In this prototype system, the morphological analyzer *Sen*[1] (*Java* port of *MeCab*[2]) is used. Currently, this system can cope with research papers written in Japanese. However, it can also support English discussion, if English

[1] https://sen.dev.java.net/

[2] http://mecab.sourceforge.net/

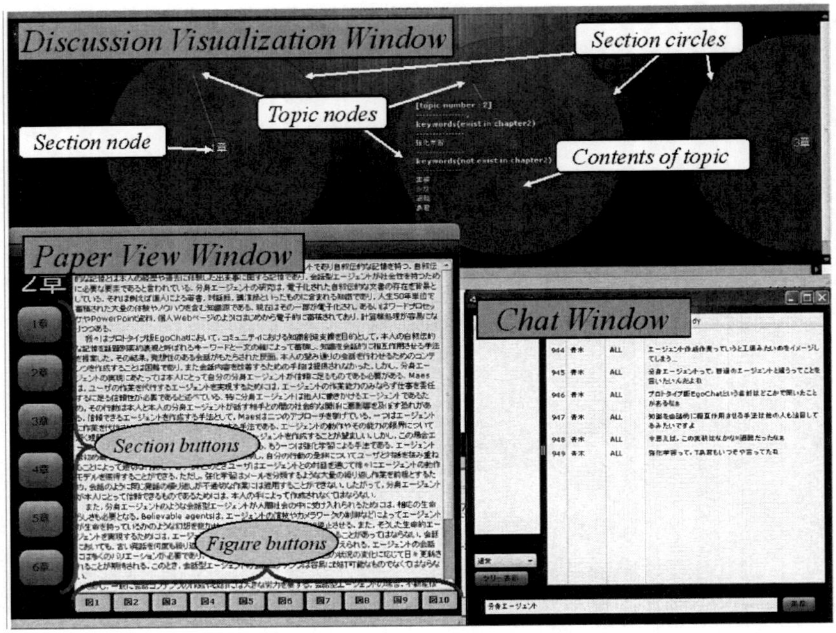

Fig. 6. Interface of prototype system

morphological analyzer is introduced. The interface of this system consists of three windows, as shown in Figure 6.

Participants make messages using the chat window. The paper view window displays the paper contents and provides identical contents to all participants. By selecting the section button, the contents of the selected section appear in all participants' windows. By selecting the figure button, the figure in the section is displayed in a separate window. When a section button is pushed, our system regards that the topic has changed, and retrieves messages that compose the topic. These messages are also analyzed by our visualization method, and the result is sent to the discussion visualization window.

In the discussion visualization window, each section is viewed as a circle. The section nodes exist in the center of each circle. Topic nodes are represented by red circles around the section node. The information of words within the topic is shown by moving the mouse cursor over a topic node (Figure 7). Words included in the topic are viewed as either keywords of the target section or other words. In addition, the topic messages are displayed by clicking a topic node (Figure 8). By clicking the inside of a section circle, keywords of the section are displayed. These keywords are arranged at the angles of appearance positions in the section (Figure 9). Discussion of a specific keyword may be encouraged by observing such keywords. By clicking on a displayed keyword, it is posted in the chat window's input area.

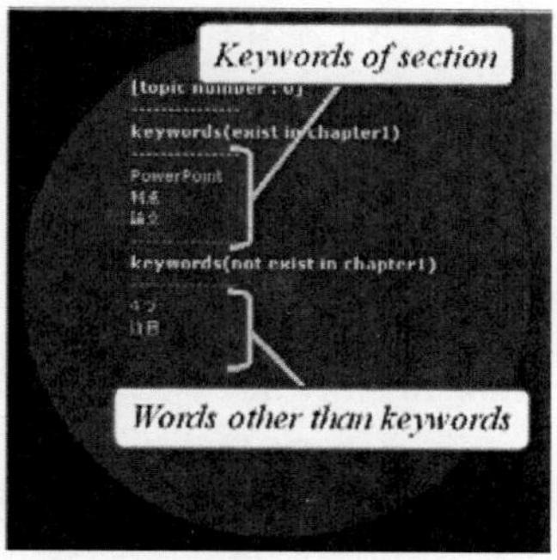

Fig. 7. Information of words in topic

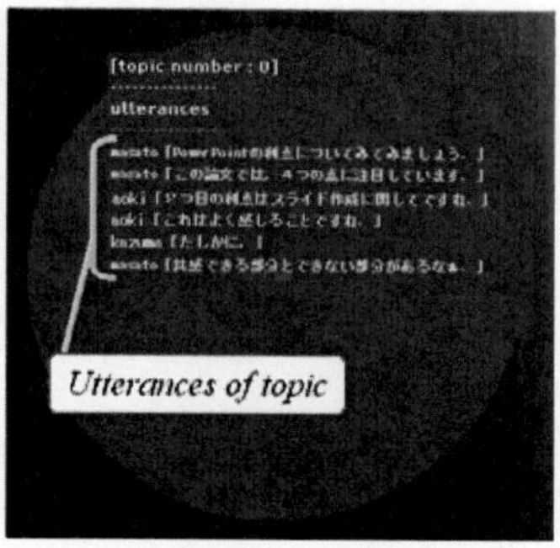

Fig. 8. Topic messages

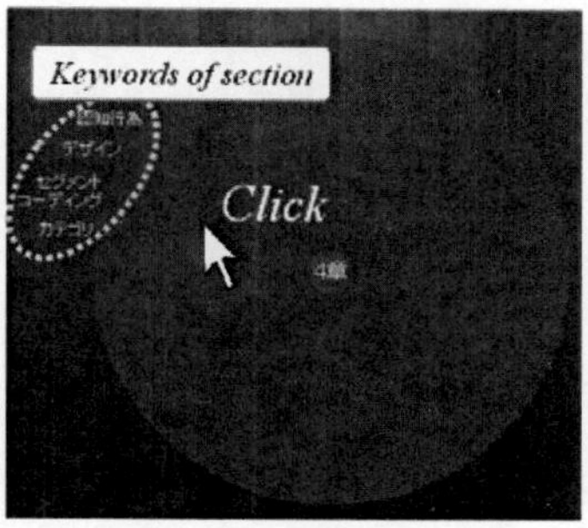

Fig. 9. Section keywords

6 Experiment

6.1 Experiment of Extracting Keywords

Validity of extracted keywords was evaluated by comparing them with manually selected keywords. The research field of the target paper is knowledge management. The paper consists of six sections. Total length of the paper is nine pages. Numbers of characters in each section are 1110, 1915, 1236, 2832, 1750, 390 and 159. In Section 1, background of the research is described. Requirement and approach are explained in Section 2, and proposed method for knowledge management is written in Section 3. In Section 4, applicability of the method is discussed. Related works are shown in Section 5 and conclusion is described in Section 6.

Correct keywords for each section were manually collected by one of the authors subjectively. The numbers of keywords for each section were 7, 9, 14, 8, 4 and 3. Since keywords with high important degrees are important for discriminating the section of the topic, extracted keywords by our method whose important degrees were within top twenty were investigated if they were included in the correct keywords.

Table 1. Result of extracted keywords

Section	1	2	3	4	5	6	total
Recall rate	71.4% (5/7)	66.7% (6/9)	42.9% (6/14)	50.0% (4/8)	75.0% (3/4)	100.0% (3/3)	60.0% (27/45)

Table 1 shows recall rate of extracted keywords. Recall rate is calculated as Equation 6.

$$Recall\,rate = \frac{number\,of\,extracted\,keywords\,in\,correct\,keywords}{number\,of\,correct\,keywords} \qquad (6)$$

In Sections 1, 5 and 6, extracted keywords covered correct keywords more than 70%. These keywords appear frequently in the target section and do not often appear in other sections. The numbers of characters in these sections are smaller than other sections. In such sections, the numbers of described topics are not many, so the appearance ratio of particular words may be large. On the contrary, if the number of topics is many, several words from different topics exist, so the ratio becomes smaller. Thus, our method could extract characteristic words for the sections of small size.

In whole paper, correctly extracted keywords are those that are defined in the paper as key concepts of the paper, such as *kihon story* (basic story). On the contrary, the incorrectly extracted keywords in these sections include some meaningless complex words, such as *kaiwa rogu bunshin agent* (conversation log avatar). Currently, all successive nouns are regarded as one complex word. The number of appearance of the complex word, such as $count(s, a)$ in Equation 1, is counted by adding the numbers of all individual nouns that compose the complex word. Since such complex word includes plural nouns, its appearance becomes large even if the complex word itself does not exist many times. Therefore, we should improve the method for counting the appearance numbers of complex words.

This experiment was executed for only one research paper. We need further evaluation using other papers.

6.2 Experimental Setting of Using System

We evaluated the adequacy and effectiveness of the visualization method using our prototype system. In this experiment, the maximum distance between section and topic node was normalized to 100, and α was set to 50 (half of the normalized maximum distance).

Groups A and B of four students in our laboratory discussed a research paper. Group A consists of one doctoral student, two graduate students and one undergraduate student. Group B consists of two doctoral students and two graduate students. The paper used in Section 6.1 was a target paper. They have knowledge for the knowledge management, but they had not read the paper before.

Table 2. Result of discussion

	Group A	Group B
Time	1:12:04	1:25:14
Number of topics	12	22
Number of messages	123	457
Largest distance	91.04 (maximum:100)	94.41 (maximum:100)
Smallest distance	52.37 (minimum:0)	48.06 (minimum:0)
Farthest angle	175.63° (maximum:180°)	177.89° (maximum:180°)
Nearest angle	6.08° (minimum:0°)	1.50° (minimum:0°)

The discussion purpose was to acquire related knowledge of the paper from various perspectives. Each examinee was asked to read the paper and understand its contents in advance. If they want to check its contents of a section which is not being discussed during the discussion, they were asked to read their own papers and to avoid using the paper view window. The paper view window was only used for changing the target of the discussion topics. One examinee of each group was asked to determine the end of the discussion. In both groups, discussion continued for more than one hour.

After the discussion, the examinees answered questionnaires about the similarities between a topic and its target section, and among topics. From the result of the questionnaires, we evaluated the effectiveness for supporting the discussion by our system. In addition, examinees observed the discussion record of the other group for each topic. To evaluate the validity of the calculated degrees of similarity between a topic and its target section, the examinees divided the topics of the other group based on the relation to the target section as either slightly or greatly related. For evaluating the validity of the calculated degrees of similarity among topics, examinees also selected combinations of similar topics of the other group and described their reasons. Moreover, the examinees answered another questionnaire about use of the whole system.

6.3 Experimental Results of Using System

Table 2 shows the result of the discussion. Table 3 is the results of the average distances of greatly or slightly related topics to the section calculated by the system (Equation 2 in Section 3). These distances are measures for the similarity between a topic and its target section. In this research, we aim to place topics which are strongly related to the section near the section. For every examinee of both groups, the average distances of the slightly related topics are larger than those of the greatly related ones. Therefore, the system adequately expressed the similarity between a topic and the target section as these distances.

Table 3. Average distances between topic and target section

	Discussion of group A				Discussion of group B			
Examinees	e	f	g	h	a	b	c	d
Topics slightly related to section	85.18	83.16	83.71	81.54	77.59	74.74	81.14	74.72
Topics greatly related to section	64.54	68.13	67.86	64.74	68.23	69.05	70.14	70.73
All topics	73.14				71.64			

Table 4. Average angles between topics

	Discussion of group A				Discussion of group B			
Examinees	e	f	g	h	a	b	c	d
Similar topics	31.80°	63.53°	82.86°	49.71°	34.94°	39.16°	38.07°	39.92°
All topics	90.62°				70.51°			

Table 5. Question scores about visualization

	Group A				Group B				Average
Examinees for answering	a	b	c	d	e	f	g	h	
a. Adequacy of distance between topic and section	4	2	4	2	4	4	4	2	3.25
b. Adequacy of distance between topics	4	5	4	2	4	4	4	4	3.88
c. Effectiveness for grasping discussion situation	4	4	4	2	4	3	4	5	3.75
d. Effectiveness for selecting target section	1	3	2	1	2	1	3	4	2.13
e. Effectiveness for selecting target location of section	1	2	4	2	2	1	3	1	2.00

Table 6. Average question scores about use of system

	Group A				Group B				Average
Examinees for answering	a	b	c	d	e	f	g	h	
Related contents about the paper can be discussed.	4	4	4	4	2	3	5	4	3.75
You want to use this system for acquiring related contents again.	3	4	3	3	4	2	4	4	3.38

The average angles between similar topics are shown in Table 4. Since we cannot directly compare angles of topics whose target sections are different, angles of topics for different sections are not considered in this experiment. In this research, we aim to place similar topics near. For all examinees, the average angles between similar topics are smaller than those of all topics. Therefore, the system properly placed similar topics in the near locations.

Table 7. Free comments about system

Target	Comments about interface
Chat window	- No problem for communication.
Paper view window	- I hesitated to push the section buttons. - I did not understand the timing for pushing the section buttons. - I felt responsible for the topic when I pushed the section button. - I think a mechanism should be prepared for obtaining the agreements of others before changing sections. - I couldn't understand when section buttons were pushed, so topics were imperceptibly changed.
Discussion visualization window	- I could roughly grasp the diversity of the discussed topics. - I was able to review the past topics. - It was easy to understand the variety of topics in each section. - Keywords around the circle overlapped, so some were difficult to recognize. - Topic changes should be estimated automatically. - I rarely used this window because I couldn't control the discussion by its information.

These results evaluate appropriateness of positions of topics relatively, such as nearer or farther. The validity of calculated distances or angles are not discussed. We should evaluate validity of these values in our further experiments.

The questionnaire results of the topic visualization method are shown in Table 5. For each question, 1 is the worst, and 5 is the best. In questions about the adequacy and effectiveness of visualizing topics *(a,b,c)*, the answers were good. Therefore, the topic visualization method is appropriate for understanding the discussion situation. In the results of questions about the effectiveness for triggering new topics *(d,e)*, it was indicated that the visualization did not lead participants to discuss specific topics. Some examinees commented that discussion topics changed based on the context, so it was difficult to change topics based on keywords in the circumference of the section circle. Therefore, a method for guiding a discussion topic is needed, so as to reflect the context of the discussion.

Table 6 shows the average scores about the use of this system. For each question, 1 is the worst, and 5 is the best. For both groups, the average scores of the two questions are greater than 3. In addition, some examinees claimed that they would like to use this system more in the future. The above results revealed that the examinees thought this system was useful for promoting various discussion.

Result of the system usability is shown in Table 7. Based on the comments about the chat window, examinees were able to communicate with others. However, based on the comments of the paper view window, the section button seems to inhibit a smooth discussion. Currently, our system detects topic changes by pushing the section button and generates a topic node when the button is pushed. Some examinees also complained about the burden of pushing the section button. Therefore, we should develop a method for identifying the target section by analyzing the words in all topic messages. For the comments of the discussion visualization window, examinees successfully grasped the discussion situation for each section. However, the keywords displayed in the window did not contribute to deriving specific topics. If keywords are displayed effectively, we believe that examinees can focus on and begin to discuss them.

7 Conclusion

We proposed a system that supports collaborative discussion for obtaining the contents related to research papers. Discussion topics are displayed for each section based on the similarity between a topic and the section, and among topics.

The experimental results showed that the visualization of topics is appropriate for grasping the discussion situation, but it does not contribute to leading a discussion for specific topics. One explanation is that the same keywords are always displayed in the discussion visualization window regardless of the discussion. For our future work, we will devise a method for leading effective discussions by showing the keywords based on the discussion progress. Appropriate keywords for the next topic may be related to previous topics. To select such keywords, their detection, which is not discussed effectively in the previous topics and is related to the current topic, needs to be developed.

This system can express relationships between topics in each section, but it cannot express topics across multiple sections. To represent such topics, we have to devise a visualization method that enables topics to indicate multiple sections. Moreover, we must devise a method for identifying multiple target sections.

In this experiment, only validity of topics is evaluated. We need additional experiment for qualitative evaluation of discussed contents.

Our collaborative discussion system focuses on the research activity of reading research papers to clarify the originality of research by evaluating the capabilities of other research. To help participants assess papers after discussion, the discussed topics should be arranged from each participant's viewpoint. In future research, we will help participants evaluate papers using the discussion results.

References

1. Looi, C.K.: Exploring the Affordances of Online Chat for Learning. International Journal of Learning Technology 1(3), 322–338 (2005)
2. Goodman, B., Geier, M., Haverty, L., Linton, F., McCready, R.: A Framework for Asynchronous Collaborative Learning and Problem Solving. In: Proc. of the 10th International Conference on Artificial Intelligence in Education 2001, vol. 68, pp. 188–199 (2001)
3. Conklin, J., Begerman, M.: glBIS: A Hypertext Tool for Team Design Deliberation. In: Proc. of Hypertext 1987, pp. 247–251 (1987)
4. Janssen, J., Erkens, G., Kirschner, P., Kanselaar, G.: Online Visualization of Agreement and Discussion During Computer-supported Collaborative Learning. In: Proc. of the 8th International Conference on Computer Supported Collaborative Learning, pp. 314–316 (2007)

5. Zhu, M., Hu, W., Wu, O.: Topic Detection and Tracking for Threaded Discussion Communities. In: IEEE/WIC/ACM International Conference on Web Intelligence and Intelligent Agent Technology, pp. 77–83 (2008)
6. Kojiri, T., Yamaguchi, K., Watanabe, T.: Topic-tree Representation of Discussion Records in a Collaborative Learning Process. The Journal of Information and Systems in Education 5(1), 29–37 (2006)
7. Leshed, G., Cosley, D., Hancock, J.T., Gay, G.: Visualizing Language Use in Team Conversations: Designing through Theory, Experiments, and Iterations. In: Proc. of the 28th of the International Conference Extended Abstracts on Human Factors in Computing Systems, pp. 4567–4582 (2010)
8. Viegas, F.B., Donath, J.S.: Chat Circles. In: Proc. of SIGCHI 1999, pp. 9–16 (1999)
9. Erickson, T., Kellogg, W.A., Laff, M., Sussman, J., Wolf, T.V., Halverson, C.A., Edwards, D.: A Persistent Chat Space for Work Groups: The Design, Evaluation and Deployment of Loops. In: Proc. of the 6th ACM conference on Designing Interactive Systems, pp. 331–340 (2006)
10. Xiong, R., Donath, J.: PeopleGarden: Creating Data Portraits for Users. In: Proc. of the 12th annual ACM Symposium on User Interface Software and Technology, pp. 37–44 (1999)
11. Tat, A., Carpendale, S.: CrystalChat: Visualizing Personal Chat History. In: Proc. of the 39th Annual Hawaii International Conference on System Sciences, vol. 3, pp. 58–68 (2006)
12. Lam, F., Donath, J.: Seascape and Volcano: Visualizing Online Discussions Using Timeless Motion. In: Proc. of CHI 2005 extended abstracts, Conference on Human factors in Computing Systems, pp. 1585–1588 (2005)
13. Inaba, A., Okamoto, T.: Negotiation Process Model for Intelligent Discussion Coordinating System on CSCL Environment. In: Proc. of the AIED, vol. 97, pp. 175–182 (1997)
14. Hayashi, Y., Kojiri, T., Watanabe, T.: Focus Support Interface Based on Actions for Collaborative Learning. International Journal of Neurocomputing 73, 669–675 (2010)

Chapter 14
The Proposal of the System That Recommends e-Learning Courses Matching the Learning Styles of the Learners

Kazunori Nishino, Toshifumi Shimoda, Yurie Iribe,
Shinji Mizuno, Kumiko Aoki, and Yoshimi Fukumura

Kyushu Institute of Technology, Faculty of Computer Science and Systems Engineering,
680-4 Kawazu, Iizuka, Fukuoka, 820-8502 Japan
nishino@lai.kyutech.ac.jp, shimoda@smile.kyutech.ac.jp,
iribe@imc.tut.ac.jp, s_mizuno@aitech.ac.jp,
kaoki@ouj.ac.jp, fukumura@oberon.nagaokaut.ac.jp

Abstract. In providing e-learning, it is desirable to build an environment that is suitable to the student's learning style. In this study, using the questionnaire to measure the student's preferences for asynchronous learning and the use of ICT in learning that has been develoed by authors, the relationship between the learning preferences of a student that have been measured before and after the course and his or her adaptability to the course is explored. The result of multiple regression analyses, excluding the changes in learning preferences that may occur duirng the course, shows that a student's learning adaptability can be estimated to some extent based on his/her learning preference measured before the course starts. Based on this result, we propose a system to recommend e-learning courses that are suitable to a student before the student takes the courses.

Keywords: e-learning, learning preferences, e-learning adaptability, multiple regression analysis, course recommendation.

1 Introduction

E-learning has been widely adopted in vocational training, higher education and life-long learning. In Japan, some higher education institutions have signed agreements to transfer credits earned through e-learning in other institutions. The scale of such credit transfer systems has increased and nowadays students can select courses they want to take among many available courses.

The advancement of information and communication technologies (ICT) allows learning management systems (LMS) with diverse functions to be developed and utilized. The research on instructional design [1] and learning technologies with regards to e-learning has flourished, and now e-learning takes various forms ranging from classes based on textual materials and classes utilizing audio-visual materials such as still images and videos to classes mainly focusing on communications between students and instructors or among students [2].

T. Watanabe and L.C. Jain (Eds.): Innovations in Intell. Machines – 2, SCI 376, pp. 203–214.
springerlink.com

E-learning allows student-centered learning in which students themselves, instead of instructors, set the time, place and pace for their study. Therefore, in e-learning it is desirable to establish a learning environment that matches the learning style of the student. The study proposes a system that suggests appropriate e-learning courses that matches the learning style of a student based on the data about the student's learning preference gathered in advance.

2 Flexibility of Learning Styles and Learning Preferences

2.1 Flexibility of Learning Styles

The research on learning styles and learning preferences has been prolific in Europe and North America. According to the Learning Skills Research Center (LSRC) in U.K., the number of journal articles on the learning styles and learning preferences has reached more than 3,800. In those articles, 71 different theories and models for learning styles and preferences have been presented. LSRC has selected 13 most prominent theories and models of learning styles and preferences from the 71 theories and models, and further studied the 13 models [3]. LSRC classified the 13 models of learning styles and preferences into five categories from the most susceptible to the least susceptible to environments based on Curry's onion model [4].

Previously some studies were conducted using the Kolb's learning style [5] in developing computer-based training (CBT) [6] and examining the influence of learning styles on the "flow" experience and learning effectiveness in e-learning [7]. Other studies used GEFT (Group Embedded Figure Text) [8] to see the influence of learning styles and learning patterns on learning performance [9] and the instrument developed by Dunn, Dunn and Price [10] to build a system which provides learning environment suitable to the student's learning style [11].

When investigating learning styles and learning preferences in e-learning, how should we consider the "flexibility of learning styles and preferences?" E-learning has the potential to provide "student-centered learning" and tends to be designed based on the pedagogy of providing learning environments according to the students' needs, abilities, preferences and styles rather than providing uniform education without any consideration of individual needs and differences. Therefore, it is meaningful to provide students and teachers with information about the students' adaptability to e-learning courses by using a questionnaire on learning preferences in e-learning. Here we use the term "learning preferences" instead of "learning styles" as the term, "preferences" connotes more flexibility than "styles."

This study looks at learning preferences of students in e-learning courses and determines if their learning preferences regarding asynchronous learning and the use of ICT of a student changes after taking an e-learning course.

2.2 Asynchronous Learning and the Use of ICT

As e-learning is usually conducted asynchronously, it requires more self-discipline of students in comparison with face-to-face classes. E-learning might be easier for students who want to learn at their own pace to continue and complete a study. However, it can be challenging for those who do not like studying on their own and prefer studying in face-to-face classes.

The use of learning management systems (LMS) can ease the distribution of course materials and the communication among students or between students and teaching staff. Some measures have been taken to help students understand the content of e-learning materials and also to motivate students in studying materials through e-mails sent by teachers and tutors of e-learning courses [12]. However, the use of ICT in e-learning tends to become complex as its functionality increases and may discourage those students who are not familiar with the ICT use.

The use of ICT and asynchronous learning is a typical characteristic of e-learning. However, as it is stated earlier, those who do not like asynchronous learning or the use of ICT may have the tendency to drop out in the middle of e-learning courses [13]. Therefore, it is desirable that students and their teachers know the students' learning preferences and their adaptability of e-learning courses in advance [14, 15].

To investigate the learning preferences in e-learning, we developed learning preference questionnaire items asking preferences in studying, understanding, questioning, and doing homework [16].

This study investigates the change in learning preferences after taking an e-learning course, using the learning preference questionnaire mentioned above. Furthermore, through multiple regression analyses the study confirms the hypothesis that the adaptability to an e-learning course can be estimated before the student's taking the course based on his/her answers to the learning preference questionnaire and proposes a system that recommends e-learning courses suitable to a student based on his/her learning preferences.

3 Survey on Learning Preferences and e-Learning Course Adaptability

3.1 Survey on Learning Preferences

The survey on learning preferences was administered to those students who enrolled in the eHELP (e-Learning for Higher Education Linkage Project) which is a credit transfer system for e-learning courses offered by multiple member universities in Japan.

In eHELP, students take one to three full online course(s) offered by other institutions in parallel to taking courses offered by their own institution. In taking an e-learning course, a student studies the content which is equivalent to 15 face-to-face classes (90 minutes per class). The majority of e-learning courses offered in eHELP is those in which students study by watching video lectures of instructors while using the downloadable text materials.

In order to improve the quality of learning experiences in e-learning, it is required to build a system in which students can have regular communication with their instructors as well as peer students using discussion boards and chat. eHELP has developed a system in which students can communicate within the LMS the students are familiar with [17,18] and provided a synchronous e-learning system utilizing Metaverse to respond to various needs of learners [19,20].

This study was conducted from the early December of 2008 to the early January of 2009 when all the e-learning courses were completed. All the items in the questionnaire were asked with the 7-point Likert-type scale; from 1 being "don't agree at all" to 7 "agree strongly," and we obtained valid responses from 53 students. We discarded responses that had marked all the same points regardless of reverse coded (i.e., negatively phrased) items. The questionnaire consists of 40 items asking preferences in studying, understanding, questioning, and doing homework in terms of asynchronous learning and the use of ICT. The questionnaire was made available online and students accessed the questionnaire online.

As the result of the factor analysis [3], we could extract three factors with eigenvalues over .07(see Appendix 1): the factor 1 being "preference for asynchronous learning," the factor 2 "preference for the use of ICTs in learning" and the factor 3 "preference for asynchronous digital communication."

3.2 The Survey on e-Learning Course Adaptability

When the learning preference questionnaire was administered, the questionnaire on e-learning course adaptability was also administered to the students who enrolled in eHELP courses. The items in the questionnaire are shown in the Table 1. The questionnaire consists of 10 items asking psychological aspects of learning such as the level of students' understanding and the level of satisfaction.

The questionnaire (see Table 1) was administered online to the students enrolled in each of the eHELP courses upon their completion of the course (i.e., between December 2008 and January 2009) and 69 completed responses were obtained. All the items in the questionnaire were asked with the 7-point Likert-type scale; from 1 being "don't agree at all" to 7 "agree strongly." The scores for the item (g) and (h) were reverse-coded. The mean score was 4.7. The mean score was calculated for the e-learning course adaptability, the factors 1, 2, and 3 respectively for each student, and the values were used in the subsequent analyses. In addition, the reverse-coded items were recoded to adjust to the other items.

Table 1. The question items in the e-learning course adaptability questionnaire

Item	Mean
(a) The content of this e-learning course is more understandable than regular class contents.	4.51
(b) The style of learning of this e-learning course is easier to learn than regular classes.	4.90
(c) The pace of this e-learning course is more suitable than regular classes.	4.91
(d) This e-learning course is more satisfying than regular classes.	4.36
(e) This e-learning course is more effective than regular classes.	4.35
(f) This e-learning course is more interesting than regular classes.	4.91
(g) This e-learning course makes me more tired than regular classes. (recoded)	4.84
(h) This e-learning course makes me more nervous than regular classes. (recoded)	5.59
(i) This e-learning course brings me more endeavor than regular classes.	4.07
(j) This e-learning course brings me more motivation than regular classes.	4.41

3.3 Correlations

Correlations between the scores of the three learning preference factors and the score for the e-learning course adaptability were analyzed among the 69 respondents who completed both of the two questionnaires. The correlation *r* is shown in Table 2.

A statistically significant (p <0.01) correlation was seen between the learning preference factor 1 (the preference for asynchronous learning) and the e-learning course adaptability and between the factor 2 (the preference for the use of ICT in learning) and the adaptability. The correlation between the learning preference factor 3 (the preference for asynchronous digital communication) and the e-learning course adaptability is not as high; however, the correlation is statistically significant at the level of p<.05.

Table 2. Correlations between the course adaptability and the leaning preference factors

	R	*p*	*N*
Adaptability - Factor 1	0.53	< 0.01	69
Adaptability - Factor 2	0.60	< 0.01	69
Adaptability - Factor 3	0.29	<.05	69

3.4 Multiple Regression Analysis

In order to further investigate the relationships between the e-learning course adaptability and each of the three factors of learning preferences, a multiple regression analysis was conducted. The results are shown in Table 3.

As shown in Table 3, the regression coefficients of the factor 1 and the factor 2 are relatively high and the p values are less than 0.01. However, as for the factor 3, the regression coefficient is low and its p value is also not significant enough. It can be suspected that the multicollinearity is high between the factor 2 and the factor 3. Therefore, another multiple regression was conducted excluding the factor 3.

As the result, the regression coefficients for the factor 1, the factor 2, and the intercept resulted in 0.23, 0.45, and 1.84 respectively. Therefore, the multiple regression equation was derived as follows,

$$\text{Adaptability to e-learning courses} = 1.84 + 0.23 \times \text{factor 1} + 0.45 \times \text{factor 2} \quad (1)$$

Table 3. The result of a multiple regression analysis

Variable Name	Regression Coefficient	*p*
Intercept	1.82	**< 0.001
Factor 1(preference for asynchronous learning)	0.23	**0.0054
Factor 2(preference for the use of ICT in learning)	0.45	**0.0003
Factor 3(preference for asynchronous digital communication)	0.01	0.938
Multiple *R*-square	0.43	** < 0.001

**significant at *p*=0.01 *significant at *p*=0.05

It is possible to predict the e-learning course adaptability for students in eHELP based on this formula.

4 Estimation of e-Learning Course Adaptability

4.1 Changes in Learning Preferences

In order to further investigate the flexibility of learning preferences in e-learning discussed in the section 2, the changes in learning preferences of students before and after taking an e-learning course in the Spring semester (from the early April, 2009, to the early July, 2009) were investigated, using the learning preference questionnaire mentioned previously.

The Figure 1 indicates the changes in learning preferences of 18 students who responded to the questionnaire both at the beginning of the course and at the end of the course. It shows the scores of responses at the beginning of the course deducted from the scores at the end of the course in bar charts. Hence, the vertical axis in the Figure 1 indicates the sum of the differences in scores before and after the course. The white bars on the left side in the Figure 1 (from q1 to q17) shows the changes in the factor 1 (preference for asynchronous learning) and the black bars on the right side indicates the changes in the factor 2 (preference for the use of ICT). As a whole, the factor 1 got positive scores and the factor got negative scores.

The paired sample t-test was conducted to the changes in scores shown in the Figure 1. As a result, q1, q2, and q26 showed significant differences ($p<0.05$) as indicated ** in the figure. In addition, the paired sample t-test of scores on q3 and q4 showed significant differences at $p<0.10$ as indicated * in the figure. Hence, it has been found that the learning preferences change after taking e-learning courses with regards to the five items indicated above.

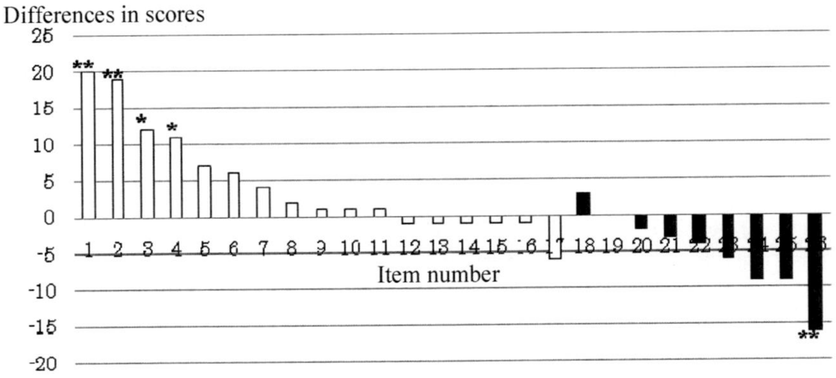

Fig. 1. The changes in learning preferences regarding asynchronous learning and the use of ICT after taking an e-learning course

By taking e-learning courses, students' preference for asynchronous learning tends to change positively, while their preference for the use of ICT tends to change negatively. Therefore, it has been found that the learning preferences for asynchronous learning and the use of ICT can change in e-learning environments.

4.2 Estimation of e-Learning Course Adaptability through Multiple Regression Analyses

Applying the scores obtained at the beginning of the course to the multiple regression formula, we attempted to predict the e-learning course adaptability at the end of the course. The Figure 2 shows the correlation between the e-learning course adaptability scores calculated by applying the item scores for the factors 1 and 2 obtained at the beginning of the Spring course and the actual scores obtained at the end of the course (n=18). In this case, the correlation coefficient is 0.01, which indicates that the e-learning course adaptability cannot be predicted.

The Figure 3 shows the correlation between the actual scores and the scores calculated using the multiple regression formula without the scores of the five items (from q1 to q4 and q26) that tend to change after experiencing e-learning as discussed in the section 4.1. As the correlation coefficient is 0.65, it can be concluded that there is a strong correlation between the calculated scores and the actual scores.

It shows the possibility of predicting the e-learning course adaptability at the end of the course by using the formula (**1**) discussed in the section 3.4 with the scores excluding those of the learning preference items that may change after taking an e-learning course.

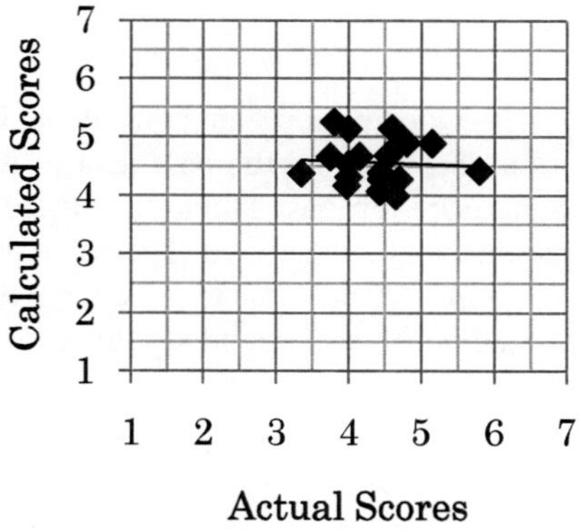

Fig. 2. The correlation between the actual scores and the calculated scores of e-learning adaptability using all the items of the factor 1 and 2 (the spring semester)

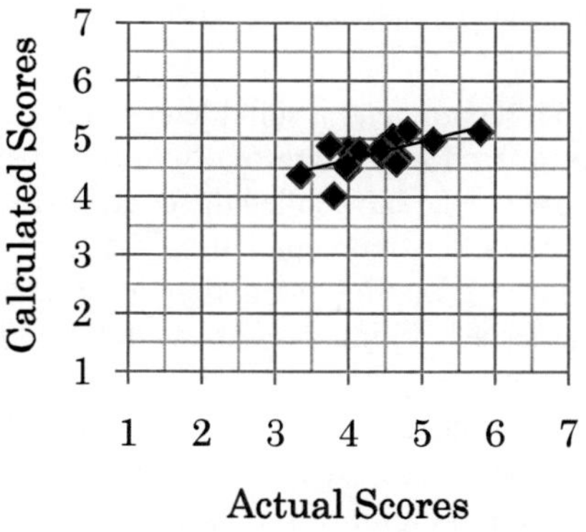

Fig. 3. The correlation between the actual scores and the calculated scores of e-learning adaptability excluding the items that are changeable after taking an e-learning course (the spring semester)

4.3 Development of a System to Recommend e-Learning Courses Suitable to a Student

The section 4.2 has shown that a student's e-learning course adaptability can be estimated before the course starts based on his/her responses to the learning preference questionnaire items on asynchronous learning and the use of ICT in learning. Thus, as shown in the Figure 4, based on the past data, the study considers the development of a system that recommends e-learning courses suitable to a student according to the result of the learning preference questionnaire of the student before the course starts.

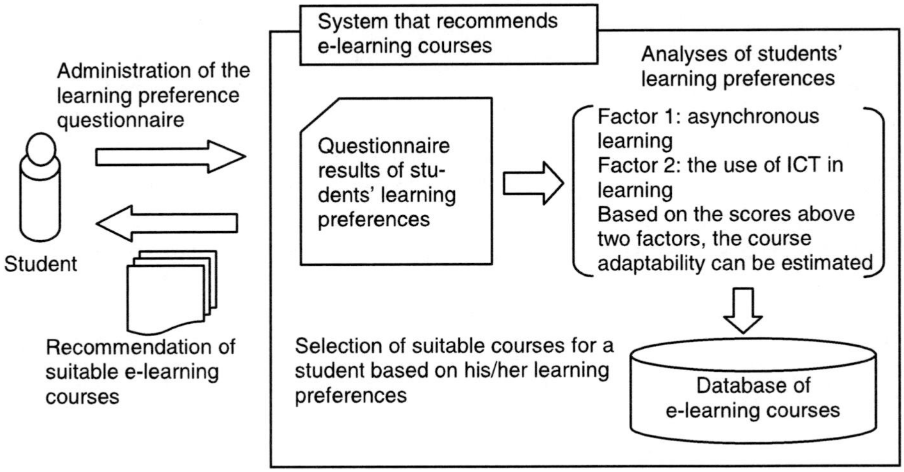

Fig. 4. The system of recommending e-learning courses suitable to a student in consideration of the student's learning preferences

To recommend a suitable course based on a learner's learning preferences, it can be considered to use the multiple regression equation derived from the past data of the learners who have taken the course. However, as there are few past data for a particular course at this moment, a multiple regression equation cannot be derived. Therefore, we use the grouping method to estimate the adaptability of the course based on the learner's learning preference for building the system to recommend a course to a learner.

The course (i)'s adaptability scores (Zi) of past students gathered after the courses were over are classified into three (high, middle and low) levels. Then the average of learning preference scores for asynchronous learning ($\overline{x}_{i,h}$, $\overline{x}_{i,m}$, $\overline{x}_{i,l}$) and for the use of ICT in learning ($\overline{y}_{i,h}$, $\overline{y}_{i,m}$, $\overline{y}_{i,l}$) are calculated for each group (i.e., high, middle and low level in course adaptability scores).

In the results of the learning preference questionnaire of the student (a) administered before the start of the course, when the average score of the preferences for asynchronous learning being x_a and the average score for the use of ICT in learning being $y_{a,}$, the deviations from each of the means of course adaptabilities at three different levels can be calculated as:

$$D_{a,h}=\sqrt{(x_a-\bar{x}_{i,h})^2-(y_a-\bar{y}_{i,h})^2}$$
$$D_{a,m}=\sqrt{(x_a-\bar{x}_{i,m})^2-(y_a-\bar{y}_{i,m})^2}$$
$$D_{a,l}=\sqrt{(x_a-\bar{x}_{i,l})^2-(y_a-\bar{y}_{i,l})^2}$$

For the eHELP course (k), the scores of adaptability to the course are grouped into three levels as $Z_h>=4.5$, $4.5>Z_m>4.0$, and $Z_l<=4.0$. Table 4 shows the means of learning preference scores for asynchronous learning (x) and for the use of ICT in learning (y) at each of the three levels.

Table 4. Means of learning preference scores for each of the three levels of the adaptability to the course (k) (*n*=13)

Course adaptability \ Learning preference	Learning preference of students in the course (k) (means)	
	Asynchronous learning ($\bar{x}_k$)	The use of ICT in learning ($\bar{y}_k$)
$Z_h>=4.5$	1.20	0.800
$4.5>Z_m>4.0$	0.329	0.178
$Z_l<=4.0$	-0.235	0.037

Figure 5 plots the learning preferences at each level of course adaptability shown in Table 4 in each axis. ● indicates the mean of the Z_h group; ▲ indicates the mean of the Z_m group; and × indicates the mean of the Z_l group. For example, if the student (a)'s preferences for asynchronous learning and the use of ICT in learning are indicated as in Figure 5, the adaptability to the course can be expected as being Z_h. In a similar vein, the student (a)'s course adaptability can be calculated for other courses.

The students of eHELP will answer the set of questions regarding the learning preferences at the beginning of the course. (See the Appendix 1.) The learning preferences of past students, who were classified into three groups based on their course

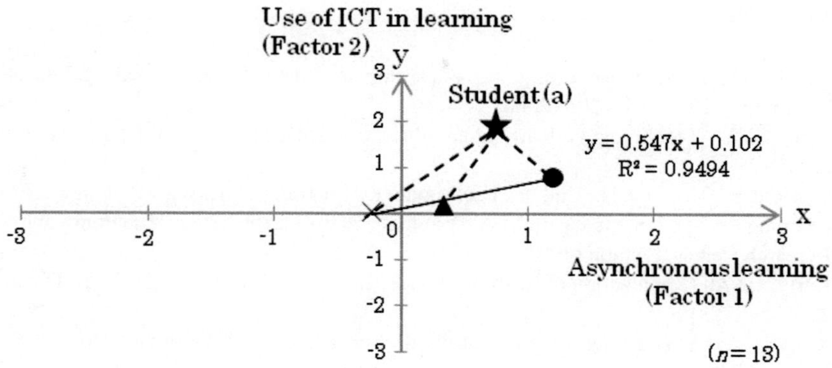

Fig. 5. The distances from the course adaptability score of the student (a) in the course (k) and the means in the three different levels

adaptability, were plotted making the average scores of preferences for asynchronous learning and the use of ICT on the x- and y- axes respectively. As indicated in the Figure 5, the correlation between the course adaptability and the learning preferences (i.e., the preference for asynchronous learning and the preference for the use of ICT) is quite high (R^2=0.95). In this way, the course adaptability can be estimated in advance to some extent if the course shows a strong correlation between the learning preferences and the course adaptability. At present, two courses including the course k among the total of 32 courses had more than 10 students in the past, both of which show strong correlations between the learning preference and the course adaptability.

5 Conclusion

This study investigated the relationship between learning preferences and e-learning course adaptability by administering questionnaires to students who were enrolled in e-learning courses at higher education institutions. The results of the study show that the learning preferences regarding asynchronous learning and the use of ICT may change after taking e-learning courses. It has been also found that there is a significant correlation between the actual e-learning course adaptability scores and the scores calculated using the multiple regression formula with the response scores of the items that do not tend to change. Based on those results, it is concluded that the e-learning course adaptability of an individual student at the end of the course can be predicted by administering the questionnaire at the beginning of the course.

Based on these results, a system recommending e-learning courses suitable to a particular student based on the student's score on the learning preference questionnaire administered before the beginning of the course is proposed.

In the future, the system to recommend suitable courses to a student is planned to be built and to be used in the operation of e-learning courses at eHELP.

In addition, we would like to continue the research on the factors influencing e-learning so that students can choose learning methods and environment that suit their learning preferences.

References

1. Suzuki, K.: Instructional design for e-learning practices. Japan Journal of Educational Technology 29(3), 197–205 (2006)
2. Bates, A.W.: Technology, e-learning and distance education, 2nd edn. Routledge, London (2005)
3. Coffield, F., Moseley, D., Hall, E., Ecclestone, K.: Learning Styles and Pedagogy in Post-16 Learning. In: Systematic, A. (ed.) A Systematic and Critical Review, Learning and Skills Research Center, London (2004)
4. Curry, L.: An Organization of Learning Styles Theory and Constructs. In: ERIC Document, 235185 (1983)
5. Kolb, D.A.: LSI Learning-Style Inventory. McBer & Company. Training Resources Group, Boston (1985)
6. Henke, H.: Learning Theory: Applying Kolb's Learning Style Inventory with Computer Based Training. In: A Project paper for A Course on Learning Theory (1996)

7. Rong, W.J., Min, Y.S.: The effects of learning style and flow experience on the effectiveness of e-learning. In: Fifth IEEE International Conference on Advanced Learning Technologies, pp. 802–805 (1996)
8. Witkin, H., Oltman, P., Raskin, E., Karp, S.: A Manual for The Group Embedded Figures Test, Palo Alto, California (1971)
9. Lu, J., Yu, C.S., Liu, C.: Learning style, learning patterns, and learning performance in a WebCT-based MIS course. Information & Management, 497-507 (2003)
10. Dunn, R., Dunn, K., Price, G.E.: The Learning Style Inventory. Price Systems, Lawrence (1989)
11. Wolf, C., Weaver, I.: Towards 'learning style'-based e-learning in computer science education. In: Australasian Computing Education Conference (ACE 2003), vol. 20 (2003)
12. Kogo, C., Nakai, A., Nojima, E.: Relationship between procrastination tendency and student dropouts in e-learning courses. Research report of JSET Conferences 2004(5), 39–44 (2004)
13. Fuwa, Y., Ushiro, M., Kunimune, H., Niimura, M.: Efforts toward the Establishment of Quality Assurances for Adults Students of Distance Learning on e-Learning System -Practice and Evaluations of Support and Advice Activities-. Journal of Multimedia Aided Education Research 3(2), 13–23 (2007)
14. Nishino, K., Ohno, T., Mizuno, S., Aoki, K., Fukumura, Y.: A Study on learning Styles of Japanese e-learning learners. In: 11th International Conference on Humans and Computers, pp. 299–302 (2008)
15. Nishino, K., Iribe, Y., Mizuno, S., Aoki, K., Fukumura, Y.: An analysis of learning preference and e-learning suitability for effective e-learning architecture. In: Intelligent Decision Technology, vol. 4, pp. 269–276. IOS Press, Amsterdam (2010)
16. Nishino, K., Toya, H., Mizuno, S., Aoki, K., Fukumura, Y.: The relationship between the learning styles of the students and their e-learning course adaptability. In: Velásquez, J.D., Ríos, S.A., Howlett, R.J., Jain, L.C. (eds.) KES 2009. LNCS, vol. 5712, pp. 539–546. Springer, Heidelberg (2009)
17. Yukawa, T., Suzuki, I., Fukumra, Y.: A cross-LMS chat system and its evaluation, ED-MEDIA 2009, pp. 1340–1345 (2009)
18. Yukawa, T., Takahashi, H., Fukumura, Y., Yamazaki, M., Miyazaki, T., Yano, S., Takeuchi, A., Miura, H., Hasegawa, N.: Online collaboration support tools for project-based learning of embedded software design. In: Velásquez, J.D., Ríos, S.A., Howlett, R.J., Jain, L.C. (eds.) KES 2009. LNCS, vol. 5712, pp. 531–538. Springer, Heidelberg (2009)
19. Barry, D.M., Kanematsu, H., Fukumura, Y., et al.: International Comparison for Problem Based Learning in Metaverse, ICEE/ICEER 2009, pp. 59–65 (2009)
20. Kanematsu, H., Fukumura, Y., et al.: Practice and evaluation of problem based learning in Metaverse, ED-MEDIA 2009, pp. 2862–2870 (2009)

Appendix 1: The Learning Preference Questionnaire

Factor 1 preference for asynchronous learning

q1	When I study through computers, I tend not to care how others study.
q2	I tend to learn more actively when I study alone than studying with others at one place.
q3	I can familiarize myself better when I study independently at my convenience than studying with others at one place.
q4	I study at my own pace and do not care how others study.
q5	I would rather study alone at the place and time convenient to me than learn in class with other people.
q6	I can concentrate better when I study independently at my convenience than studying with others at one place.
q7	I can learn better when I study at the time I decide than when I study at the time decided by others.
q8	I would rather do group learning through computers than face-to-face.
q9	I would rather follow the computer instruction rather than study reading textbooks.
q10	I understand better when I study at my convenient time rather than learning in class with other people.
q11	I feel more motivated when I study using computers than learning from teachers in person.
q12	It is easier for me to take test individually than to take one in a place with others.
q13	I feel less tired looking at a computer screen than looking at a blackboard or a large screen in a classroom.
q14	I want to study at my own pace.
q15	I can be more creative when I study alone than studying with others at one place.
q16	I feel more motivated when I study at my convenience than learning in class with other people.
q17	I feel less tired when I study independently at my convenience than studying with others at one place.

Factor 2 preference for the use of ICT in learning

q18	I understand better when I learn through computers than when I learn by reading books.
q19	I tend to learn more actively using computers than studying in class.
q20	I prefer learning through computers to learning by reading books.
q21	I am familiar with computers.
q22	It is easier for me to take test on a computer than on paper.
q23	I would rather submit my report in an electronic format than in a paper and pencil format.
q24	I prefer taking notes using a computer than writing on paper.
q25	I can concentrate better looking at a computer screen than looking at a blackboard or a large screen in a classroom.
q26	It is easier for me to memorize what is on a computer rather than to review printed materials.

Factor 3 preference for asynchronous digital communication

q27	I would rather receive answers later from teachers via mail than asking questions in person or through chat.
q28	I prefer communicating via email to communicating through telephones.
q29	I would rather ask questions using email or bulletin boards than asking teachers in person.
q30	It is easier for me to communicate through computers or cell phones than to communicate face-to-face.
q31	I can be more creative when I think on paper than using computers.

Chapter 15
Design of the Community Site for Supporting Multiple Motor-Skill Development

Kenji Matsuura[1], Naka Gotoda[2], Tetsushi Ueta[1], and Yoneo Yano[2]

[1] Center for Administration of Information Technology, The University of Tokushima, Minamijosanjima 2-1, Tokushima, 770-8506, Japan
{matsuura,tetsushi}@ait.tokushima-u.ac.jp
[2] Faculty of Engineering, The University of Tokushima, Minamijosanjima 2-1, Tokushima, 770-8506, Japan
{gotoda,yano}@is.tokushima-u.ac.jp

Abstract. Conventional approaches for computational support on complex motor skill acquisition and development are usually domain specific. However, today's social web services allow members to facilitate the skill development from one to many types based on the detected shared parameters. This paper describes the design and development on a community site that supports motor-skill learning. Commitments to a community have strong advantage on both self-assessment and detection of the relative level about a skill. Moreover, some parameters among different skills might be the same in assessment of the acquired skills of individuals. Therefore, we design and implement the web-community system that integrates different skill-communities to interact each other. With respect to the technical interest, our environment offers users an authoring environment where they can adopt different types of modules to embed at the same time.

Keywords: motor skill, skill development, online community, and media type.

1 Introduction

A motor-skill is one of very complex targets for learning science and research domains on technology-enhanced learning [1]. There are several definitions and typical taxonomies based on both scientific and phenomenal achievements. A well-known motor skill is a "gross motor skill" which especially draws attention of field researchers about educational technology. Gross motor-skills are regarded as the human ability of physical fruition that integrates partial body-movements precisely.

Motor skill learning, which includes both acquisition and development, is a change of individual ability or potential change of interaction ways in the real world. The conditions for human beings who try to activate learnt skills sometimes change depending on the skill type, environmental context, objects, collaborators and so forth. Therefore, the empirical knowledge or overtimes experience itself becomes very important in this domain.

T. Watanabe and L.C. Jain (Eds.): Innovations in Intell. Machines – 2, SCI 376, pp. 215–224.
springerlink.com

Human computer interaction is also a key concept to solve this problem. Hollan et al. [2] suggested that the theory of distributed cognition has a special role to play in understanding interactions between people and technologies. Bansler and Havn surveyed the shift of working and researching style with computer-mediated interaction [3]. Computer supported skill learning brings us a new environment which has a potential to provide vicarious experience of training by other people through watching video image or visiting articles.

In a sense, human cognition in the real world varies depending on the skill. Though skill learning is individual activities, current researchers about CSCL, standing for Computer Supported Collaborative Learning, focus on its strong advantage in the interaction with other learners. Sharing knowledge is a core method to circulating best practices among members of a community of interest [4]. Fischer et al. [5] also reported theoretical framework of socio-cultural approach for learning in a context of collaborative settings. The theory on communities of practice (CoP) is also the key issue in computer supported collaborative learning. Technology stewarding for communities is well organized and summarized in "Digital Habitats" by Wenger et al [6]. Many community-based learning styles are discussed in the context of technology oriented. We believe that verbal communication with other people on the web is actually a key factor for collaborative learning but adopting other media for nonverbal things enriches learning and development of physical skills [7].

Today's technologies dealing in learning science are not only networked PCs but also other devices such as sensor, touch-screen, cell-phones, and whatever available for learning and communication. Further, some researchers adopt haptic devices for supporting motor-skill domain [8]. They made us possible to monitor physical activities at real time or asynchronously. The concrete domain of this study is sports including multiple types of repeating actions such as jumping, running and so forth. We designed the supporting environment with various kinds of media in a social networking service for encouraging communication with other members.

This environment has strong advantages from a technical viewpoint. At first, learners can get opportunities to know relative skills of other learners beyond the different categories of sports. At second, they can also find a new skill in a sub-community. As a result, we expect better influence for users to accelerate the development of standing skill while they would get new opportunities of challenging unknown skills. As the third, the architecture does not limit its target domain. It covers lots of motor skills in the same system.

2 Motor-Skill Learning

2.1 Preliminary Discussion

Studies on motor skill have the long history as one of difficult domains [8]. According to well-known taxonomy for motor skill by Gentile [9], human skill could be divided into two-dimensional space by the axis of environmental-context and action function. There are still four sub spaces in each space by classifying regulatory conditions, intertribal variability, body orientation, and manipulation.

In regard to the process of skill development, we should take care which phase is the focused subjects in considering the support. For example, in early stages of skill training, we can often find the rapid improvement in skill performance but hardly do that in the latter stages. Gupta and Noelle summarized the similar condition in training [10].

A lot of approaches for this domain are mainly based on the analytical view. If successful or failure performance of a certain action is captured by sensors, movies and so forth, researchers try to analyze the reasons why they could or could not. The analysis is carried out quantitatively and qualitatively. However, the analyzed reasons in some cases would be difficult to suggest improvement for the next time. It is because either environmental conditions or physical conditions will change from those in the past. Though people of skill learning cannot expect the detail conditions at the next time of trial, they should correspond to a new situation based on the acquired skill. Furthermore, it is difficult to suggest an adaptive way of improvement in a precise and accurate manner for individuals.

2.2 Open and Closed Skill

Another presumable discussion for complex motor skill arose by Allard and Starkes about the difference between "Open skill" and "Closed skill" [8]. Closed skill assumes predictable conditions in a stable environment. Once a performer developed the skill, s/he can do it again in the same situation next time. For instance, a professional football player can perform juggling many times under the same condition. On the other hand, open skill cannot assume such predictions. Regarding open skills, a player tries something in facing different objects, conditions, and environmental factors at every trial time. Since the environmental factors, i.e. weather conditions, always change, most of field sports or trainings are open skill. As another instance, a professional football player can recognize the best way at every time in facing opponent in the field. Therefore it is necessary to acquire an ability to accept environmental change or potential change even if the physical or mental conditions are different each time.

We believe that it is better to know or to train various kinds of skill than mastering only single skill. However, from the design perspective of supporting systems for skills, most systems are based on a domain dependent concept. For example, the system designer usually setup specialized function to control the skill development based on a certain model. Typical examples are seen in music [11] and cooking. In this study, we try to design the supporting system that aims at community-based training for players of multiple sports. For instance, to deal with swimming and running as targets, some analyzing parameters in the system are almost the same, "distance", "speed" and "time". The performance of body movement is surely different between them. The difference for them is only the scale or range from systematic viewpoint. Therefore, we believe we can handle the parameters in a same way in the same system. Then we can design a bridging function of users in this context.

2.3 Media Type

Activities in training can be monitored and stored by the system in a various way. Some typical skills are available to be monitored and represented in a video and/or sensors while others use text if they can be represent in a verbal way. This study designs an authoring environment where each community-author of system users can freely customize the training record space for a skill by way of combining some modules. An example is "rope skipping", which integrates video and text media for inputting and representing.

If a potential user is interested in a community room, s/he can register oneself to this room. Then, as a community member, s/he stores her/his daily exercising records to this room. In such a situation, we have to take into account of input and output for each module. Though current policy about the input/output of a module from designing viewpoint is one by one, we will extend the variety of combining input and output for improving flexibility and scalability.

2.4 Process of a Skill Development

Fig.1 indicates a human process of a skill learning that is assumed in this study. If learners want to learn a skill listed in a community space, they can join the community to recognize their stage of current skill. The recognition is carried out in both ways; the self-recognition by capturing their action with a video, sensors or text as a meta-recognition, or the detection of a current status for the skill in a comparison among community articles. Then, they train themselves with recording something in a community space. During the training, they make trial and error about the skill including off-line activities. After its process, they can evaluate their article by own to decide the next action whether continuing training or completing it. If they satisfy their skill represented in an article, they can complete and proceed to next skill or keep training for refinement according to their decision. However, if not, they should go back to the recognition process in this figure. Though a sophisticated method for the next skill challenge is not implemented concretely so far, we can implement the recommendation function of the next item in case the system as one of implications.

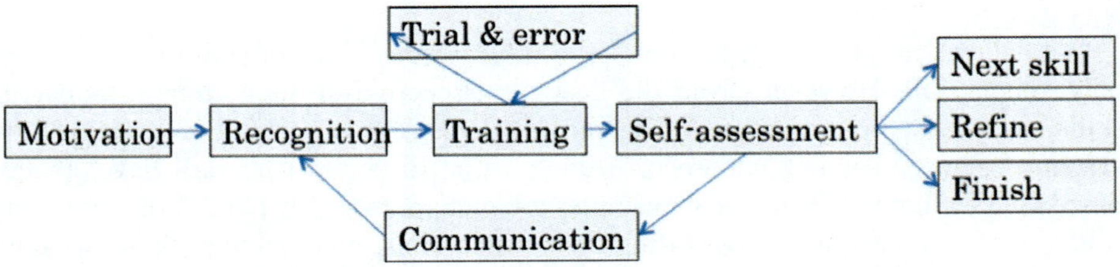

Fig. 1. General process of a motor-skill development

We designed supporting functions based on this assumed process. The first one is a description space that makes members know the target skill, methodology in a practice, criterion for achievement, evaluation and so forth before enrolling to the community. The next one is a view space of an article in a comfortable way. The recognition and self-evaluation process need this function. The third one is flexible input method combining various media. During training, learners sometimes input their data of practice to the system. The forth one is communication that enables learners to interact each other by text or multimedia to know the better way for a skill if needed. At last, as an optional function, the automatic analysis function will support the skill-development activities with this system. However, it will be domain dependent technology is necessary. Therefore, we made a few functions for this purpose at this studying stage.

3 Design and Development

3.1 Framework of the Architecture

Based on the principal discussed in the former section, we developed the prototype that enabled users to create a new community space for a skill. The architecture of the system is illustrated in Fig.2. The prototype was developed on "OpenPNE" platform (http://www.openpne.jp/) that is an open source application of "symphony" PHP framework (http://www.symfony-project.org/). The description of symphony in the official website is "*Symfony is a full-stack framework, a library of cohesive classes written in PHP.*" The system is a kind of social-networking service, sometimes called "social software" on a web.

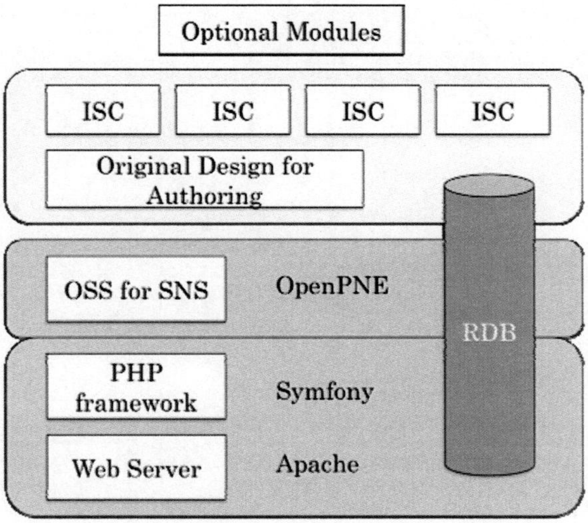

Fig. 2. System architecture

The prototype provides some templates for each module that is, for example, a video module for inputting and showing in an embedded html. "ISC" in the figure is an "Implemented item of a Skill Community". If an author creates a community space in the system, it means that a new ISC is created. The prototype uses a database to store platform data for OpenPNE and the original data by our proposal altogether. If an advanced module is developed for analysis of an article and so forth, it is called from outside. For instance, an image processing module for movie data is a typical module that should be separated from the server as it needs high performance. In this way, the prototype has several modules and the organization is flexible from a technical viewpoint.

3.2 Authoring Environment on the Web

The target user of the authoring environment is a sub-community owner who intends to found the specific space for the community of interest. The system provides an interactive interface for the user to select and combine some abstract modules, which are previously designed and developed, into a unified space.

Fig.3 indicates the configuration process and the snapshots in authoring interface. To create new sub-community, four steps are required for the user. At the first step, the user set the name and detail description of the skill community. For example, the objective of the skill community and the training method are described in this process. The user has to select action modules adjusting to the skill community. The snapshot of step 3 is an interface for selecting and configuring the parameters. In this case, the user creates an environment that presents members "text area" for a title, "text area" for description of blog-body, "point" for the number of repeating actions.

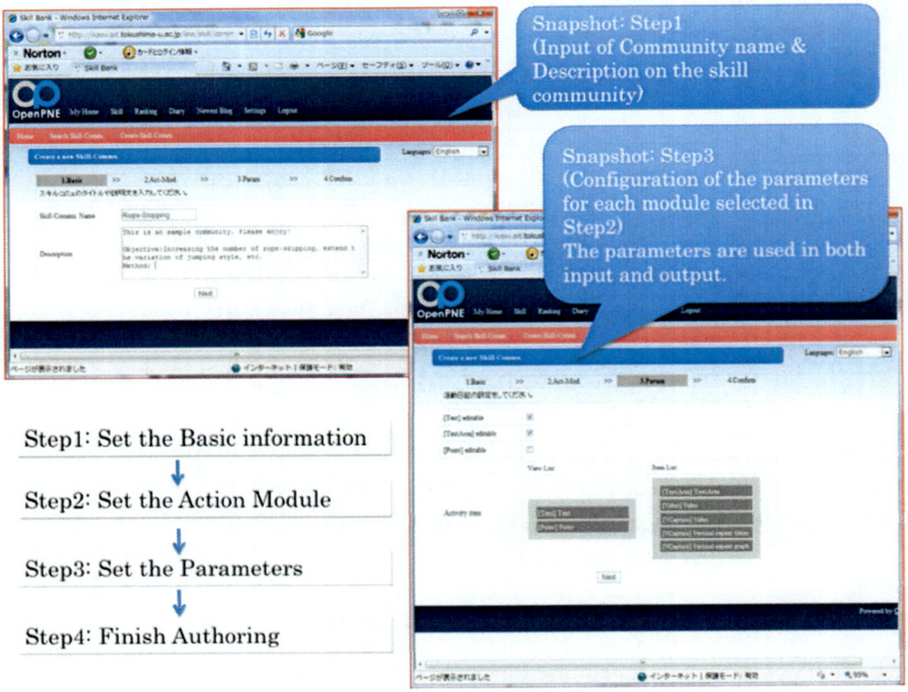

Fig. 3. Snapshot and the process of authoring a community space

The drag and drop operation by a mouse device is available at this phase. If a user finds requisite elements on the right of the interface, s/he can move them to the left. The order of embedding position for each module is configurable freely. However, members have to input information of prior-specified to each module assigned to a motor skill for the community theme. We have developed six types of modules for the prototype so far; 1)Text (Single-line / Multi-line available), 2)Score, 3)Image, 4)Video, 5)HRM(Heart Rate Monitor), 6)GPS(Global Positioning System).

From the operational viewpoint of community members, those modules offer an easy interface to input data. For example, module 1) and 2) provide a traditional web-form that accepts text from users. At a form for the "Score", if the description suggests a certain range of values, community members have to obey the instruction.

As for module 3), 4), 5) and 6) users also easily use it with an accustomed operation that is associated with a file uploading function on a web. The module 3) offers only one file at a time as a default. Therefore, if plural number of image files is needed, a community owner has to set up several modules of 3) at the same time. A video for module 4) sometimes becomes large in size. The maximum size of one file depends on the configured settings of the system. In viewing them as a community member, the video file has been converted in a Flash streaming media. Therefore, we believe they don't feel stress about the media type such as 3gp, Quicktime and Windows Media. In order to handle the last two modules for HRM and GPS data, the community owner supposes that community members have sensors for them and they can get the data from sensors of HRM and GPS to their computer. The communication method between a computer and sensors is often wired but sometimes wireless in some cases using ANT software of Garmin products. The data should be a text file so as to parse them at the server side.

3.3 Displaying Environment

Fig.4 is a snapshot of a user interface of a sub-community space that aims at sharing information for developing rope-skipping skill (ISC in Fig.3). It is one of the several communities created on the same SNS platform. This instance combines several modules such as "text" and "video". In addition, as a pluggable module for optional function, a video processing technique is installed in this case. The purpose of the integration of the image processing technology is to detect wave motion of the gross motor skill. The detection is carried out based on the sequential movement about the crown of a head. Wave motion is useful to count actions. It is also available to compare the stability of the waveform among each trial time. Stability is originally defined by a calculation using variance of local-max and local-min. Learners in a state of plateau may need good examples of skill images in such an easy-to-understanding way.

In this sub-community, the image processing requires high performance of a computer. Therefore, the pluggable module for this purpose was developed outside of the main system. The displaying frame embedded the video that has been converted from any kinds of conceivable media to the Flash-video because this media type seems comfortable for cross platform clients. The conversion function is carried out using our originally developed application that uses FFmpeg tool (See http://www.ffmpeg.org/). In the created Flash-video, the jumping wave form is over-wrapped onto the captured image. It was useful for community members to understand the stability and the counting adequacy by the system.

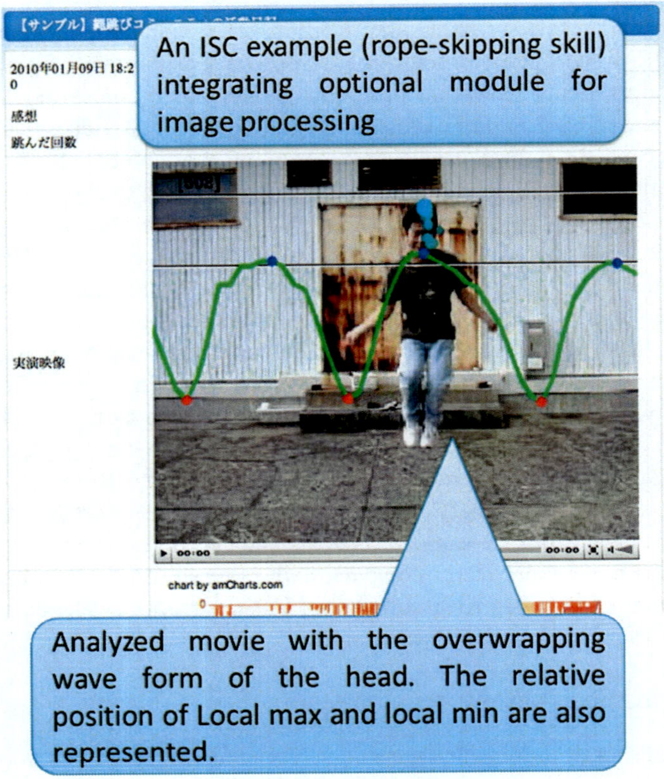

Fig. 4. A snapshot of a rope-skipping community

4 Trial Use

4.1 Organization of Participants

A present experiment of academic work is usually designed to test the predictions of the controlled performance and experimental performance [12]. However, our proposal cannot be estimated upon the comparison against the relative work but should be observed the practical use of community members.

Twenty students of our university have used this system for four weeks. Since the system was a prototype, we attempted to investigate how many skill-communities the system allowed community members to create at first. We also wanted to survey the number of combination patterns of module they can create.

4.2 Findings

Subjects created 28 different communities for motor skills such as jogging, playing pencils, Hula-Hoop, football juggling. These communities were created by fifteen different owners that means 75 % of members could done individually. Only in a few weeks trial term, total 82 articles were created in these communities. Therefore, three articles were included in each community space in average.

Table 1. All patterns of modules in created communities

Pattern	Included Modules
1	Text area (Multi-line)
2	Movie
3	Picture-Image
4	Text, Movie
5	Text, Text area (Multi-line), HRM
6	Text, Text area (Multi-line), Movie
7	Text, Text area (Multi-line), Score, Movie
8	Text, Text area (Multi-line), Score, Picture-Image
9	Text, Text area (Multi-line), Score, Picture-Image, Movie
10	Text area (Multi-line), Movie
11	Text area (Multi-line), Score, Movie
12	Text area (Multi-line), Score, Picture-Image, Movie

In terms of module combination, there were twelve patterns we found through this trial (See table 1). For instance, "text", "text area", "score" and "video" were combined at the community of Hula-Hoop. Some skill communities shared the same combination pattern of module even though the target skill was different. It implicates that there is a potential extension to make the linkage between these different skill-communities. For example, jogging community and swimming community are completely different each other on skill category perspective. However, the combination of media types and the parameters that are dealt with them, i.e. distance, time and speed, can be the same in the system. The community member in each space may mutually contribute in order to keep their motivation highly.

Even in a very short period of four weeks use, a small number of community members could try to improve the same skill. However, they visited many communities to expand their skills that have been created by other owners. The evaluation on these activities from quantitative approach is hard at this stage but we will keep observation on mutual influences and what really happen in their activities in both in-personal and inter-communities.

5 Summary and Future Implications

This article touched upon the design issue of an SNS system that promotes skill-development of human beings. To know various kinds of different motor skill is important and required for improving and refinement for skill acquisition and development in a community. From a technical viewpoint, several types of input and output methods should be integrated into the unified SNS so as to combine them in a flexible way for a motor skill according to the sub-community author. This framework offers additional possibilities to support linking relevant skills due to the integrated media types, analyzing method, and human decision. The system helps learners to transfer possible skills among community-members, to understand a skill deeply, and to discover new skills or articles relevant to the history of their own training record.

We conducted the trial use though the research project is in an ongoing stage. Subjects were pleased to create various kinds of sub-community. We found some interesting implications for the future work. However, we have to go further investigations on motivation of users. We also believe the principal subject on detecting methodology for similar skill-community and sometimes contrasting ones will be interesting topic for another studying stream.

Acknowledgments. This work was partially supported by Grant-in-Aid for Young Scientists (B) No.20700641. In terms of the system development, an alumnus of our laboratory, Mr. Toyoaki Nabeshima, made a profound contribution to this work.

References

1. Higgins, S.: Motor Skill Acquisition. Physical Therapy 71, 123–139 (1991)
2. Hollan, E., Hutchings, E., Kirsh, D.: Distributed cognition: toward a new foundation for human-computer interaction research. ACM transactions on Computer-Human Interaction 7(2), 174–196 (2000)
3. Bansler, J.P., Havn, E.C.: Sharing best practices: An Empirical Study of IT-Support for Knowledge Sharing. In: Proceedings of The 9th European Conference on Information Systems, pp. 653–664 (2001)
4. Fischer, G.: Communities of Interest: Learning through the Interaction of Multiple Knowledge Systems. In: Proceedings of the 24th IRIS Conference, pp. 1–14 (2001)
5. Fischer, G., Rohde, M., Wulf, V.: Community-based learning: The core competency of residential, research-based universities. International Journal of Computer-Supported Collaborative Learning 2(1), 9–40 (2007)
6. Wenger, E., White, N., Smith, J.D.: Digital Habitats. CPsquare (2009)
7. Matsuura, K., Gotoda, N., Ueta, T., Yano, Y.: Bridging Multiple Motor-Skills in a Community Site. In: Setchi, R., Jordanov, I., Howlett, R.J., Jain, L.C. (eds.) KES 2010. LNCS(LNAI), vol. 6279, pp. 145–152. Springer, Heidelberg (2010)
8. Feygin, D., Keehner, M., Tendick, F.: Haptic Guidance: Experimental Evaluation of a Haptic Training Method for a Perceptual Motor Skill. In: Proceedings of 10th International Symposium on Haptic Interfaces for Virtual Environment and Teleoperator Systems, pp. 40–47 (2002)
9. Allard, F., Starkes, J.L.: Motor-skill experts in sports, dance, and other domains. In: Ericsson, K.A., Smith, J. (eds.) Toward a General Theory of Expertise, pp. 126–150. Cambridge University Press, Cambridge (1991)
10. Gentile, A.M.: Skill acquisition: Action, movement, and neuromotor processes. In: Carr, J.H., Shepherd, R.B. (eds.) Movement science: Foundations for Physical Therapy in Rehabilitation, pp. 111–187 (2000)
11. Gupta, A., Noelle, D.C.: A Dual-Pathway Neural Network Model of Control Relinguishment in Motor Skill Learning. In: International Joint Conference on Artificial Intelligence 2007, pp. 405–410 (2007)
12. Heuser, F.: A theoretical framework for examining foundational instructional materials supporting the acquisition of performance skills. In: International Symposium on Performance Science, pp. 385–390 (2007)
13. Wulf, G., McNevin, N., Shea, C.H.: The automaticity of complex motor skill learning as a function of attentional focus. The Quarterly Journal of Experimental Psychology 54(4), 1143–1154 (2001)

Chapter 16
Community Size Estimation of Internet Forum by Posted Article Distribution

Masao Kubo[1], Keitaro Naruse[2], and Hiroshi Sato[1]

[1] National Defense Academy of Japan, 1 Hashirimizu, Yokosuka, Kanagawa 239-8686, Japan
{masaok,hsato}@nda.ac.jp
http://www.nda.ac.jp/cc/cs

[2] University of Aizu, Aizu-Wakamatsu, Fukushima-ken 965-8580, Japan
naruse@u-aizu.ac.jp
http://web-ext.u-aizu.ac.jp/official/index_e.html

Abstract. Anyone can easily know the number of people who submit messages, articles, and comments to a social media web site, for example, blogs, Internet forums, and bulletin boards, by enumerating the IDs. However, it is very difficult to detect the population of a community that includes lurkers who are interested in such a web site but have not submitted any messages. It is not simple task, even if one refers to the log data on the servers for social media. The population of lurkers is very important information for many applications. In this paper, a new method, based on the dynamics of complex systems, is proposed to estimate the population of a social media site that includes lurkers. As shown in recent research of Internet forums, the message distribution tends to follow a power law and the proposed method utilizes this fact. This method only needs the number of messages posted per user, and, therefore, all readers can use this method. Moreover, one does not need any language-specific knowledge. In this paper, the proposed method is confirmed by two practical experiments.

Keywords: Internet community, social dynamics, social intelligence, visualization of Internet community, power law, web intelligence, web mining.

1 Introduction

In this paper, a new method is proposed to estimate the population of an Internet community of social media web sites, such as Internet forums, bulletin boards and blogs [17][14]. Usually, anyone who can read messages on an Internet media web site can also post their own messages. One example of desirable information for these users is its population because it helps them to select a forum. However, it is nearly impossible to know the population unless he or she has the authority to log on to its server. Therefore, questioning whether we are able to estimate its population by using only published information is both natural and important.

T. Watanabe and L.C. Jain (Eds.): Innovations in Intell. Machines – 2, SCI 376, pp. 225–239.
springerlink.com

An Internet community can be divided into two groups: people who have posted at least one message and people who have not yet posted [12][16][3]. The population of the posting group is found by enumerating a type of ID for each message, but the population of the non-posting group cannot be counted in the same manner. We believe that it is possible to estimate the population by using published information and usually observable data, and so we focus on the regularity of the frequency of the number of posts per member [11]. The relation between the population of people who have posted messages and the number of messages per person forms a line in a log-log plot [18][4][1]. By using this regularity, we propose methods to estimate the population, including the lurkers, who are people who have not yet posted. Two estimation methods are proposed: the first uses the ID information and the another method utilizes the ID and time stamp of each message.These methods utilize only the posted messages that anyone can read.

It is important to choose a good forum on which user can expect to obtain sufficient number of responses when it posts its message. For selecting an active forum, the number of people who are willing to submit their messages is important. However, it is very difficult to know exact number of the population. Generally speaking, it is said that there are a several times larger number of reader than poster of forum. For example, 20:80 law is very famous. Ogawa introduces the following related data about the population of reader and poster of forum in her paper [13]: In 2002, "@cosme"(http://www.cosme.net) which is a web site of user evaluation on cosmetic products has 14822 Silent users and 1570 Active users. Also, in 1995, internet provider's forum, "Nifty forum" had (S/A)=(56.8%/43.2%). 89.9% of user in 15 mailing lists on "egroup" and "freeml" which are free internet mailing list sites are silent users. The balance of reader and poster of Internet community seems to be fluctuated case by case as these data said. Therefore a new technology to estimate the population is required.

Roughly speaking, 2 types of information are available. The first is information which anyone can access which includes messages, IDs, and submission date, etc. On the other hand, the second is information which administrator only can use. For example, the number of access to Internet forum is included in the second type of information. There is no guarantee that active communication still continues consecutively after members submitted a large amount of messages. Generally, the number of posted messages per forum also follows power law. Therefore, it is not adequate to estimate the characteristics of forum by using foundamental statistics valus, for example, average. So, some new estimation methods which deduce from each communication dynamics are prefered.

The second type of information which user without any authority can not utilize tends to be disclosed by amiable administrator. However, the access count is heavily influenced by access of software-bots. Also, the access goes up and down according to our life rhythm in spite of goodness of communication on each forum[4]. Therefore, a technology which can examine whether the disclosure information is reasonable are required.

Related method is proposed by Firth et.al. [7] which estimates the population of internet forum by Bass model. Bass model which discusses diffusion of innovation applied to 3 forums for freshmen at university for estimation of maximum population of each communities. It worked well but user without any authority can not use their method because logged data of server are necessary.

Note that there are no clear definitions of a community of an Internet forum. Traditionally, people who have posted are admitted as members of that community [16][6]. This definition is appropriate but inconvenient because the traditional definition does not consider prospective new posters of a forum. Actually, the question of whether a person who reads a forum for a few seconds/hours should be included as a member is a difficult one. Therefore, in this paper, we define a community as a set of people who definitely submit messages when sufficient time has passed. The proposed method deduces the population under this assumption. Therefore, the estimation may be different from the actual logged data results on a server, and so it is meaningful to determine the difference between the logged information and the estimations.

The remainder of the paper is organized as follows. In the next section, we discuss whether preferential attachment is a reasonable mechanism for generating the regularity of posting activity. In the third section, two estimation methods of the population of a community are proposed. Finally, we discuss the differences between the estimation and the actual data. In this paper, we describe two practical experiments: (1) a comparison with the access count of a server of a forum and (2) a comparison of a television drama program and the community size of the related forum.

2 Characteristics of the Posting Activity of an Internet Forum

2.1 The Data Source

We collected data from http://www.2ch.net, the biggest Japanese Internet bulletin board system, which has over 700 categories and 11.7 million people. Each category exhibits a few hundred bulletin boards with a total of approximately 100 million page views per day. We collected data from this web site from June 2005 to August 2005 and from December 2006 to March 2007, and a total of 327,333 forums were obtained.

2.2 Characteristics of the Posting Activity

Let us show a few examples of the frequency of contributions of an Internet forum. Figure 1 illustrates the numbers of users and submissions for three different forums. The x axis is the number of messages per user, and the y axis is the number of corresponding users. For example, point (x, y)=(1, 45) means that 45 people submitted one message. On the logarithmic scale, all of the distributions tend to form a line. Figure 2 shows the result of a regression analysis of the

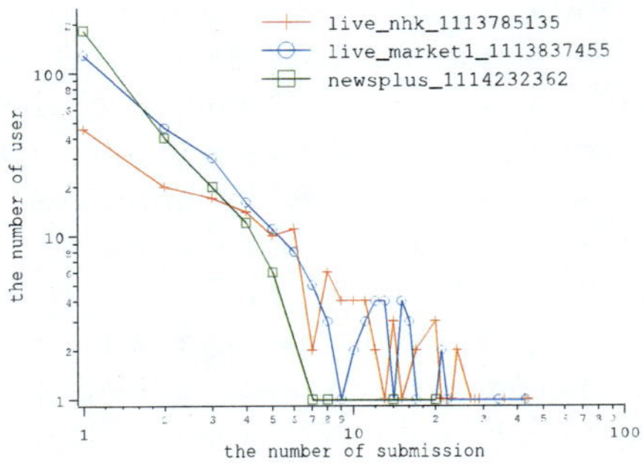

Fig. 1. Frequency distribution of the posting activity of three forums.

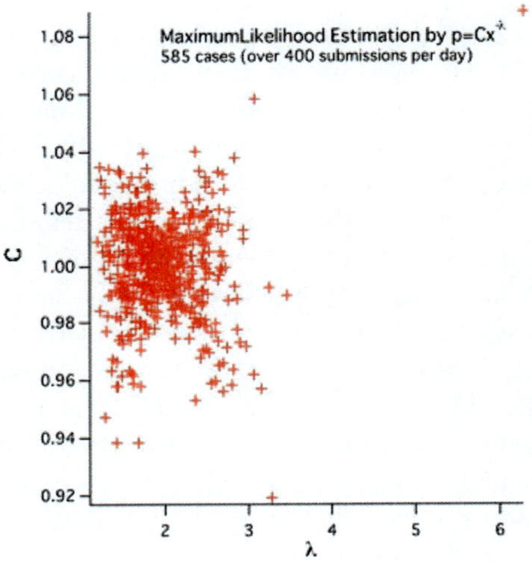

Fig. 2. Result of regression analysis of 584 bulletin boards that include over 400 messages.

collected data. Here, 584 bulletin boards having over 400 submissions per day are selected from the collected data and the coefficients of Cx^{λ} are estimated. The two coefficients of the 584 cases are shown. In this figure, many of them are concentrated around (C, λ)=(1, 2.0).

2.3 Preferential Attachment as a Generating Mechanism of the Power-Law-Like Trend of an Internet Forum's Posting Activity

In this paper, we assume that all members of a community eventual submit a message. If the usersf purpose is communication, they will submit again in response to the comment messages of other users. Actually, Joyce et al. [9] reported that the frequency of posting an additional message increases by 12% when any comments occur.

Therefore, we propose that preferential attachment [5] is a good candidate for generating the power-law-like trend of Internet forums because we presume that this reciprocity promotes dynamics similar to "a rich man gets richer"; that is, a user who frequently posts messages will submit more because he or she has a much larger chance to elicit comment messages from other users.

If the generating mechanism of this posting activity is preferential attachment, the probability of user i submitting a message is $\Pi(k)$, which is written as

$$\Pi(k) = \frac{k^{\alpha}}{\Sigma_{k'} k'^{\alpha}} \tag{1}$$

where k is the number of messages posted by user i and $\alpha = 1$.

Coefficient α of the collected data is investigated by the method of Massen and Doye [10]. In this study, the α of 496 of 584 cases (84.9%) is obtained

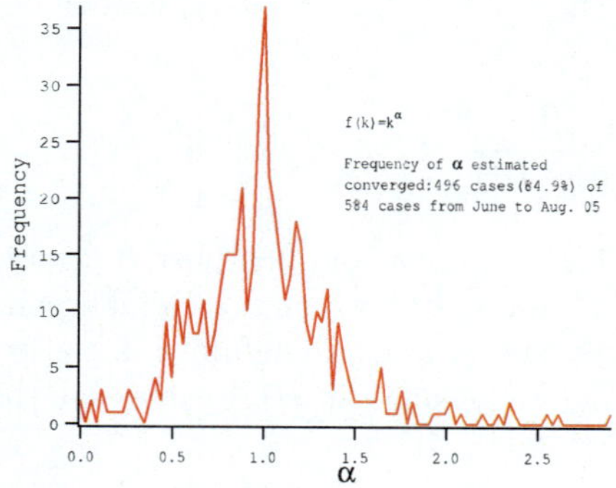

Fig. 3. Coefficient α of the preferential attachment estimation.

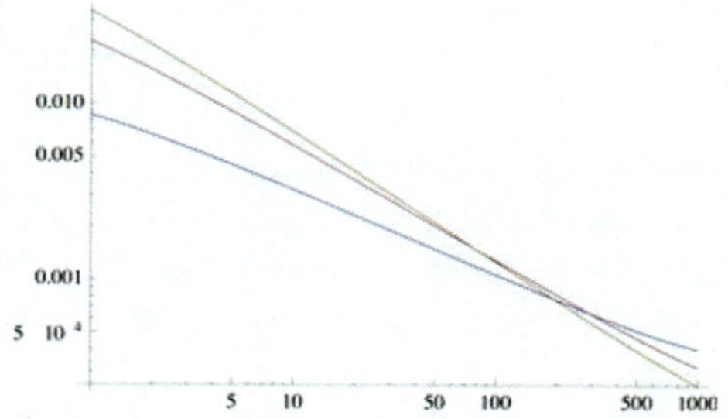

Fig. 4. Example of the proposed model (Eq. 4).

successfully. Figure 3 shows the results of the estimation. As shown, a high peak occurs at $\alpha = 1$. Therefore we can conclude that preferential attachment is one of the primary mechanisms of this posting activity.

3 The Proposed Method

3.1 The Community Model

So far, we have shown that it is reasonable to use preferential attachment as a generating mechanism for the power-law-like regularity of the posting activity of an Internet forum. We can estimate the coefficient α of each set of messages of a forum. However, it is still difficult to estimate the population of people who have not posted. The reasons are as follows. (1) The initial population of a community is not small, but the number of initial nodes in the original preferential attachment model is very small. (2) The population of a whole community should be constant, but the number of nodes in the preferential attachment grows continuously. This second assumption may sound illogical. Actually, the population of the community is changing continuously because people leave and join. However, these individual changes are too difficult to estimate. Therefore, we consider the community to be stable over a sufficiently short period used for collecting data. In this paper, we assume that the community only changes at midnight. We think that this assumption is convenient for a longer-term analysis because it is easy to measure and is not too long for detecting their attitudes. By these assumptions, the following mechanism based on preferential attachment is proposed.

First, we suppose that there is one forum with its own community. Hereafter, we collectively use the term "forum" for similar social media. Let N be the population of this community. Only one message is posted per time step and the forum includes a total of t messages posted at time step t. Also, let $n_i(t)$ be the number of messages posted by member i. Then the probability $\Pi(i)$ of member i posting an additional message is

$$\Pi(i) = \frac{wn_i + 1}{\sum_{j=1}^{N}(wn_j + 1)} \tag{2}$$

where w is a weight parameter and $w \geq 0$. A member who has posted a large number of messages submits again with high probability, whereas a member who has not yet posted any messages (n_i=0) still has a chance to submit a message. If w=0, then member posting activity is random because all $\Pi(i)$ values are the same.

In this community model, the frequency distribution $P(k)$ of the number of messages posted per member is

$$P(k) = \frac{\partial}{\partial k}(1 - \int_0^{(t+\frac{N}{w})\frac{1+\frac{1}{w}}{k+\frac{1}{w}} - \frac{N}{w}} P_{first}(t)dt) \tag{3}$$

$$= \frac{\Gamma(\frac{(1+w)(N+tw)}{w(1+kw)} - \frac{1}{w})}{(1+kw)\Gamma(\frac{(1+w)(N+tw)}{w(1+kw)})} \tag{4}$$

where $P_{first}(t_i)$ is the probability of member i first submitting a message at time step t_i.

$$P_{first}(t_i) = (\prod_{t=0}^{t_i-1}(1 - \frac{1}{wt+N}))\frac{1}{wt_i+N}. \tag{5}$$

Figure 4 shows several $P(k)$ values with different w values (=2, 3, 4). The x axis is the number of postings per member and the y axis is the corresponding probability. For example, let $N = 1000$ and $t = 1000$. As can be seen, the community dynamics generate a similar line-like distribution to that of the actual data. More detail information about this modeling is shown by [11].

3.2 The Proposed Community Population Estimation Methods

The Principle to Estimate the Population When Sufficient Time Has Passed. The population N can be obtained if the parameters of $P(k)$ in Eq. 4 are estimated by fitting the actual data. However, this equation includes a Γ function, which causes a serious error tolerance.

Therefore, other approaches should be considered. First, if sufficient time has passed since the forum started, the population is estimated by the limit of Eq. 4.

Let $Q(k)$ denote the ratio of a pair of consecutive posting probabilities. That is,

$$Q(k) = \frac{P(k)}{P(k+1)}. \tag{6}$$

The limit of $Q(k)$ at $t = \infty$ is

$$\lim_{t\to\infty} Q(k) = (\frac{1+(k+1)w}{1+kw})^{\frac{w-1}{w}}. \tag{7}$$

Therefore, given the actual data of r_k, which is the ratio of of k-times posters and $(k+1)$-times posters,

$$r_k = (\frac{1+(k+1)w}{1+kw})^{\frac{w-1}{w}} \tag{8}$$

where w can be estimated numerically. Note that the maximum possible r_k is

$$\lim_{w\to\infty} (\frac{1+(k+1)w}{1+kw})^{\frac{w-1}{w}} = 1 + 1/k. \tag{9}$$

Therefore, extra caution is needed if the actual value is larger than this bound. Once w is determined, $Post_0$, which is the population of people who have not posted yet, can be found by $Post_1$, for example,

$$Post_0 = Post_1 \lim_{t\to\infty} Q(0) = Post_1(1+w)^{\frac{w-1}{w}}. \tag{10}$$

So far, the proposed population estimation method assumes that sufficient time has passed.

The estimation Method When the Observation Time Is Known. The method proposed in the last section can be applied only when sufficient time has passed. In this section, we propose another estimation method based on the observation time t.

Let $Post_x(t)$ be the population of people who have posted x times until time step t. In the following, "$^{-}$" (e.g.,$\overline{Post_0}(t+1)$) denotes the expectation.

The group of 0-time posters, which is the lurker group, gets smaller with probability $\frac{Post_0(t)}{tw+N}$. Therefore, the expectation at $t+1$ is as follows.

$$\overline{Post_0}(t+1) = \overline{Post_0}(t) - \frac{\overline{Post_0}(t)}{tw+N}. \tag{11}$$

From this recurrence formula,

$$\overline{Post_0}(t) = \frac{N\Gamma(t - \frac{1}{w} + \frac{N}{w})\Gamma(\frac{N}{w})}{\Gamma(t + \frac{N}{w})\Gamma(-\frac{1}{w} + \frac{N}{w})}. \tag{12}$$

In the same manner as $Post_0(t)$, the recurrence formula for $Post_1(t)$ is

$$\overline{Post_1}(t+1) = \overline{Post_1}(t) + \frac{\overline{Post_0}(t)}{tw+N} - \frac{\overline{Post_1}(t)(w+1)}{tw+N}. \tag{13}$$

Then, we obtain

$$\overline{Post_1}(t) = \frac{t\Gamma(1+\frac{N}{w})\Gamma(-1+t-\frac{1}{w}+\frac{N}{w})}{\Gamma(t+\frac{N}{w})\Gamma(-\frac{1}{w}+\frac{N}{w})}. \quad (14)$$

Generally, $Q_{x/x+1}(t)$, the ratio of x-times posters and $(x+1)$-times posters, is

$$Q_{x/x+1}(t) = \frac{\overline{Post_x}(t)}{\overline{Post_{x+1}}(t)} = \frac{(x+1)(-1+N+(-(x+1)+t)w)}{(-x+t)(1+xw)}. \quad (15)$$

Therefore, given a trinomial of consecutive population data, for example, $Post_1(t)$, $Post_2(t)$, and $Post_3(t)$, N and w are found by simultaneous equations.

Let $Post_{x-1}(t)$, $Post_x(t)$, and $Post_{x+1}(t)$ be given. In this case, $\hat{w}$, the expectation of w, is

$$\hat{w} = Post_{x-1}(t)Post_{x+1}(t)(-1-t+x)(1+x) + Post_x(t)^2(-x+t)/D. \quad (16)$$

And $\hat{N}$, the expectation of N, is

$$\begin{aligned}\hat{N} = xPost_x(t)(t(-t+x)Post_x(t) - (1+x)Post_{x+1}(t)) \\ +(1+t-x)Post_{x-1}(t)((-t+x)Post_x(t) \\ +(-2+t)(1+x)Post_{x+1}(t))/D,\end{aligned} \quad (17)$$

where

$$\begin{aligned}D = (x^2(-t+x)Post_x(t)^2 + (1+t-x)(-1+x^2)Post_{x-1}(t)Post_{x+1}(t) \\ -x(1+x)Post_x(t)Post_{x+1}(t)).\end{aligned} \quad (18)$$

Example. Let me show you an example. In one Internet forum, 449 messages were submitted on a day. There were 75 users who posted once. The number of users who submitted twice and 3 times were 24, 20, respectively. When its fluctuation is sufficient small, $Post_1$=75,$Post_2$=24,$Post_3$=20, t=449 in equation 17 . Then we get the total population is 314 and w=1.526. Indeed, on this day, 156 users posted at least once. Therefore, this estimation suggests 158 users had not submitted their messages.

Analysis of the Proposed Method. In this section, we give an analysis of Eq. 16. Figure 5 shows a contour map of $\hat{w}$ when $Q_{1/2}$ and $Q_{2/3}$ are given at $t = 100$. This map can be divided into the following four regions: $\hat{w}$ is positive (**So**), $\hat{w}$ is negative (**S1**), $\hat{w}$ is not derived (**S2**), and $\hat{w}$ in the region around $D = 0$ is extremely large (**S3**).

Equation 16 does not consider any fluctuation. Therefore, it sometimes happens that an estimated $\hat{w}$ is out of the **So** region by fluctuation and in that case this method cannot estimate the population.

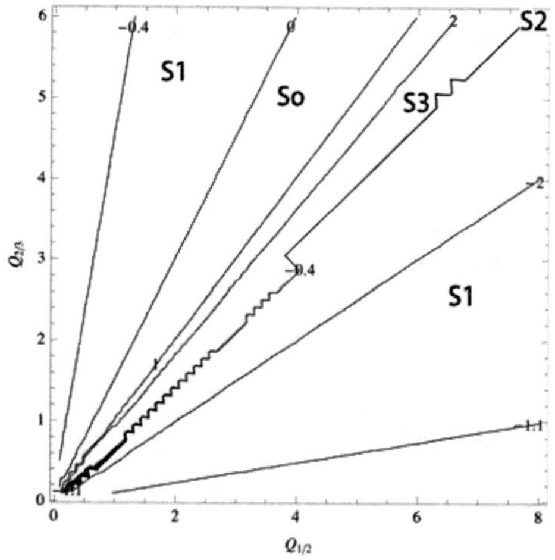

Fig. 5. Map of $\hat{w}$, where $Q_{1/2}(t)$ and $Q_{2/3}(t)$ are given at $t = 100$.

Therefore, we introduce additional heuristic estimation rules, shown in the following, when the method of Eq. 16 does not work.

S1 $w < 0$
$Post_0 \leftarrow Post_1$ because we suppose $Post_0$ is at least same as $Post_1$

S2, S3 $Post_0 \leftarrow 0$ because we cannot obtain any meaningful information when D=0

S4 $w \gg 1$
$Post_0 \leftarrow \frac{Post_1}{Post_2} Post_1$ because this value suggests the preferential attachment model is not adequate, when **S4** includes the case in which $Q_{1/2}$ and $Q_{2/3}$ are not obtained.

4 Experiments

In the preceding section, the population estimation methods are proposed. In this section, these methods are evaluated by using actual data.

Generally speaking, if logged data are not utilized, evaluation of the methods is extremely difficult, as in other computational social science studies. In this paper, we evaluate our proposed methods by calculating the correlation [8] of indices that have strong relations within the population. For example, the number of votes and the related community size have a strong relation.

In this paper, we evaluate our proposed methods by comparison with (1) the access count of the forum, (2) the viewing rate of a television drama program and the community size of the related forum. Fortunately, the viewing rate of many television programs is published by many companies. If the proposed estimation is good, we assume it makes the correlation stronger.

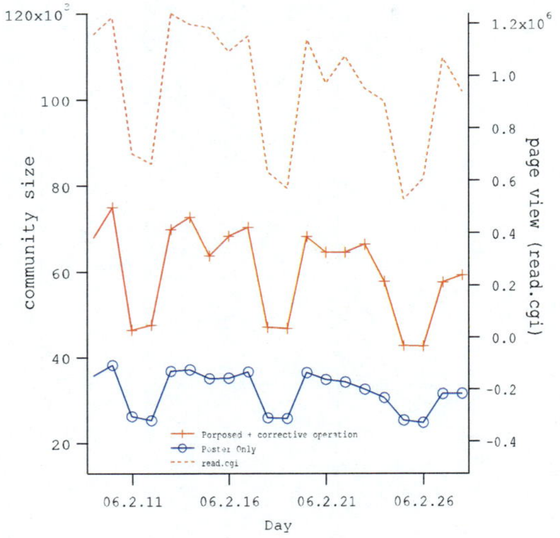

Fig. 6. Access count of the server "ex13.2ch.net" and its estimated total population.

4.1 Correlation of the Number of Access Counts in a Web Server Log

The access log for each Internet forum is not open to the public, and the access requests to these forums are divided across several server computers. However, a part of the log information of these servers is available to the public. Therefore, a comparison of server levels is possible: namely, the correlation between the sum of the estimated population of each forum on a particular server and the number of access requests to the server. Although the sizes of both are usually different, it is obvious that both have a strong relation. Therefore, we expect that a strong correlation.

The Data Set. Only the daily log data of the server are available. Therefore, a forum in which users are in discussions after midnight is not suitable. Hence, server "ex13.2ch.net" is adopted for this verification because this server includes only a relatively small number of unsuitable forums. The posted comments and messages were collected from February 2006 to March 2006. The data of 33,069 forums were collected, which included 3,128,155 messages. Next, the sum of the populations of these forums from the second week to the fourth week of February is estimated by the proposed method. We use Eq. 17 and the heuristics. Finally, the cross-correlation between the estimation and the number of requests to access the server is calculated.

The server log data released for "pv.kakiko.com" is adopted for this experiment. This site's log data are classified into the data types to be accessed, namely, "html", "cgi", "dat", "text", and "picture".

Table 1. Correlation of the access count to the server: Pearson product-moment correlation.

	Poster population	Estimated community size
Total	0.9765	0.9781
read.cgi	0.9433	0.9533
html	0.9461	0.9521
dat	0.9811	0.9820

Result of the Estimation. In Fig. 6, 3 graphs are depicted. The line with the "+" mark represents the estimation by the proposed method. A poster submits at least one message. The population of the posters can be obtained by enumerating the IDs on the forums. The graph with "" represents the population of the posters only. The dotted line represents the access count of users to the cgi file. The x axis indicates the date. The result of the Pearson product-moment correlation is shown in Table 1. For "read.cgi," the correlation of the proposed method is 0.9533, whereas that of the poster population is 0.9433.

Summary of Experiment 1. In this section, the correlation between the estimated population by the proposed method and the access count to the server is examined. We expected that if our method is suitable, the correlation should be similar to the correlation between the population of the poster and the access count because the access count and the population of the community have a close relation. The results are interesting from two viewpoints. The first is the very high correlation of the population of the poster. This suggests that the server load behavior is very similar to the poster population. The second is our estimation is similar to the correlation of the poster population, as we expected. The results imply that our method is relevant because the correlation becomes a very low value if the estimation is meaningless, that is, if the estimation contains noise.

4.2 Correlation with Viewing Rate and the Estimated Population of an Internet Forum Related to a TV Drama

Second, to evaluate the proposed method, we collect seven television drama viewing rates and their corresponding forums (see Table 2) from December 2006 to March 2007. Then, the following two types of correlations are calculated.

1. The correlation between the viewing rate and the number of people who have posted (number of posters). The number of posters can be obtained by enumerating the IDs on the forums.
2. The correlation between the viewing rate and the population of the community that includes people who have not posted yet (lurkers). The community size is estimated by the proposed method.

Table 2. Number of forums for each TV drama.

Program title	Number of forums
K1	5
H1	24
H2	13
K2	30
K3	6
T1	9
H1	7

Table 3. Comparison with the viewing rate: Pearson product-moment correlation coefficient r (Eq. 20).

	Poster population by ID enumeration	Proposed
correlation coefficient	0.7565	0.7987

Summary of Experiment 2. It is an advantage of an Internet forum that people can read or write a post anytime. Some people may access an Internet forum when a TV drama program is broadcasted while other people may read or write to the same forum after a few days. Therefore, we suppose that the viewing rate of a weekly TV program is a good criteria for verification of the estimated population. First, we estimate the population $\hat{N}$ of a forum for each day from December 2006 to March 2007. We use Eq. 17 and the heuristics. As Table 2 shows, each TV program has more than one forum. Next, the correlation between the sum of the estimated populations for a week of all related forums and the viewing rate is calculated as follows. Let $\hat{N}_x^{day}$ be the population of community $x \in B_d$ on a *day*. The sum of the population for a week, $pop_{B_d}^{week}$, is

$$pop_{B_d}^{week} = \sum_{x \in B_d} \sum_{day \in DAY} (\hat{N}_x^{day}) \tag{19}$$

where $day \in DAY = \{day1, ..., day7\}$ is each day after the broadcast. The Pearson product-moment correlation coefficient r is calculated as follows.

$$r = \frac{\frac{1}{M}\sum_i (pop_i - \overline{pop})(v_i - \overline{v})}{\sqrt{\frac{1}{M}\sum_i (pop_i - \overline{pop})^2}\sqrt{\frac{1}{M}\sum_i (v_i - \overline{v})^2}}. \tag{20}$$

We suppose that there are M related forums.

Results. The estimation results are shown in Figs. 7 and 8 and Table 4.2. Figure 7 shows an example of the estimated daily population transition of Internet communities for a TV program. The x axis is the day, and the y axis is the sum

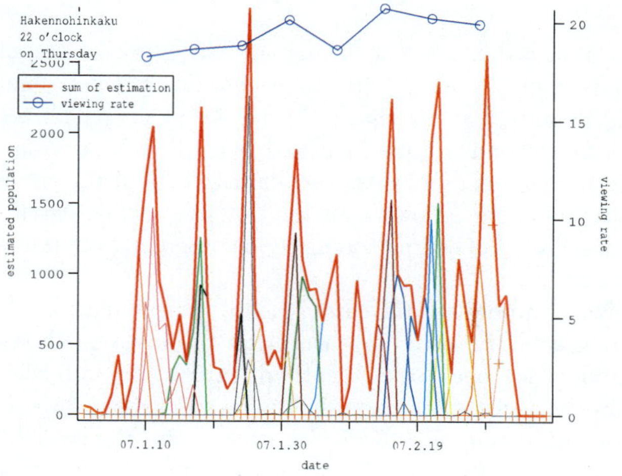

Fig. 7. Example of the estimated population and the viewing rate of a TV program.

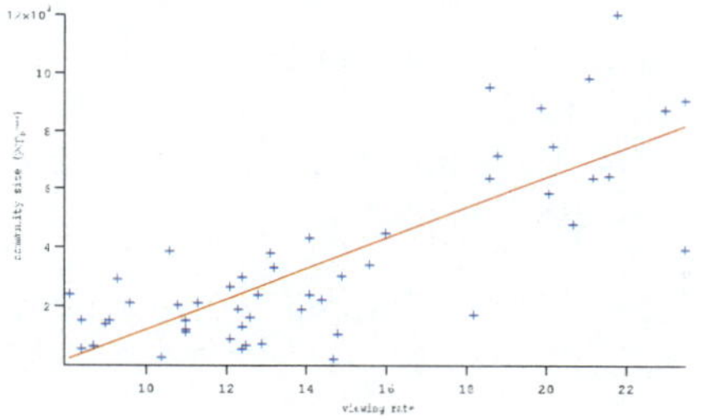

Fig. 8. Correlation between the estimated population and the viewing rate.

of the population (solid line) and its corresponding viewing rate (). The bold line indicates the sum of the population of the related forums. Figure 8 plots the viewing rate and $pop_{B_d}^{week}$. The x axis is the viewing rate and the y axis is $pop_{B_d}^{week}$. As one can see, these form a line. Table 4.2 shows the result of the comparison of the correlation between the population of posters and the estimated population, which includes the lurkers. The poster population is obtained by enumeration of the IDs of the posted messages. By comparison, the score of the poster population has a high correlation (0.7565) with the viewing rate. However, the proposed population estimation methods has a higher correlation (0.7987) when the number of samples is 51 and the significance levels, 5% and 1%, are approximately 0.28 and 0.37, respectively.

This does not contradict our expectation. We conclude that this series of experiments proves the validity of our proposed population estimation methods.

5 Conclusion

In this paper, a new concept which anyone can estimate community size of internet forum by using collective emergent characteristics of communication is proposed. Traditionally, it seems that there are no typical patterns or regularities on an internet forum and a bulletin board because it is a result of voluntary behavior of an individual user. However, distribution of the number of posted articles per user on them follows as a kind of power law distribution. In this paper, we proposed a set of estimation methods based on preferential attachment by using this trait.

Usually, only the administrator can estimate the population of a community of a social media site. However, by using the proposed methods, anyone can estimate the population because our methods utilize only published information and usually observable data, for example, the ID and time stamp of the messages.

With small modification of preferential attachment, the model for generating the posting activity of an Internet forum is proposed. Then, two estimation methods are proposed: the first uses the ID information and the another method utilizes the ID and time stamp of each message.

Finally, we verified our methods by using the correlation between (1) the access count to the server of a forum and (2) the viewing rate of a television program and the estimated population of the related Internet forum. The correlations of both experiments are similar to that of the population of posters by enumeration of the IDs. We think these results confirm the validity of our methods because the correlation is strong.

Acknowledgments. This paper is supported by a Grant-in-Aid for Scientific Research of Japan.

References

1. Albert, R., László Barabási, A.: Statistical mechanics of complex networks. Reviews of Modern Physics 74(1), 47–97 (2002)
2. Amaral, L.A.N., Scala, A., Barthélémy, M., Stanley, H.E.: Classes of small-world networks. Proc. of the National Academy of Sciences 97(21), 11149–11152 (2000)
3. `http://www.tlsoft.com/arbitron/jul95/arbitron.summary.txt`
4. Baldi, P., Frasconi, P., Smyth, P.: Modeling the Internet and the Web: probabilistic methods and algorithms. Wiley, Chichester (2003)
5. Barabási, A.L., Albert, R., Jeong, H.: Mean-field theory for scale-free random networks. Physica A 272, 173–187 (1999)
6. Fisher, D.: Studying Social Information Spaces. In: From Usenet to Cowebs: Interacting With Social Information Spaces, pp.3–19 (2003)
7. Firth, D.R., Lawrence, C., Clouse, S.F.: Predicting Internet-based Online Community Size and Time to Peak Membership Using the Bass Model of New Product Growth. Interdisciplinary Journal of Information, Knowledge, and Management Volume 1, 1–12 (2006)
8. Gruhl, D., Guha, R.V., Liben-Nowell, D., Tomkins, A.: Information diffusion through blogspace. In: WWW 2004, pp. 491–501 (2004)

9. Joyce, E., Kraut, R.E.: Predicting Continued Participation in Newsgroups. Journal of Computer-Mediated Communication 11(3), 723–747(25) (2006)
10. Massen, C.P., Doye, J.P.K.: A self-consistent approach to measure preferential attachment in networks and its application to an inherent structure network. Physica A: Statistical Mechanics and its Applications 377(1), 351–362 (2007)
11. Keitaro, N., Kubo, M.: Lognormal Distribution of BBS Articles and its Social and Generative Mechanism, Web Intelligence. In: Proceedings of the 2006 IEEE/WIC/ACM International Conference on Web Intelligence, pp. 103–112 (2006)
12. Nonnecke, B., Preece, J.: Lurker demographics:counting the silent, Conference on Human Factors in Computing Systems. In: Proceedings of the SIGCHI conference on Human factors in computing systems, pp. 73–80 (2000)
13. Ogawa, M., Sasaki, Y., Tsuda, H., Yoshimatsu, T., Kokuryo, J.: Silent Members (SM): Their Communication Behaviour and Influence on Purchases of Others. In: PACIS 2003 Proceedings, pp. 108–121 (2003), `http://aisel.aisnet.org/pacis2003/8`
14. Killworth, P.D., McCarty, C., Russell Bernard, H., House, M.: The accuracy of small world chains in social networks. Social Networks 28, 85–96 (2006)
15. Smith, M.A.: Invisible Crowds in Cyberspace: Mapping the Social Structure of the Usenet, Communities in Cyberspace: Perspectives on New Forms of Social Organization. Routledge Press, London (1999)
16. Mark, A.: Smith, Measures and Maps of Usenet, in From Usenet to Cowebs: Interacting With Social Information Spaces, pp. 47–78 (2003)
17. White, N.: Community Member Roles and Types (2001), `http://www.fullcirc.com/community/memberroles.htm`
18. `http://www.nslij-genetics.org/wli/zipf/`

Chapter 17
A Design of Lightweight Reprogramming for Wireless Sensor Networks

Aoi Hashizume[1], Hiroshi Mineno[2], and Tadanori Mizuno[1]

[1] Graduate School of Science and Technology, Shizuoka University, Japan
[2] Faculty of Informatics, Shizuoka University, Japan
3–5–1 Johoku, Naka–ku, Hamamatsu, Shizuoka 432-8011, Japan
aoi@minelab.jp, {mineno,mizuno}@inf.shizuoka.ac.jp

Abstract. Considering a maintenance of sensor networks, wireless reprogramming techniques are required. Particularly, when sensor networks are introduced to targeted environment, several parameters of sensor nodes must be calibrated, for example, sensing interval, data sending interval. To change these parameters, we don't have to reprogram the whole of program. It is needed to update only program sections that include targeted variables in our sensor network. We designed and implemented a lightweight reprogramming scheme. This scheme doesn't require reboot of sensor nodes, therefore we don't have to stop the services in long term. Our proposal makes reprogramming efficient in respect of service availability and energy consumption.

1 Introduction

Advances in microelectromechanical system and low power wireless communication technology have led to the development of wireless sensor networks (WSN). A typical WSN consists of a number of small battery-powered sensor nodes that sense, collect, and transfer various data autonomously. There are many WSN applications and services, including structural monitoring, security, and position tracking. These networks include state-of-the-art technologies (ad-hoc network routing, data processing, position estimation, etc.), and these technologies are implemented as specific code on the sensor nodes. These technologies are highly advanced and still developing. Therefore, these codes will be modified or extended in the future for long-running applications using WSNs. Thus, a method to efficiently reprogram many deployed sensor nodes is necessary.

Wireless reprogramming has been extensively researched [1][2][3][4]. Wireless reprogramming distributes new code to many sensor nodes using wireless multihop communication. Then, sensor nodes reprogram themselves with the received data. Reprogramming protocols now mainly focus on software data dissemination and discuss little about reprogramming [5]. Therefore, we have to address not only data dissemination but also reprogramming. We designed and implemented a reprogramming scheme on real sensor nodes.

T. Watanabe and L.C. Jain (Eds.): Innovations in Intell. Machines – 2, SCI 376, pp. 241–249.
springerlink.com

This paper is organized as follows. In Section 2, we explain several issues related to wireless reprogramming. Our reprogramming design and implementation is introduced in Section 3. We describe our experimental results in Section 4. Finally, Section 5 concludes the paper and mentions future work.

2 Reprogramming in Sensor Networks

Many wireless reprogramming protocols share design challenges [1]. The first challenge is completion time. The reprogramming completion time affects services using WSNs. When we reprogram a network, we have to stop the services and wait until the code update is completed. Thus, we have to minimize the reprogramming completion time. The second challenge is energy efficiency. Sensor nodes are usually battery-powered, and the sensor node battery provides the energy used in reprogramming. This battery also supplies energy for computing, communication, and sensing functions. Therefore, reprogramming must be energy-efficient. The third challenge is reliability. Reprogramming requires the new code to be delivered throughout the entire network, and the delivered code must be executed correctly on the sensor nodes.

Reprogramming protocols use several techniques to resolve the challenges listed above. Pipelining was developed to accelerate reprogramming in multihop networks [6] [7]. Pipelining allows parallel data transfer in networks. In pipelining, a program is divided into several segments, and each segment contains a fixed number of packets. Instead of receiving a whole program, a node becomes a source node after receiving only one segment. Particularly for transmitting a large program, pipelining can reduce the completion time significantly. However, pipelining has detrimental characteristics. Pipelining works well when there is a large number of hops between base stations and the farthest destination nodes. In contrast, having a small number of hops causes delay in code distribution. Negotiation is used to avoid data redundancy and improve reprogramming reliability. As explained above, pipelining is done through segmentation. After the data are segmented, it is necessary to avoid broadcast storms that are caused by dealing with a large number of segments. A negotiation scheme was developed in SPIN [8]. In this scheme, the source node knows which segment is requested before sending it out. This reduces data redundancy. Hierarchical reprogramming [9] [10] accelerates software distribution and reduces the number of control packets. First, the base station sends program codes to nodes in the upper layer of the node hierarchy (i.e., pseudo-base stations). Then the pseudo-base stations distribute the codes to other nodes in their local areas. Hierarchical reprogramming protocols improve reprogramming efficiency. Except for Firecracker and Sprinkler, most reprogramming protocols start distributing software from a single base station in the network and assume no hierarchy. Hierarchical reprogramming protocols improve reprogramming efficiency. Deploying base stations in suitable places greatly improves the efficiency of the reprogramming [11].

3 Proposed Reprogramming Scheme

3.1 Targeted Environment

As described in Section 2, most existing protocols focus on software data dissemination and hardly consider the reprogramming scheme on the sensor node. In the following sections, we consider a reprogramming scheme that includes hardware, network topology, program sections, and algorithms.

The development of this scheme concerns the construction of a sensor network utilizing a ZigBee node produced by Renesas Solutions Corporation (RSO) [12]. The first is the battery-operated Renesas sensor board in Fig. 1 which has three kinds of sensors: motion, temperature, and light. It can be operated either with four AAA batteries or an AC adaptor. This board is used as the sensor node.

The other board is a ZigBee evaluation board, shown in Fig. 2. It has no sensors or batteries, and therefore a power supply cable is required. This board is equipped with an RS-232C interface which can be connected with a PC. The board can be used in three ways, one board each: the first as a ZigBee router node, the second as a coordinator node, and the third as a sink node. Specified node works as a base station node (coordinator node).

The assumed network topology is shown in Fig. 3. It requires several PCs, PC-connected base station nodes, and sensor nodes. A PC sends new code to a base station, and the base station sends the code to neighboring sensor nodes. Then, sensor nodes update their firmware (FW) with the receipt code.

Fig. 1. Sensor Board

Fig. 2. ZigBee Evaluation Board

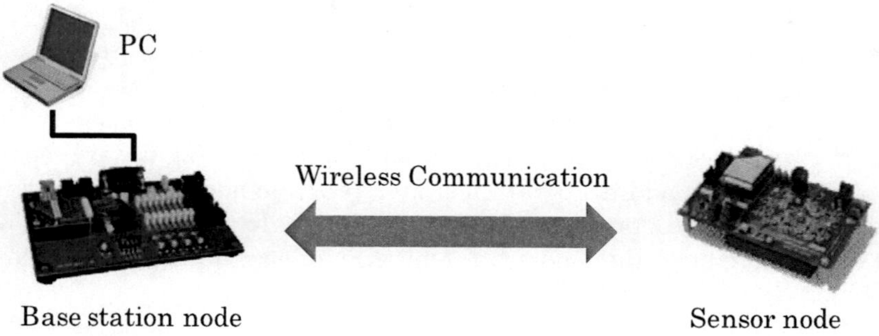

Fig. 3. Network Topology

Table 1. Program data allocation of microcomputer

Data	Section code	Allocation
Automatic variable	stack	RAM
Static variable (has no initial value)	base	
Static variable (has initial value)	data	
Initial value of static variable	data_I	ROM
String, constant number	rom	
Program code	program	

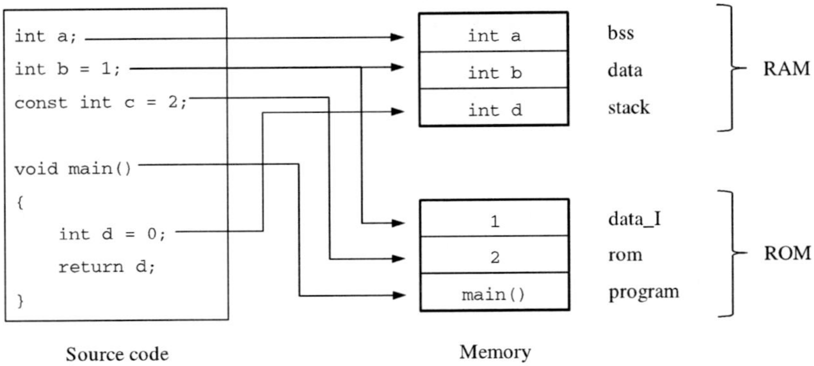

Fig. 4. Example of allocation

3.2 Design and Implementation

We designed the reprogramming scheme to solve several of the problems described in Section 2.

Here, we describe the memory allocation of the program sections. Table 1 and Fig. 4 show the program data allocation of a microcomputer in our targeted environment.

We designed several program sections not only for general use but for reprogramming use in the memory. Table 2 and Fig. 5 show address allocation of program sections. Program data are allocated by unit of the sections in the memory area. Program sections for general use and reprogramming use are separated to avoid collisions of address parts when we reprogram nodes. Then, we have to update only the reprogramming section and do not have to change the whole program. This reduces the amount of reprogramming data. Less data shortens completion time and reduce energy consumption.

Table 2. Address allocation of program sections

Section name	Section code	Address allocation
update_data	data	0x003000 – 0x0030FF
update_bss	bss	
update_bss_I	bss_I	0x0F8000 – 0x0F80FF
update_rom	rom	0x0F9000 – 0x0F90FF
update_prog	program	0x0FA000 – 0x0FA1FF

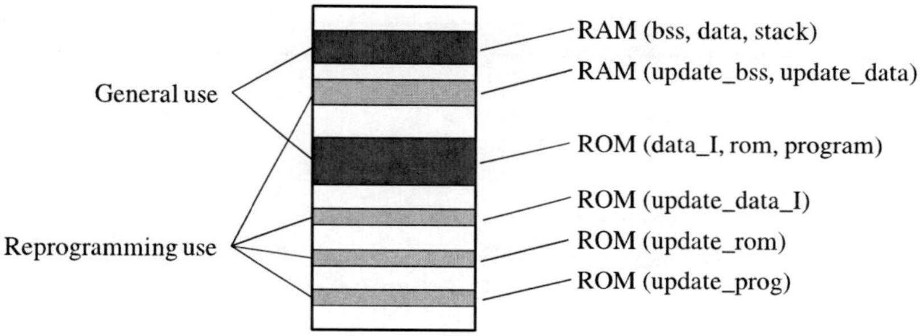

Fig. 5. Program Section

3.3 Message Format

After a base station node received new program data from PC, it starts to send receipt data to neighboring nodes by a FW data message. A FW data message includes a sequence number and FW data. Fig. 6 and Table 3 show a message format of a FW data message. A base station node waits an arrival of ACKs from target nodes after sending a FW data message. In this duration, all messages are rejected without ACKs. A base station node sends or resends a FW data message based on an ACK information. If sensor nodes received FW data message competely, they notices it to a base station node by ACKs. Fig. 7 and Table 4 presents details of a FW data ACK message. Additionally, one of five values represented in Table 5 is set to ACK parameter of a FW data ACK message. After a sensor node received a FW data message, it sends back a FW data ACK message and waits an arrival of next FW data message. In this duration,

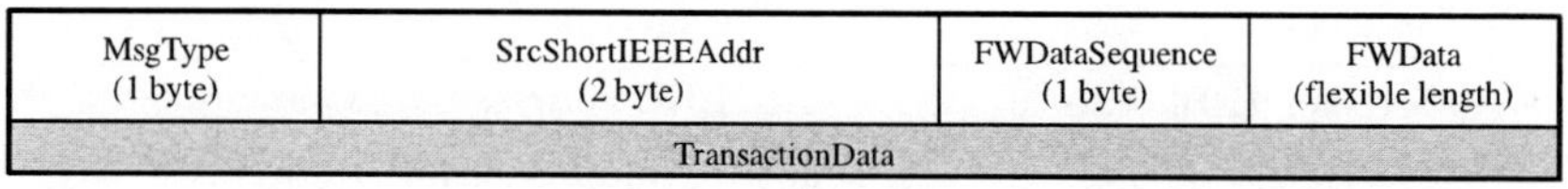

Fig. 6. Message format of a FW data message

Table 3. Explanations of a FW data message

Parameter	Explanation
MsgType	Message type
SrcShortIEEEAdrr	A short IEEE address of a source node
FWDataSequence	A sequence number of FW data
FWData	FW data

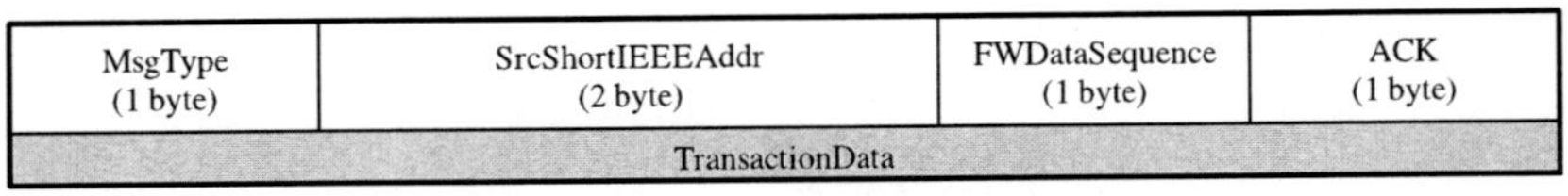

Fig. 7. Message format of a FW data ACK message

all messages are rejected without FW data messages. All of FW data messages is received, then a sensor node updates itself. If FW udpate succeeded, the node sends a FW data ACK message that includes FWDATA_ACK_UPDATE_OK prameter. Failure results make the sensor node sends a FW data ACK message that includes FWDATA_ACK_UPDATE_NG prameter.

3.4 Reprogramming Algorithm

Base Station Node. Here, we explain the reprogramming algorithms. Algorithms 1 and 2 are firmware update algorithms on the base station nodes.

Algorithm 1 shows the initialization process of the firmware update. If the process mode is a FW update, the base station node sets a FW update flag and calls a FW update function (*appFWupdate*). After *appFWupdate* is returned, the return value of it is set to process mode. Furthermore, the node resets the FW update flag and the FW data receipt flag.

Algorithm 2 shows the main firmware update process. The base station calls the FW data get function (*appFWdataGet*) and then gets the FW data from the PC by calling this function. If the base station gets the FW data, it enters a

Table 4. Explanations of a FW data ACK message

Parameter	Explanation
MsgType	Message type
SrcShortIEEEAdrr	A short IEEE address of a source node
FWDataSequence	A sequence number of FW data
ACK	ACK information

Table 5. Explanations of ACK parameters

ACK	Explanation
FWDATA_ACK_OK	Request to send next FW data
FWDATA_ACK_NG	Request to resend current FW data
FWDATA_ACK_UPDATE_OK	Notice a success of FW update
FWDATA_ACK_UPDATE_NG	Notice a failure of FW update
FWDATA_ACK_UPDATE_TIMEOUT	Notice a timeout occurred

Algorithm 1. Initialization process of firmware update (Base station node)

```
if process mode = FW update then
  set FW udpate flag
  call FW update function (appFWupdate)
  set the return value of appFWupdate as process mode
  reset FW udpate flag
  reset FW data receipt flag
end if
```

loop process. If FW data ACK is FWDATA_ACK_UPDATE_OK/_NG or a time out occurs, this loop is broken.

Sensor Node. Next, Algorithms 3 and 4 are firmware update algorithms on the sensor nodes.

Algorithm 3 shows the initialization process of the firmware update. If the process mode is a FW update, the sensor node analyzes the FW data message and sets the FW data size to a FW data structure. Furthermore, the node initializes a sequence number and a FW data counter. After the initialization of the FW data structure, the base station calls a FW update function (*appFWupdate*). After *appFWupdate* is returned, the return value of it is set to process mode.

Algorithm 2. Main process of firmware update (Base station node)

```
call FW data get function (appFWdataGet)
if appFWdataGet returned FW_DATA_OK then
  repeat
    call FW data send function (appFWdataSend)
    wait FW data ACK
    if get FW data ACK then
      call ACK check function (appFWdataRepCheck)
    end if
  until ACK = FWDATA_ACK_UPDATE_OK || ACK = FW-
  DATA_ACK_UPDATE_NG || TIMEOUT
end if
```

Algorithm 3. Initialization process of firmware update (Sensor node)

```
if process mode = FW update then
  analyze FW data message
  set FW data size to FW data structure
  initialize sequence number and FW data counter
  call FW update function (appFWupdate)
  set the return value of appFWupdate as process mode
end if
```

Algorithm 4. Main process of firmware update (Sensor node)

```
call FW data message check function (appFWdataChk)
if appFWdataGet returned FW_DATA_OK then
  repeat
    call FW data send function (appFWdataSend)
    wait FW data ACK
    if get FW data ACK then
      call ACK check function (appFWdataRepCheck)
    end if
  until ACK = FWDATA_ACK_UPDATE_OK || ACK = FW-
  DATA_ACK_UPDATE_NG || TIMEOUT
end if
```

Algorithm 4 shows the main firmware update process. The sensor node calls the message check function (*appFWdataChk*). If *appFWdataChk* returns FWDATA_GET_OK the node sends a FW data ACK message to the base station node to request the next FW data. If *appFWdataChk* returns FWDATA_GET_NG, the node sends a FW data NACK message to the base station node to request the lost FW data. After sending the FW data ACK/NACK message, the sensor node receives a FW data message and then calls *appFWdataChk*. If *appFWdataChk* returns FWDATA_GET_COMP, the node calls a ROM update function (*appFWupdateRom*). Finally, this node sends a FW data ACK and finishes the operation. In this instance, the return value is "No Operation."

4 Experimental Results

We allocated sensing interval and LED blink interval variables on the program section for reprogramming use. Then, we sent new variables from the PC to the sensor nodes through the base station node. Result showed that the sensor nodes were correctly reprogrammed.

5 Conclusion

A reprogramming scheme was implemented in a wireless sensor network. First, we described reprogramming challenges and approaches. Then, we explained the design of the reprogramming scheme, the allocation of program sections and reprogramming algorithms on the sensor nodes and the base station nodes. In addition, we demonstrated that our proposed scheme worked correctly.

In future work, we will try to implement the latest reprogramming protocols in real sensor networks. Then we will research several problems and solutions in actual environments.

Acknowledgments

This work is partially supported by the Knowledge Cluster Initiative of the Ministry of Education, Culture, Sports, Science and Technology.

References

1. Wang, Q., Zhu, Y., Cheng, L.: Reprogramming wireless sensor networks: challenges and approaches. IEEE Network 20(3), 48–55 (2006)
2. Thanos, S., Tyler, M., John, H., et al.: A remote code update mechanism for wireless sensor networks. CENS Technical Report (2003)
3. Pradip, D., Yonghe, L., Sajal, K.D.: ReMo: An Energy Efficient Reprogramming Protocol for Mobile Sensor Networks. In: Proc. IEEE PerCom, pp. 60–69 (2008)
4. Leijun, H., Sanjeev, S.: CORD: Energy-efficient Reliable Bulk Data Dissemination in Sensor Networks. In: Proc. IEEE INFOCOM, pp. 574–582 (2008)
5. Pascal, R., Roger, W.: Decoding Code on a Sensor Node. In: Proc. IEEE DCOSS, pp. 400–414 (2008)
6. Jonathan, W.H., David, C.: The Dynamic Behavior of a Data Dissemination Protocol for Network Programming at Scale. In: Proc. ACM SenSys, pp. 81–94 (2004)
7. Sandeep, S.K., Limin, W.: MNP: Multihop Network Reprogramming Service for Sensor Networks. In: Proc. IEEE ICDCS, pp. 7–16 (2005)
8. Kulik, J., Heinzelman, R.W., Balakrishnan, H.: Negotiation-Based Protocols for Disseminating Information in Wireless Sensor Networks. Wireless Networks 8(2–3), 169–185 (2002)
9. Philip, L., David, C.: The Firecracker Protocol. In: Proc. ACM SIGOPS European Workshop (2004)
10. Vinayak, N., Anish, A., Prasun, S.: Sprinkler: A Reliable and Energy Efficient Data Dissemination Service for Wireless Embedded Devices. In: Proc. IEEE RTSS, pp. 277–286 (2005)
11. Aoi, H., Hiroshi, M., Tadanori, M.: Base Station Placement for Effective Data Dissemination in Sensor Networks. Journal of Information Processing 18, 88–95 (2010)
12. Renesas Solutions Corporation, `http://www.rso.renesas.com/`

Chapter 18
Simulation Evaluation for Traffic Signal Control Based on Expected Traffic Congestion by AVENUE

Naoto Mukai[1] and Hiroyasu Ezawa[2]

[1] Dept. of Culture-Information Studies, School of Culture-Information Studies, Sugiyama Jogakuen University
17-3, Hoshigaoka-motomachi, Chikusa-ku, Naogya, Aichi, 464-8662, Japan
nmukai@sugiyama-u.ac.jp

[2] Dept. of Electrical Engineering, Graduate School of Engineering, Tokyo University of Science, Kudankita, Chiyoda-ku, Tokyo, 102-0073, Japan
j4309617@ed.kagu.tus.ac.jp

Abstract. In these years, the traffic jam becomes a serious problem according to the increasing of vehicle holders in Japan. One of the key issues to ease the traffic jam is a traffic signal control, i.e., optimization of traffic signal parameters (cycle, split, and offset). The information technologies such as probe car and road-to-vehicle communication enable to progress to the next stage for the traffic signal control. In this paper, we focus on "expected traffic congestion (ETC)" which is a simple indicator for traffic jam. The value of ETC is based on the shared probe data (i.e., path information) among vehicles by road-to-vehicle communication. We apply the ETC to the optimization of traffic signal parameters. Moreover, in order to evaluate the effectiveness of the optimization, we introduce "AVENUE" which is a popular traffic stream simulator. We developed an outer module to calculate the ETC and update the traffic signal parameters for AVENUE. The experimental results using the outer module and AVENUE indicate that our traffic signal control can reduce the traveling time of vehicles.

1 Introduction

There are significant traffic problems (e.g., traffic jam) according to the increasing of vehicle holders in Japan. Intelligent Transportation System (ITS) is a system to resolve the various traffic problems by using computers, electronics, and advanced sensing technologies. One of the important subjects of the ITS is a traffic jam problem. A traffic signal control (i.e., optimization of traffic signal parameters) is a typical and effective approach to ease the traffic jam. Previously, some patterns of traffic parameters (cycle, split, and offset) are prepared, and a suitable pattern is selected from the prepared patterns depending on the traffic situations. However, this method is not so flexible, and inadaptable to a unexpected traffic situation (e.g., traffic accidents). Thus, adaptive methods of traffic signal control have been studied for a number of years [1,2,3,4,5].

T. Watanabe and L.C. Jain (Eds.): Innovations in Intell. Machines – 2, SCI 376, pp. 251–263.
springerlink.com

On the other hand, in these years, the latest information technologies such as "probe car" and "road-to-vehicle communication" can be available for general users. For example, some of taxi companies adopt probe cars as taxis to identify the customer trends in Japan. The probe data of taxis are called taxi probe which include customer's get on/off positions and time[1]. Moreover, the famous application of road-to-vehicle communication is Vehicle Information and Communication System (VICS) which provides traffic information to car navigation system by FM broadcast and radio beacon[2]. Such information technologies enable to progress to the next stage for the traffic signal control. In [6,7,8], vehicles release pheromone to road link depending on their probe data (i.e., position and speed), and the vehicles can avoid traffic jam according to the accumulated pheromone on the road link. In [9,10,11], an on-demand traffic control scheme called Advanced Demand Signals scheme (ADS) and its extended version are proposed. In the schemes, traffic signals are controlled by demands of vehicles (or pedestrians) without traditional signal parameters (i.e., request for keep indication or change indication). In [12], a delay time at a red light is estimated by using probe cars instead of vehicle detectors, and traffic signal parameters are optimized to minimize the estimated delay time.

In this paper, we focus on a simple indicator for traffic jam which is called "Expected Traffic Congestion (ETC)". The ETC is proposed by Yamashita et al. [13], and used for path selection of vehicles. There are some traditional indicators such as delay time (i.e., the wasted waiting time of vehicles) and queue length (i.e.,the length of vehicles in front of traffic signals). However, such indicators need to fixed sensors (e.g., vehicle detectors). On the other hand, the ETC can be calculated easily by probe cars and road-to-vehicle communication. In order to calculate the ETC, vehicles periodically send their probe data (i.e., positions and scheduled paths) to a server by road-to-vehicle communication, and the server estimates the degree of traffic jam as ETC on the basis of shared probe data. Moreover, we propose a method of traffic signal control which optimizes the cycle and split according to the ETC. The ETC is an indicator for not only current traffic jam but also future traffic jam. Thus, we expect that our method can avoid the aged deterioration of traffic trend (i.e., building the new road) and the sudden change of traffic flow (i.e., road accidents). We implemented our method as an outer module of "AVENUE"[3]. The AVENUE is one of the popular traffic stream simulators, and used for various traffic evaluations. We performed simulation experiments on AVENUE, and the results indicate that our method can reduce the traveling time of vehicles.

The remainder of this paper is organized as follows. Section 2 defines the expected traffic congestion. Section 3 describes the traffic signal model of AVENUE. Section 4 presents our traffic control method based on the expected traffic congestion. Section 5 shows experimental results and discusses it. Finally, Section 6 offers conclusions and future works.

---

[1] Japan Society of Civil Engineers "http://www.jsce-int.org/"
[2] Vehicle Information and Communication System Center "http://www.vics.or.jp/"
[3] i-Transport Lab "http://www.i-transportlab.jp/"

2 Definition of Expected Traffic Congestion

In this section, we explain the definition of expected traffic congestion (ETC) which is proposed by Yamashita [13]. The ETC represents the amount of shared paths (a path from present position to destination) among vehicles. In this paper, the ETC is used to optimize the parameters of traffic signals (i.e., cycle and split).

2.1 Representation of Path

In order to calculate the ETC, all vehicles send their probe data (i.e., positions and scheduled paths) periodically to a neighbor signal control machine (i.e., a server) by road-to-vehicle communication. The signal control machine broadcasts the received probe data to other signal control machines by wide-area wireless communication, and calculates ETC from the collected probe data. A path of vehicle is represented by a sequence of links as Equation (1) ($l(s_0, s_1)$ is a link between signals s_0 and s_1). For simplicity, each vehicle always selects the path of the shortest distance.

$$R = (l(s_0, s_1), l(s_1, s_2), l(s_2, s_3), l(s_4, s_5)) \tag{1}$$

2.2 Expected Traffic Congestion

Here, we explain how to calculate the expected traffic congestion (ETC) by collected probe data. First, a path weight (PW) of link $l(s, s') \in R$ is defined as Equation (2). $|R|$ is the path size of R, and $N(l(s, s'))$ is the descending sequence number of link $l(s, s')$ in R. Thus, the values from $|R|$ to 1 are assigned to each link from current link to destination link, respectively. Consequently, the path weight of nearby links will be high, and the path weight of far links will be low. For example, $|R|$ and N for $l(s_2, s_3)$ in Equation (1) are 4 and 2, thus the path weight is 2/4.

$$PW(l(s, s')) = \left(\frac{N(l(s, s'))}{|R|}\right) \tag{2}$$

Next, the ETC of link $l(s, s')$ is defined as Equation (3). The ETC is the sum of the path weights for all collected paths. For example, in Equation (4), there are three collected paths (R_0,R_1, and R_2). The ETCs of links in the collected paths are calculated as follows. A link $l(s_0, s_1)$ is contained in the both paths R_0 and R_1, thus, the ETC of $l(s_0, s_1)$ is the sum of $PW_0(s_0, s_1) = 4/4$ and $PW_1(s_0, s_1) = 2/2$. Moreover, a link $l(s_1, s_2)$ is contained in the all paths, thus, the ETC of $l(s_1, s_2)$ is the sum of $PW_0(s_1, s_2) = 3/4$, $PW_1(s_1, s_2) = 1/2$, and $PW_1(s_1, s_2) = 3/3$. The ETCs of the all links are summarized in Table 1.

$$ETC(l(s, s')) = \sum_i PW_i(l(s, s')) \tag{3}$$

$$\begin{aligned} R_0 &= (l(s_0, s_1), l(s_1, s_2), l(s_2, s_3), l(s_4, s_5)) \\ R_1 &= (l(s_0, s_1), l(s_1, s_2)) \\ R_2 &= (l(s_1, s_2), l(s_2, s_3), l(s_4, s_5)) \end{aligned} \tag{4}$$

Table 1. Expected Traffic Congestion

Function	$l(s_0, s_1)$	$l(s_1, s_2)$	$l(s_2, s_3)$	$l(s_3, s_4)$
PW_0	1	0.75	0.5	0.25
PW_1	1	0.5	0	0
PW_2	0	1	0.67	0.33
ETC	2	2.25	1.17	0.58

3 Traffic Signal Model

In this section, we describe the model of traffic signal for a traffic simulator AVENUE. The indication of traffic signals consists of some phases which represent the indication time (i.e., "green", "red", and "yellow"). Moreover, the indication time depends of three control parameters: cycle, split, and offset.

3.1 Signal Indication Phases

In the traffic simulator AVENUE, the indication of traffic signals is defined as the combination of some phases. Figure 1 shows an example of phases for a crossroad, and these six phases are repeated (i.e., the next of no.6 is no.1). In the no.1, the indications for vertical traffic signals are "green" lights, and the indications for horizontal traffic signals are "red" lights. The blue arrows represent vehicular trajectories (i.e., straight, left turn, and right turn). In the no.2, the "green" lights are changed to "yellow" lights. In the no.3, all indications are changed to "red" lights. Other phases no.4, no.5, and no.6 are the reverse phases of no.1, no.2, and no.3. The indication time of each phase depends on three control parameters: cycle, split, and offset. We will explain the detail of the three control parameters in the next part.

3.2 Traffic Signal Parameters

Here, we explain control parameters of traffic signals (i.e., cycle, split, and offset) by using an example shown in Table 2. In the table, there are three traffic signals s_0, s_1, and s_2.

Cycle. The cycle is a time interval that an signal indication loops back (i.e., the total indication time of all phases). In our model, the value of cycle is set between 90 seconds and 180 seconds (the range is based on a typical cycle of traffic signals). For simplicity, in our model, the indication time of phases which include yellow lights (i.e., no.2 and no.5) is fixed to "3" seconds, and the indication time of all red phases (i.e., no.3 and no.6) is fixed to "2" seconds. Therefore, the indication time of phases which include green lights (i.e., no.1 and no.4) are only adjustable according to cycle. For example, in table 2, the cycle of s_0 is 150.

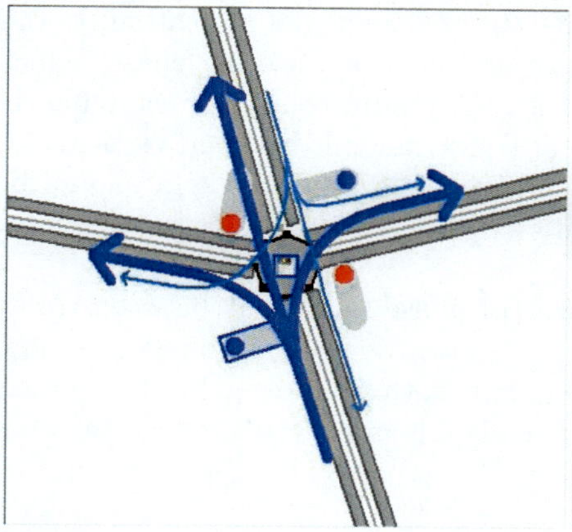

No.1

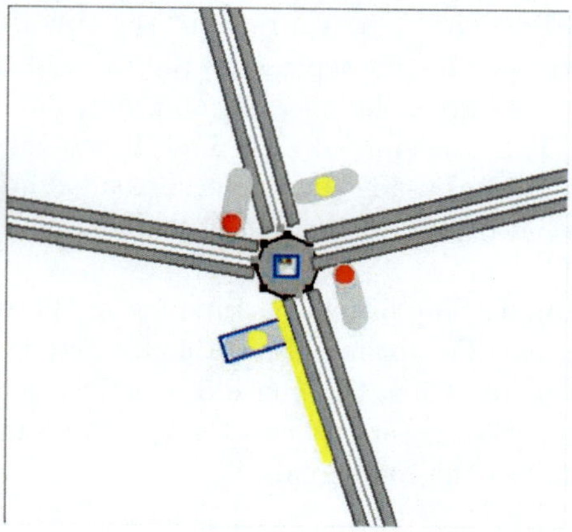

No.2

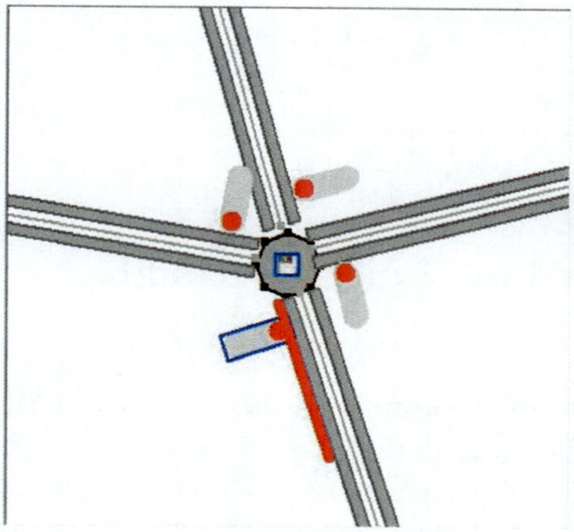

No.3

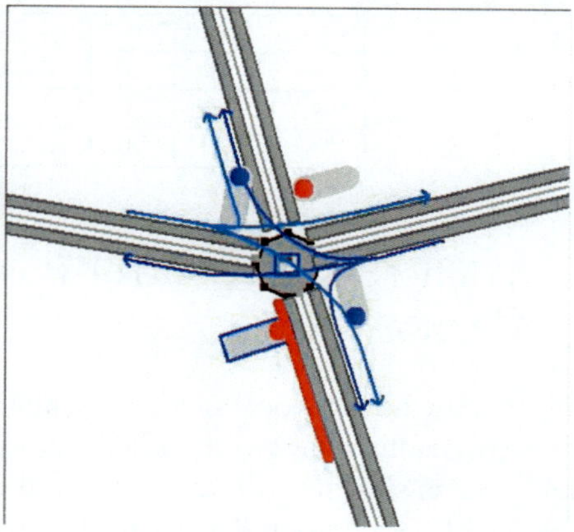

No.4

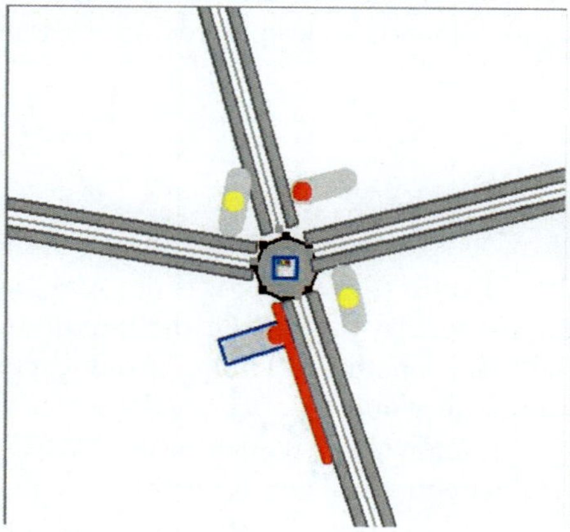

No.5

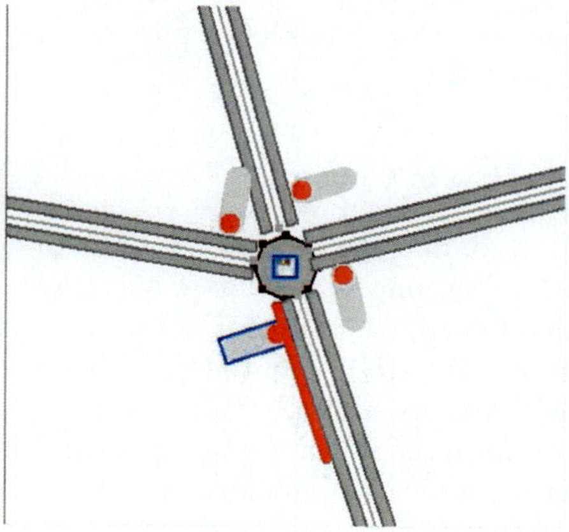

No.6

Fig. 1. Phase Numbers of Traffic Signal

Split. The split is a ratio of the time interval of each signal indication. In our model, the split represents the ratio of no.1 and no.4 because the phases which include green lights (i.e., no.1 and no.4) are only adjustable. For example, in table 2, the split of s_1 is $3:4$. If the gap of traffic amount between vertical and horizontal roads (i.e., arterial roads and side road) is wide, the split should be set according to the gap of traffic amount.

Offset. The offset is a time lag of the start of phase (i.e., no.1) between traffic signals. For example, in table 2, the offset of s_0 and s_1 is 0, and offset of s_0 and s_2 is 10. Thus, the phase of s_2 starts later than both s_0 and s_1. In many cases, the offset is set to transit time between two signals in order to reduce the stop time at the red signal.

Table 2. Indication Time (seconds)

Signal	No.1	No.2	No.3	No.4	No.5	No.6	Offset
s_0	70	3	2	70	3	2	0
s_1	60	3	2	80	3	2	0
s_2	70	3	2	80	3	2	10

4 Traffic Signal Control Based on Expected Traffic Congestion

We explain how to optimize the traffic signal parameters on the basis of the expected traffic congestion (ETC) as follows. Although there are six phases in one cycle as shown in Figure 1, the indication time of no.2, no.3, no.5, and no.6 is so small as compared with no.1 and no.4. Hence, we consider the values of ETC at no.1 and no4 as the key amount of traffic. Figure 2 shows an example of ETC at no.1 and no.4. The ETC is updated by a signal control machine all of the time. Thus, the ETC represents the total demands to keep green indications in the future.

4.1 Cycle Control

First, we find the target cycle. The target cycle is based on the maximum value of ETC in one cycle. The maximum value of no.1 is $ETC_1(l_2) = 30$, and the maximum value of no.4 is $ETC_4(l_2) = 20$. We regard the sum of the maximum value ($ETC_1(l_2) + ETC_4(l_2) = 50$) as the total amount of traffic in one cycle. Moreover, we multiply the sum of the maximum value by a weight factor α to transform the ETC to time scale (i.e., if $\alpha = 1$, one vehicle passes an intersection per second). Consequently, the target cycle is defined as Equation (5).

$$\alpha \times (max(ETC_1(l)) + max(ETC_4(l))) \tag{5}$$

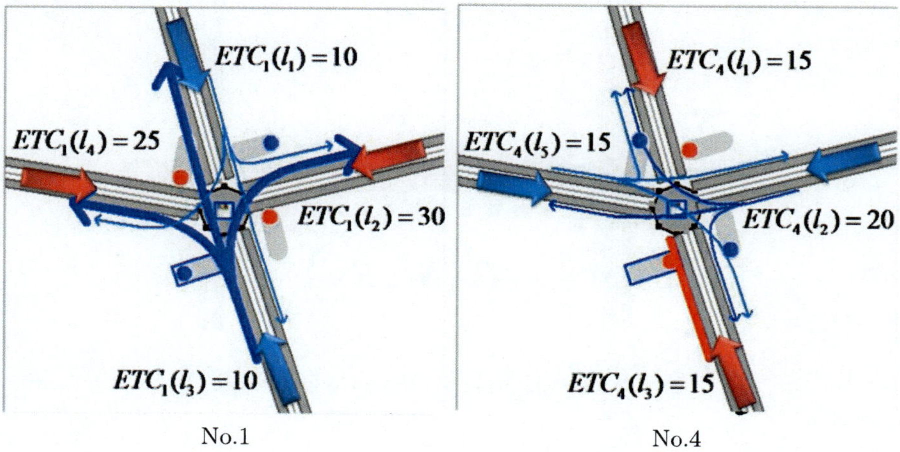

Fig. 2. An Example of ETC at No.1 and No.4

Next, we increment or decrement the current cycle. We calculate the difference between the current cycle and target cycle. If the difference is more than a threshold β, the current cycle is incremented or decremented by γ toward the target cycle. The rapid change of the current cycle may cause a traffic disturbance, thus, the γ should be set a small value, but the small value cannot follow the rapid change of traffic amount. Hence, we must find the best value of γ, empirically.

4.2 Split Control

The target split is based on the ratio of ETC in one cycle. We calculate the sum of ETC on a straight link pair in one cycle. In the Figure 2, there are two straight link pairs (i.e., (l_1, l_3) and (l_2, l_4)). The sum of (l_1, l_3) is $ETC_1(l_1) + ETC_1(l_3) + ETC_4(l_1) + ETC_4(l_3) = 50$, and the sum of (l_2, l_4) is $ETC_1(l_2) + ETC_1(l_4) + ETC_4(l_2) + ETC_4(l_4) = 90$. We regard the rate of the sum (i.e., $5 : 9$) as the target split. Consequently, the target split is defined as Equation (6) (L and L' are sets of straight link pairs).

$$\sum_{l \in L}(ETC_1(l) + ETC_4(l)) : \sum_{l' \in L'}(ETC_1(l') + ETC_4(l')) \qquad (6)$$

4.3 Offset Control

The offset is based on the distance between adjacent signals. For example, there are two traffic signals s_1 and s_2 as shown in Figure 3. The offset value of s_2 is set by Equation (7), where $d(s_1, s_2)$ is the distance between s_1 and s_2, and v is the average speed of vehicles. The cycle and split of s_2 are forcibly synchronized with s_1 to keep the offset (the cycle and split of s_1 are optimized by the above

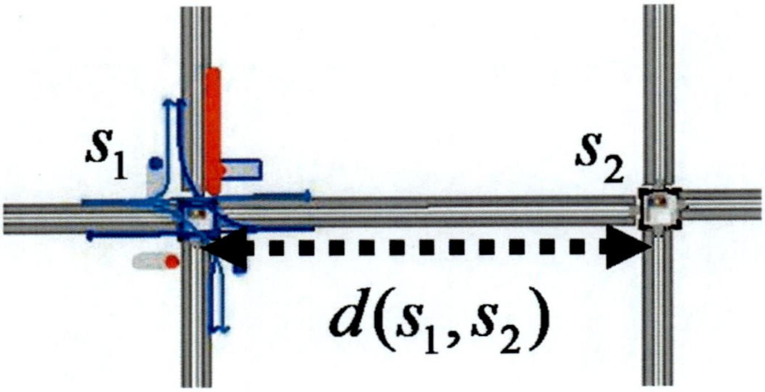

Fig. 3. An Example of Offset Cooperation

method). This way enables to reduce the amount of stop at the s_2 for vehicles which start from s_1. The operating condition for the offset control is very limited because the offset control is often useless for complex traffic flows. In fact, the offset control is mostly adopted at traffic signals on arterial highways. Thus, in this paper, we only adopt the offset control to specified traffic signals in advance. The offset control for complex traffic flow is our future work.

$$\frac{d(s_1, s_2)}{v} \tag{7}$$

5 Simulation Experiments

We have already reported the effect of our method in the past international conference[14], but its simulation environment by artisoc, which is a multi agent simulator[4], is very simple. Thus, in this paper, we adopt a traffic stream simulator "AVENUE". We developed an outer module of AVENUE to update cycle and split on the basis of ETC, but offset is fixed in advance because the offset is independent of ETC. Here, we report our experimental results.

5.1 Traffic Environment

We performed simulation experiments to evaluate "cycle&split controls" and "offset control" by using two types of road networks. Figure 4(a) is a real network topology at Kudan-shita, Tokyo for cycle&split evaluation. Moreover, Figure 4(b) is a virtual network topology for offset evaluation. A road network consists of nodes and links (i.e., intersections and road segments), and traffic signals are placed at the nodes. Moreover, there are some circles at nodes. The circles called zones represent origin and destination of vehicles. The origin and destination of vehicles are summarized in Tables 3 and 4. In Table 3, there are 10

[4] MAS Community "http://mas.kke.co.jp/"

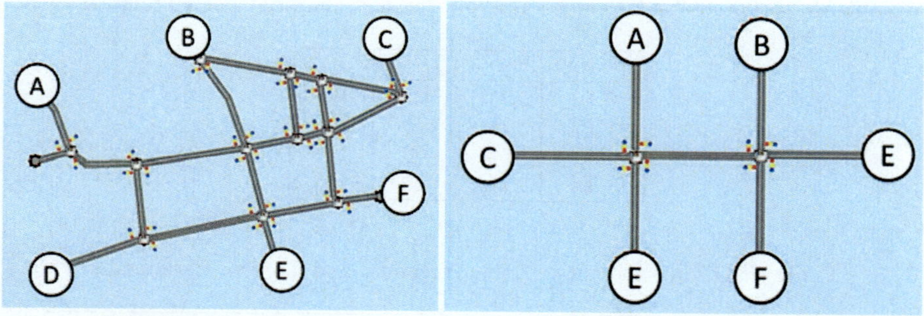

(a)Cycle&Split Evaluation (b)Offset Evaluation

Fig. 4. Road Network

Table 3. Origin-Destination Pairs for Cycle&Split Evaluation (per 15 minutes)

From → To	Numbers
A → F	100
B → D	100
B → E	100
C → D	100
D → B	100
D → C	100
D → F	100
E → F	100
F → A	100
F → D	100

origin-destination pairs, and the inflow number per 15 minutes is 100 vehicles. In Table 4, there are 3 two-way origin-destination pairs, and there are 4 inflow number patterns: PT1, PT2, PT3, and PT4. All vehicles select the shortest path from origin node to destination node. The default indication time of signals and control parameters are summarized in Table 5.

5.2 Cycle&Split Evaluation

Figure 5 is the average traveling time of cycle&split evaluation by using the road network in Figure 4(a). The simulation time of this evaluation is 15 minutes. Here, we compared three signal controls: "fix", "cycle&split", and "cycle&split(step)". The "fix" keeps the initial indication time of traffic signals shown in Table 5. The "cycle&split" replaces the current cycle with the target cycle immediately independent of the the increment value γ. The "cycle&split(step)" increases or decreases the current cycle by γ toward the target cycle ($\gamma = 5, 10, 15$). The experimental result indicates that the cycle&split controls can reduce the traveling time compared to the "fix". The direct control of cycle and split is the best result of them, but the rapidly change of cycle and split may confuse car drivers,

Table 4. Origin-Destination Pairs for Offset Evaluation (per 15 minutes)

Pattern	A ↔ E	B ↔ F	C ↔ E
PT1	100	100	100
PT2	75	75	100
PT3	50	50	100
PT4	25	25	100

Table 5. Default Indication Time (seconds) and Control Parameters

No.1	No.2	No.3	No.4	No.5	No.6	Weight α	Threshold β	Increment γ
70	3	2	70	3	2	2	10	10

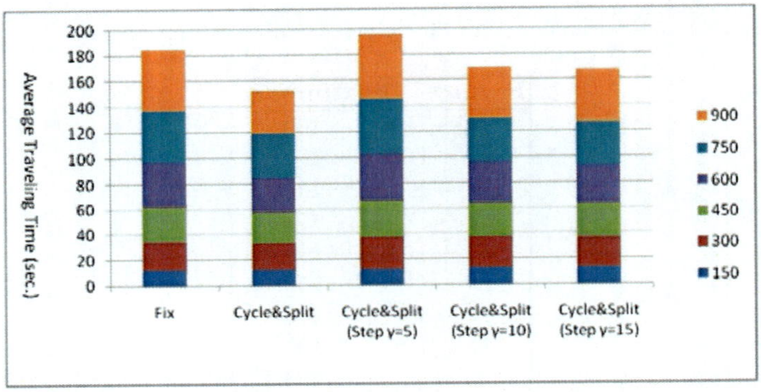

Fig. 5. Average Traveling Time of Cycle&Split Evaluation

and leads to traffic accidents. Moreover, the small increment value ($\gamma = 5$) is worse than the "fix" because it takes long time to reach the target cycle, and the long increment value ($\gamma = 15$) is almost the same with the direct control. It seems that the optimal value of γ depends on various traffic environments (e.g., network topology, vehicle speed, and so on). Therefore, we must find the optimal value in advance to keep the effect of cycle&split controls.

5.3 Offset Evaluation

Figure 6 is the traveling time of offset evaluation by using the road network in Figure 4(b). The simulation time of this evaluation is 60 minutes, and the pattern of inflow number is PT1 in Table 4. Here, we compared three signal controls: "fix", "cycle&split", and "cycle&split+offset". The "fix" and "cycle&split" are the same of the preceding section. The "cycle&split+offset" synchronizes two signals of the horizontal traffic lane (i.e., C ↔ F). The experimental result indicates that the offset control improves the performance of cycle&split controls. However, we must say that the effect of our offset control is limited because the

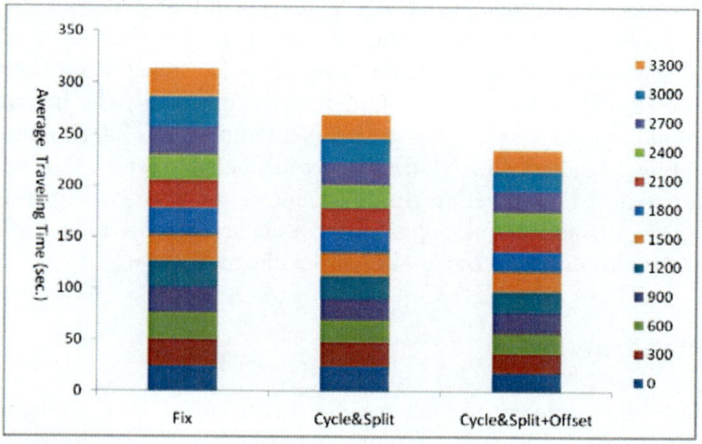

Fig. 6. Average Traveling Time of Offset Evaluation

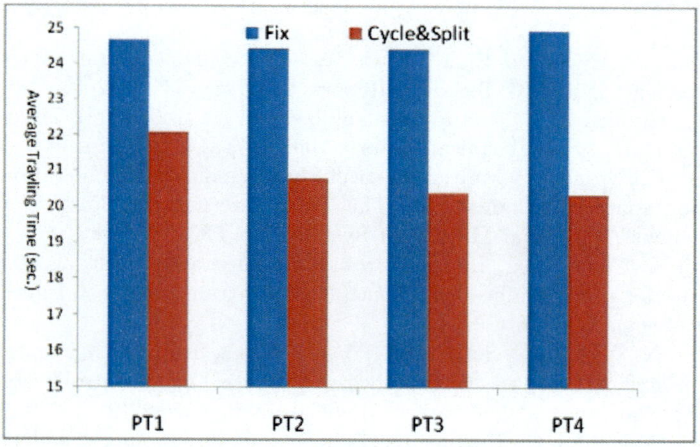

Fig. 7. Average Traveling Time of Time Series Variation

network condition for this evaluation is very simple. Thus, we must find the appropriate operating condition of offset control for complex road network. Figure 6 is the traveling time by changing the inflow patterns from PT1 to PT4. The experimental result indicates that cycle&split is adaptable to time series variation. Moreover, when the inflow number is remarkable one-side (i.e., PT4), the split control can decrease the traveling time effectively.

6 Conclusion

The most of current traffic control is not able to respond the rapidly change of traffic situation. Thus, adaptive controls of traffic signals by using probe data and road-to-vehicle communication are getting a lot of attention. In this paper, we proposed a method of traffic signal control on the basis of expected traffic congestion. The expected traffic congestion is a congestion barometer, in dynamic and decentralized manner. Our method optimizes the indication time of traffic signals in each cycle. We performed simulation experiments by using a traffic simulator called AVENUE. The experimental results showed that our method can reduce the traveling time of vehicles, effectively. In the future, we intend to develop the dynamic offset control according to traffic conditions. In addition, we try to evaluate our method in wide traffic areas.

Acknowledgment

We gratefully acknowledge the advice and criticism of Prof. Naoyuki Uchida of Tokyo University of Science, Japan. Moreover, we thank the advice of Mr. Hisamoto Hanabusa for developing an outer module of AVENUE.

References

1. Miyanishi, Y., Miyamoto, E., Maekawa, S.: A proposal of traffic signal control using its technology. IPSJ SIG Technical Reports 2001(83), 53–60 (2001)
2. Sun, X., Kusakabe, T.: A study on traffic signal control for prevention of traffic congestion. IPSJ SIG Technical Reports 2005(89), 31–34 (2005)
3. Dresner, K., Stone, P.: Multiagent traffic management: A reservation-based intersection control mechanism. In: The Third International Joint Conference on Autonomous Agents and Multiagent Systems, pp. 530–537 (2004)
4. Chen, R.S., Chen, D.K., Lin, S.Y.: Actam: Cooperative multi-agent system architecture for urban traffic signal control. IEICE transactions on Information and Systems 88(1), 119–126 (2005)
5. Jyunichi, K., Yasuo, K., Souichi, T., Yoshitaka, K., Taisuke, S.: Applying genetic programming with substructure discovery to a traffic signal control problem. Transactions of the Japanese Society for Artificial Intelligence 22(2), 127–139 (2007)
6. Ando, Y., Masutani, O., Honiden, S.: Performance of pheromone model for predicting traffic congestion. In: Proceeding of Autonomus Agents & Multi Agent Systems (2006)
7. Ando, Y., Masutani, O., Honiden, S.: Pheromone model: Appplication to traffic congestion prediction. In: Proceeding of Autonomus Agents & Multi Agent Systems, pp. 1287–1298 (2005)
8. Ando, Y., Masutani, O., Sasaki, H., Motoida, S.: Pheromone model: Application to traffic congestion prediction. The Institute of Electronics, Information and Communication Engeneers j88-D-1(9), 1287–1298 (2005)
9. Kato, Y., Hasegawa, T.: Traffic signals controlled by vehicles: Paradigm shift of traffic signals. IEICE Technical Reports ITS 100(284), 67–71 (2000)
10. Aso, T., Hasegawa, T.: Full automated advanvced demand signals ii scheme. In: Proceedings of the 12th International IEEE Conference on Intelligent Transportation Systems, pp. 522–527 (2009)

11. Aso, T., Hasegawa, T.: Traffic signal control schemes for the ubiquitous sensor network period. In: Proceedings of ITS Symposium 2009, pp. 67–72 (2009)
12. Hanabusa, H., Iijima, M., Horiguchi, R.: Development of delay estimation method using real time probe data for adaptive signal control algorithm. In: Proceedings of ITS Symposium 2009, pp. 207–212 (2009)
13. Yamashita, T., Kurumatani, K., Nakashita, H.: Approach to smooth traffic flow by a cooperative car navigation system. Transactions of Information Processing Society of Japan 49(1), 177–188 (2008)
14. Ezawa, H., Mukai, N.: Adaptive traffic signal control based on vehicle route sharing by wireless communication. In: Setchi, R., Jordanov, I., Howlett, R.J., Jain, L.C. (eds.) KES 2010. LNCS, vol. 6279, pp. 280–289. Springer, Heidelberg (2010)

Chapter 19
A Comparative Study on Communication Protocols in Disaster Areas with Virtual Disaster Simulation Systems

Koichi Asakura[1] and Toyohide Watanabe[2]

[1] School of Informatics, Daido University, 10-3 Takiharu-cho, Minami-ku, Nagoya, 457-8530 Japan
asakura@daido-it.ac.jp
[2] Department of Systems and Social Informatics, Graduate School of Information Science, Nagoya University, Furo-cho, Chikusa-ku, Nagoya, 464-8603 Japan
watanabe@is.nagoya-u.ac.jp

Abstract. In this paper, we describe a comparative study on communication protocols based on ad-hoc network technologies in disaster situations. In disaster situations, mobility and density of persons are different from those in usual situations. Thus, in order to evaluate communication protocols, we propose a virtual disaster simulation system. In this simulation system, virtual disaster areas are constructed based on hazard maps which are provided for predicting damage of disaster. Virtual disaster areas enables us to conduct experiments for communication systems effectively since we cannot conduct experiments in real disaster situations. By using this simulation system, we compare three protocols from the viewpoints of data gathering performance and power consumption. Experimental results show that our proposed ad-hoc unicursal protocol shows good performance with respect to both viewpoints in disaster situations.

1 Introduction

In disaster situations such as a big earthquake, a robust communication system is important for gathering and sharing information on disaster areas[1,2]. Although an information sharing system in disaster situations is very essential, it is difficult to provide such a system effectively. This is because communication infrastructures such as base stations for mobile phone, WiFi access points for wireless LANs and so on, may be broken and malfunction. Thus, in order to realize robust and useful communication systems in disaster situations, we have to develop communication methods which do not rely on such infrastructures.

Recently, it is well known that ad-hoc network technologies are suitable for confused situations described above[3,4,5]. In ad-hoc networks, no communication infrastructures are required. Namely, in ad-hoc networks, terminals communicate with neighboring terminals and generate communication paths cooperatively without communication infrastructures. Each terminal restructures a network autonomously according to topology change which is caused by mobility of terminals.

In order to develop effective communication systems for disaster situations, we have proposed a communication system based on ad-hoc network technologies[6]. In this

T. Watanabe and L.C. Jain (Eds.): Innovations in Intell. Machines – 2, SCI 376, pp. 265–279.
springerlink.com

system, we adopt an ad-hoc unicursal protocol for generating communication paths among terminals. In this protocol, unicursal paths are generated in order to decrease the frequency of communication, which leads the longevity of networks and terminals because of less consumption of batteries. In order to evaluate effectiveness of the ad-hoc unicursal protocol, experiments which conform to actual conditions of disaster areas must be conducted.

Our research goal is to develop an effective information sharing system suitable for disaster areas. For this purpose, we have to adopt an appropriate communication protocol. Thus, in this paper, we give a comparative study on communication protocols based on ad-hoc network technologies for information sharing systems in disaster situations. For this evaluation, we develop a simulation system of disaster areas, especially for a big earthquake. In disaster areas, buildings and roads are destructed. Collapsed buildings make persons immobile, and damaged roads make passing cost of the roads increase. Thus, for simulation systems of disaster areas, we have to realize damaged roads which persons cannot pass through or take much time to pass through, and collapsed buildings which unmovable rescuees are blocked in. In our system, disaster areas are realized based on information in hazard maps.

The rest of this paper is organized as follows. Section 2 expresses our communication protocol and movement of refugees for preparation. Section 3 describes our approach for constructing virtual disaster areas. Section 4 gives algorithms for virtual disaster areas. Section 5 shows our experiments. Finally, Section 6 concludes this paper and gives our future work.

2 Preliminaries

In this section, we describe our ad-hoc unicursal protocol and movement of refugees as preliminaries. Section 2.1 expresses an ad-hoc unicursal protocol for communication systems in disaster situations. Section 2.2 describes the movement patterns of refugees in disaster areas.

2.1 Ad-Hoc Unicursal Protocol

Infrastructures for communication systems such as base stations for mobile phones, access points for wireless networks and so on, may be broken or malfunction in disaster situations although these mechanisms are fully functional at ordinary times. Additionally, such communication systems cannot work correctly in disaster situations even if infrastructures are still alive, since frequency of communication becomes very high at short time and thus communication congestion occurs. Therefore, we have to utilize other communication mechanisms which do not depend on the communication infrastructures for communication systems in disaster situations.

Under the above circumstances, we adopt ad-hoc network technologies for communication systems. In the ad-hoc network, communication paths are generated among communication terminals with no communication infrastructures[3,4,5]. Namely, in the ad-hoc network, terminals can communicate with each other without using communication infrastructures. This property is very important for communication systems in disaster situations[7].

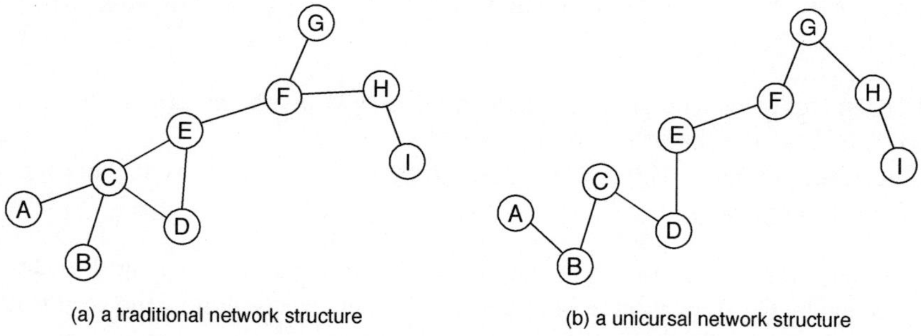

Fig. 1. Network structures

For effective and robust communication systems, longevity of network is very important. For longevity of network, we have to pay attentions to power consumption of communication activities[8,9]. This is because terminals work with limited batteries. In a wireless network environment, packet sending is the most power-consumed operation and thus frequency of packet sending must be kept low for realizing longevity of network. In our protocol, we adopt unicursal paths for constructing communication routes. In unicursal paths, each terminal communicates with at most two terminals generally. Thus, frequency of packet sending is decreased and communication loads are balanced among terminals, which leads longevity of terminals and network.

Figure 1 shows a unicursal path. Figure 1(a) shows a traditional network structures in which terminals connects with all terminals within the communication range. Thus, the terminals C or F have much effort for sending packets than other terminals. This makes power consumption of the terminals C or F higher. If these terminals cannot work due to lack of batteries, network is divided and information sharing cannot be achieved appropriately. On the other hand, as shown in Figure 1(b), our protocol adopts unicursal network structure, which enables power consumption of terminals to be balanced uniformly. Thus, information sharing can be kept in long time.

With this protocol, we develop a system for gathering rescuees' locational information. In this system, rescuees' locational information is shared among rescuees by our protocol. Refugees send a search packet periodically in order to acquire locational information of neighboring rescuees. With this system, rescuees' locational information can be collected by refugees and gathered in shelters.

2.2 Movement of Refugees

In simulation systems for the earthquake situations, we have two types of person objects: static objects and moving objects. Static objects refer to persons who cannot change the locations like rescuees in collapsed buildings or injured persons. Moving objects refer to persons who move in the area freely like refugees. In addition, even in refugees, we have many types of persons: persons who are familiar with the area and know locations of shelters, persons who do not know where shelters exist and so on. Thus, in order to realize disaster situations in simulation systems, several movement patterns of refugees have to be provided.

In our simulation systems, we distinguish refugees into the following two types: locals and outsiders.

- **Locals:** refugees who are familiar with the area know the nearest shelter and are willing to use the shortest path to the shelter.
- **Outsiders:** refugees who are stranger to the area do not know any shelters and thus follow other persons.

For movement of locals, we adopt the Dijkstra's algorithm to calculate the shortest path to shelters[10]. Locals move in the disaster areas according to the result of the Dijkstra's algorithm. Therefore, locals select roads with knowledge at ordinary times. This is because we cannot achieve precise information of road conditions in the area at the time of disaster.

On the other hand, for movement of outsiders, we adopt the following algorithm. When an outsider reaches to an intersection, the number of refugees on the roads is counted for all roads which are connected to the intersection. Then, an outsider selects the road on which the most refugees move.

3 Virtual Disaster Areas

In this section, we describe a simulation system for generating virtual disaster areas. In our system, virtual disaster areas are constructed mainly based on information in hazard maps and information on buildings. First, we describe hazard maps and information on buildings in Section 3.1 and 3.2, respectively. Then, we express requirements for such simulation systems in Section 3.3. Finally, we give related work on simulation systems for evaluating communication protocols in Section 3.4.

3.1 Hazard Maps

Japan is known as an earthquake-prone country. It is forecasted that big earthquakes occur at urban areas such as Tokyo and Nagoya in the near future. In order to predict damage to urban areas, hazard maps are provided. Hazard maps describe potential damage to disaster areas. In hazard maps, we can observe what kind of damage assumes to occur in which part of disaster areas when disaster strikes. Generally, hazard maps are published by urban governments[11,12]. There are many kinds of hazard maps for disasters: earthquake, tsunami and so on. In this paper, we focus on hazard maps for earthquakes.

In hazard maps for earthquake, the following information is included:

- **seismic intensity:** seismic intensity on the Japanese scale is described for every parts of the map.
- **ground liquefaction:** the degree of ground liquefaction is described for every parts of the map.
- **landslide:** the areas where landslide occurs are described.
- **shelters:** shelters where refugees can stay with safe are described.

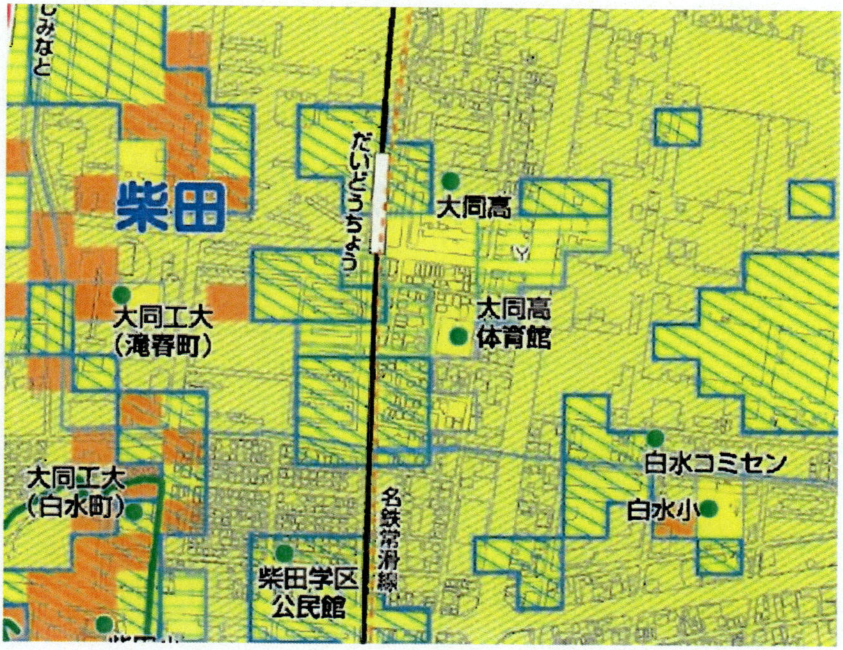

Fig. 2. An example of hazard maps

Figure 2 shows an example of hazard maps. This hazard map is provided for the earthquake which is predicted to occur in the near future on Tokai region in Japan[13]. In this map, the above information is expressed as tiled mesh. Seismic intensity is denoted as tile color and the degree of ground liquefaction is denoted as hatch of tiles. Detailed information is shown in Table 1. The positions where landslide is predicted to occur are denoted as polygons. Figure 3(a) shows such information. As shown in Figure 3(b), the positions of shelters are denoted as bullets.

As described above, hazard maps consist of predicted damage information. Thus, we can achieve realistic simulation experiments by constructing virtual disaster areas based on hazard maps.

3.2 Information on Buildings

In order to evaluate damage to disaster areas correctly, information on buildings is also important. Namely, we have to take the kind of buildings, the kind of materials of buildings, the age of construction and so on, into consideration. This is because aseismic performance of buildings clearly depends on materials of buildings, such as wood and reinforced concrete, and used earthquake-resistant regulations are different on age of construction.

3.3 Requirements

A simulation system for virtual disaster areas is used for evaluating communication protocols in the situation where big earthquake occurs and many refugees and rescuees

Table 1. Information of seismic intensity and ground liquefaction

(a) Seismic intensity

Seismic Intensity		Situation
	7	A big shake prevents us from act at our will.
	6-high	It is hard to stand. We can only crawl.
	6-low	It is hard to stand. We need a support for walk.
	5-high	It is hard to move. We may feel a fear.
	5-low	It is a little difficult to move. We act for our safety.

(b) Ground liquefaction

Degree of Risk		Predicted Situation
	very high	The degree of risk for ground liquefaction is very high. Ground liquefaction occurs in about 18-35% of the area. Sand and muddy water may be squirted out. Buildings and Bridges may slant. Sagging roads may exist.
	high	The degree of risk is high. Ground liquefaction occurs in about 5% of the area. Sand and muddy water may be squirted out. Buildings and Bridges may slant.
	low	The degree of risk is low. Ground liquefaction occurs in about 2% of the area. Sand and muddy water may be squirted out.
No Hatch	very little	There is very little risk for ground liquefaction.

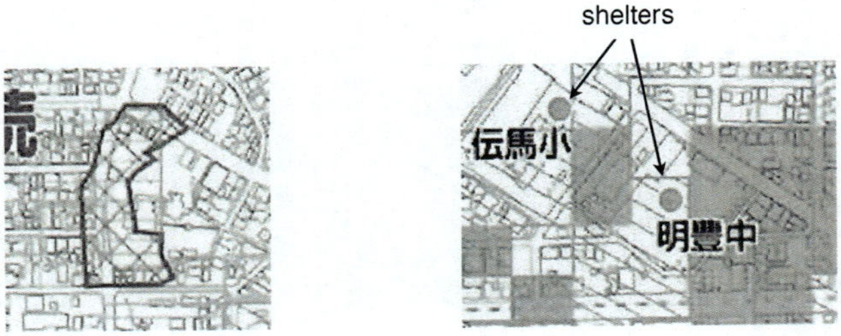

(a) predicted area for landslide (b) shelters

Fig. 3. Landslide and shelters

appear. In order to evaluate communication protocols used in disaster areas, it is important to simulate activities of refugees and rescuees realistically. For refugees, condition of passing roads has to be taken into account. In other words, it is important to calculate where road destruction occurs and how much damage roads are suffered from the earthquake. On the other hand, for rescuees, condition of buildings is important information. Distribution of rescuees in disaster areas depends on distribution of damage for buildings. Therefore, from the above consideration, estimation of damage for roads and buildings is essential for simulation systems of virtual disaster areas.

3.4 Related Work

There are many research projects on simulating communication systems. The Network Simulator ns-2 is one of the most famous projects[14]. In ns-2, users can evaluate communication protocols with various scenarios. The GloMoSim project[15] and the JiST/SWANS project[16,17] are also the same type of research projects for wireless network. However, these systems mainly focus on evaluating communication protocols in ordinary times; namely, these systems do not pay much attention to situations which are changed drastically. Thus, in these systems, we cannot evaluate communication systems under the complicated situations such as disaster areas.

4 Algorithms

In this section, we provide algorithms for constructing disaster areas. First, we describe data model for hazard maps and information on buildings in Section 4.1. Next, we give algorithms for constructing virtual disaster areas in Section 4.2.

4.1 Data Model

As shown in Section 3.1, hazard maps include information of seismic intensity and the ground liquefaction for each tile. In our system, the following data is provided for a tile t.

- $SI(t)$: it describes information of seismic intensity for the tile t.
- $GL(t)$: it describes information of the degree of ground liquefaction for the tile t. This value is normalized to 1.
- $LS(t)$: it describes information on landslide for the tile t. If a marker for landslide includes in the tile t, $LS(t)$ is defined as 1 (true). Otherwise, $LS(t)$ is defined as 0 (false).

Additionally, information of shelter locations is modeled as an intersection of roads in maps.

For information on buildings, as shown in Section 3.2, the kind of buildings, the kind of materials of buildings and the age of buildings are assigned for each tile t.

- $BT(t)$: it describes the kind of buildings. In our system, we use two types: residential type and industrial type.
- $BM(t)$: it describes the kind of materials of buildings: a wooden house or a reinforced concrete building.
- $BA(t)$: it describes the age of buildings. In accordance with [18], three ranges of the age are used: before 1971, from 1972 to 1981 and after 1982.

4.2 Calculation Methods

Figure 4 shows processing flow for constructing virtual disaster areas. This processing flow is based on an earthquake damage evaluation scheme published by the government[18]. Input data of the system, hazard maps and information on buildings, are denoted as upper circles. In this scheme, damage of buildings and damage of roads are calculated.

First, data for seismic intensity in hazard maps are translated into numerical values, maximum peak ground velocity (kine)[1]. Then, ratio of total collapse of building is calculated by using the value of maximum peak ground velocity, information on buildings and information on ground liquefaction. This ratio and information on landslide are used for deployment of rescuees since rescuees may appear where damage of buildings is extensive and landslides occur. Furthermore, the amount of rubbles caused by building destruction is achieved by the ratio of total collapse and the kind of buildings. Information on landslide is also used for calculating the amount of rubbles.

[1] 1 kine = 1 cm/s.

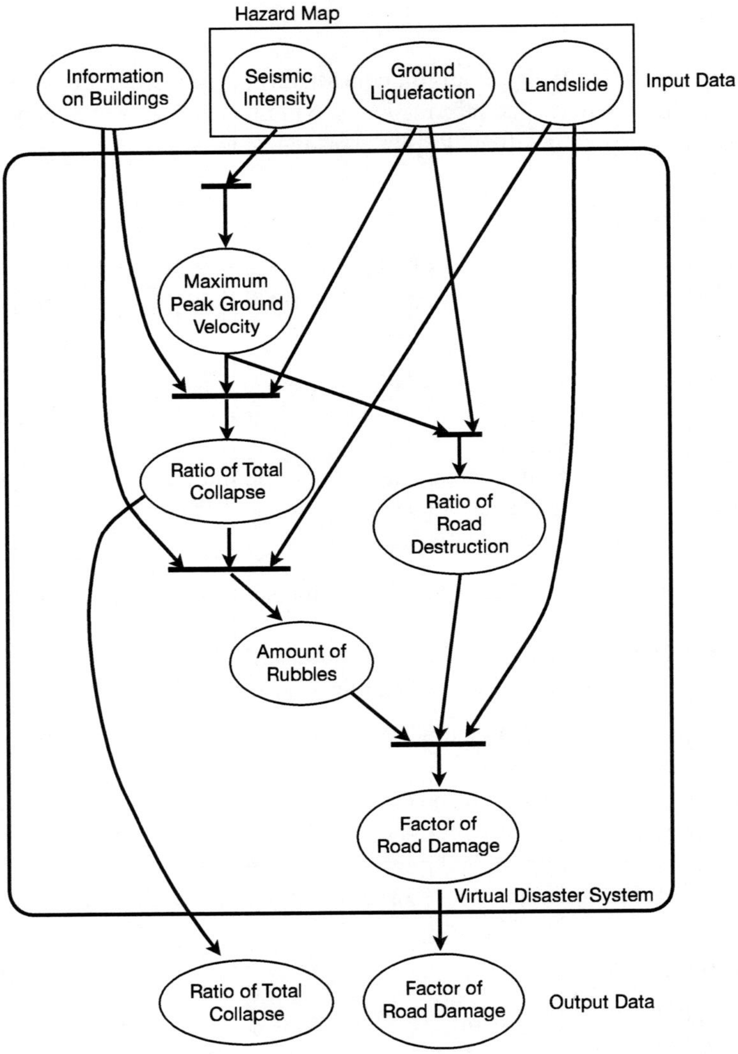

Fig. 4. Processing flow for constructing virtual disaster areas

On the other hand, road destruction is invoked by quake itself, ground liquefaction and landslide. Thus, ratio of road destruction is calculated by three parameters: maximum ground velocity and information on ground liquefaction and landslide. For evaluation of road damage, road passing costs have to be calculated. In our system, ratio of road damage which describes additional passing cost of roads than usual, is calculated. This ratio is calculated according to the amount of rubbles and the ratio of road destruction.

From now on, we describe calculation methods for each processing step.

Maximum Peak Ground Velocity. The value of maximum peak ground velocity is calculated from data of seismic intensity. A translation table is shown in Table 2. The value is determined by uniform random number generator with the given numerical range. Hereafter, the values of maximum peak ground velocity of a tile t is denoted as $PGV(t)$.

Table 2. Seismic intensity and maximum peak ground velocity

Seismic Intensity: $SI(t)$	4	5-low	5-high	6-low	6-high	7
Maximum Peak Ground Velocity (kine)	4–10	10–20	20–40	40–60	60–100	100–200

Ratio of Total Collapse of Buildings. The ratio of total collapse of buildings describes what percentage of the buildings in the tile are destructed by earthquake. This value is calculated by using the maximum peak ground velocity, the degree of ground liquefaction, the kind of materials of buildings and the age of buildings.

Figure 5 shows an algorithm for calculating the ratio. In [18], the values are provided as reference tables. Thus, in our system, the ratio is calculated by an interpolation method with the reference table. Table 3 shows a part of the reference table. For example, if the maximum peak ground velocity is 75 kine, the ratio of total collapse of wooden houses built in 1975 is 5%.

ALGORITHM: RTC
  Input: $PGV(t), BM(t), BA(t)$.
  Output: $RTC(t)$.
BEGIN
  $Table$:= Reference Table for $BM(t)$ and $BA(t)$.
  $PGA_{min} := \lfloor \frac{PGA(t)}{10} \rfloor \times 10.$
  $PGA_{max} := \lceil \frac{PGA(t)}{10} \rceil \times 10.$
  $r_{min} := \frac{PGA_{max} - PGA(t)}{10}.$
  $r_{max} := \frac{PGA(t) - PGA_{min}}{10}.$
  $RTC_{min} := Table(PGA_{min}).$
  $RTC_{max} := Table(PGA_{max}).$
  $RTC := RTC_{min} \times RTC_{min} + RTC_{max} \times RTC_{max}.$
  return RTC.
END

Fig. 5. An algorithm for ratio of total collapse of buildings

Amount of Rubbles. In a big earthquake, many rubbles from buildings break out and affect surrounding roads. Also, earth and sands by landslides occupy roads and prevents refugees from passing through the roads. Thus, in order to evaluate availability of roads in disaster areas, it is important to calculate the amount of rubbles. The amount of rubbles is calculated as shown in Figure 6. First, landslide makes damage of the roads increase drastically. In our system, roads is unavailable completely where landslide

Table 3. A reference table for ratio of total collapse of wooden buildings

$PGV(t)$ (kine)	Age of buildings		
	before 1971	from 1972 to 1981	after 1982
	...		
60	5%	2%	0%
70	8%	4%	1%
80	13%	6%	1%
90	18%	9%	2%
	...		

```
ALGORITHM: RUBBLE
   Input: LS(t), RTC(t), BT(t).
   Output: Rubble(t) [t].
BEGIN
   if LS(t) is true then return ∞.
   if BT(t) is "resident" then
      Rubble := 41.25 × RTC(t).
   else
      Rubble := 957 × RTC(t).
   endif
   return Rubble.
END
```

Fig. 6. An algorithm for total amount of rubbles

occurs. Then, the amount of rubbles is calculated from the ratio of total collapse of buildings and the kind of buildings.

Ratio of Road Destruction. Roads themselves are damaged by shake of earthquakes and ground liquefaction. We calculate the ratio of road destruction which expresses how many destruction spots exist in a 1 kilometer-long road. This value is also provided as a reference table in [18].

Factor of Road Damage. The factor of road damage describes how many passing costs of roads are increased due to earthquakes. In a usual situation, passing costs of roads are proportional to the length of roads. In our system, passing costs are computed by multiplying road length by the factor of road damage.

The factor of road damage is calculated by the amount of rubbles, the ratio of road destruction and information on landslide.

5 Experiments

We conduct an experiment for evaluating communication protocols with our simulation system. This section describes the experimental results.

5.1 Simulation Settings

In the experiments, we provide a virtual disaster area with 2.0-kilometer-width and 1.5-kilometer-height. Four shelters where refugees are moving toward are provided in the area.

The number of rescuees and refugees is 250 each. Rescuees are deployed according to the ratio of total collapse of buildings and information on landslide. In other words, many rescuees appear in the places where the ratio of total collapse is high or landslide occurs. Refugees are deployed on the road randomly. In the experiments, it is assumed that all the refugees know the locations of shelters. Namely, refugees can move to one of the shelters without losing their way. Development of realistic movement algorithms for refugees is our future work.

In our communication systems, rescuees who cannot move in collapsed buildings, send their locational information to the surrounding persons periodically. On the other hand, refugees communicate with neighboring rescuees on the way to shelters. Thus, locational information on rescuees can be collected in shelters naturally by movement of refugees.

5.2 Communication Protocols

In the experiments, we evaluate how much information on rescuees can be collected with the communication systems based on mobile ad-hoc networks. We compare three communication protocols: a unihop protocol, a flooding protocol and our ad-hoc unicursal protocol.

In the unihop protocol, rescuees attempt to send their locational information directly to moving refugees. Namely, other rescuees do not relay the messages. In this protocol, the number of sending packets can be reduced, which saves power consumption of communication systems, while their locational information cannot be delivered to the shelters sufficiently.

On the other hand, in the flooding protocol, when a rescuee receives a packet for locational information on other rescuees, the rescuee retransmits the packet to other rescuees. Thus, the packet can be reached to all rescuees in their connected network. This communication scheme makes the possibility to deliver the packets to refugees increase. However, communication frequency among rescuees also increases, which leads high consumption of batteries and thus makes the network short-lived.

5.3 Experimental Condition

In the experiments, we compare the number of information on rescuees which is received by moving refugees. The higher number denotes the higher performance of data gathering in disaster areas. For communication systems in disaster areas, it is also important to reduce power consumption of batteries, as well as to capture much information on rescuees. In wireless network environments, packet sending is the most power consumed operations. Thus, we have to adjust the interval of packet sending properly in communication systems. In the experiments, we vary the value of packet sending interval parameter.

5.4 Experimental Results

Figure 7 shows the performance of data gathering for each protocol. The horizontal axis shows the interval of packet sending, and the vertical axis shows the ratio of gathering data on rescuees. This figure denotes that performances of our ad-hoc unicursal protocol and the flooding protocol are almost the same in short packet sending intervals, while the performance of the unihop protocol is relatively low in all ranges. This is because that packet retransmission is achieved in both our protocol and the flooding protocol, which increases the chance of packet receiving of moving refugees.

Next, Figure 8 compares the frequency of packet sending for each protocol. The vertical axis shows the number of packet sending for rescuees. From this figure, we can conclude that the frequency of packet sending in our ad-hoc protocol and the unihop protocol is almost the same, while the frequency for the flooding protocol is very high. The frequency of packet sending in the flooding protocol is about 10 times of those in ad-hoc protocol.

Figure 9 shows the relationship between the ratio of data gathering and the frequency of packet sending. In this figure, plot points which locate nearer the origin shows the better total performance with respect to the data gathering and power consumption. Clearly, plot points for our proposed ad-hoc unicursal protocol locate the nearer than those for both the flooding and unihop protocols. From this experiments, we can clarify that our proposed ad-hoc unicursal protocol can achieve better performance for communication systems in disaster situations.

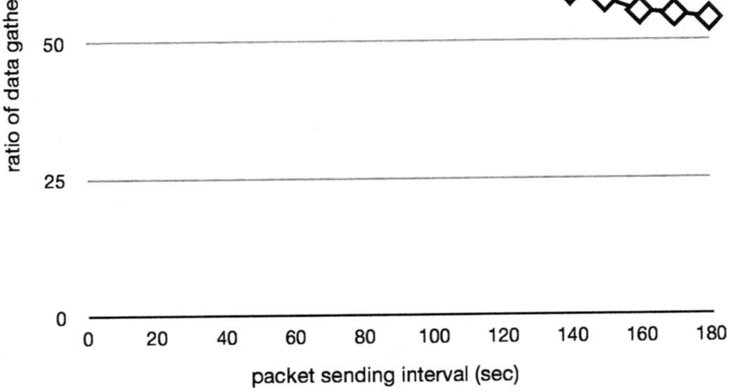

Fig. 7. The ratio of data gathering

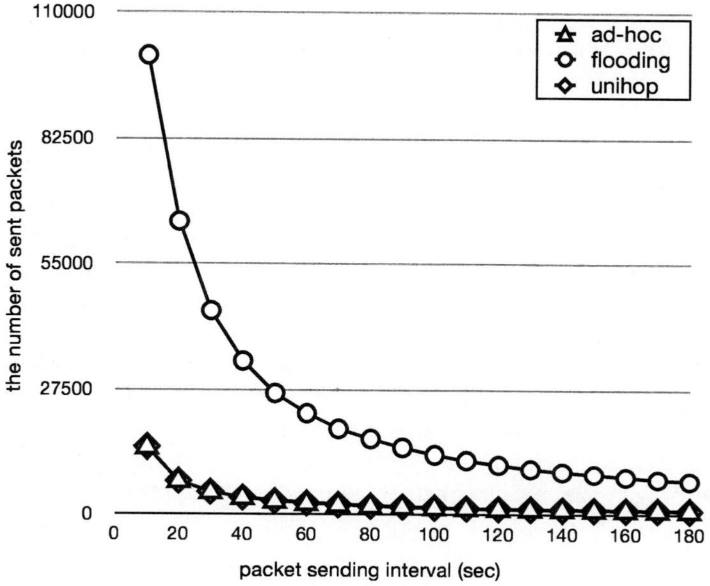

Fig. 8. The number of sent packets

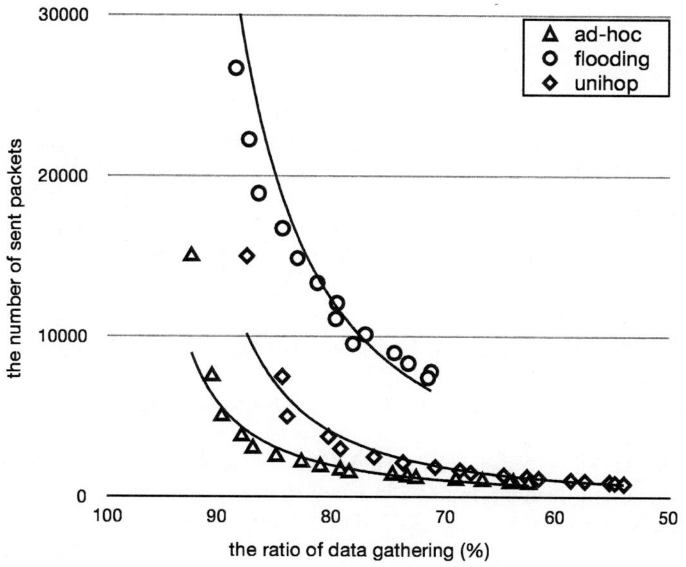

Fig. 9. The relationship between ratio of data gathering and the number of packets

6 Conclusion

In this paper, we perform a comparative study on communication protocols in disaster situations. For this evaluation, we develop a simulation system for constructing virtual disaster areas. In this simulation system, a virtual disaster area is constructed based on information in hazard maps. Since we cannot conduct experiments in real disaster areas, such a simulation system is very important for developing communication systems in disaster situations. Experimental results show that our proposed communication system based on ad-hoc unicursal protocol can work effectively for collecting information of rescuees in disaster areas.

For our future work, we have to pay much attention to constructing precise disaster areas. In disaster areas, damage for roads and buildings may increase as time passes such as spread of fire. We have to take such time-depend events into consideration. Additionally, we have to develop natural movement patterns for refugees in disaster areas. Furthermore, since an information gathering system with ad-hoc unicursal protocol enables one-way communication, we have to enhance our system and protocol so as to provide two-way communication, namely to transfer useful information from shelters to rescuees.

Acknowledgement

This research has been supported by the Kayamori Foundation of Informational Science Advancement.

References

1. Midkiff, S.F., Bostian, C.W.: Rapidly-Deployable Broadband Wireless Networks for Disaster and Emergency Response. In: The 1st IEEE Workshop on Disaster Recovery Networks, DIREN 2002 (2002)
2. Meissner, A., Luckenbach, T., Risse, T., Kirste, T., Kirchner, H.: Design Challenges for an Integrated Disaster Management Communication and Information System. In: The 1st IEEE Workshop on Disaster Recovery Networks, DIREN 2002 (2002)
3. Toh, C.-K.: Ad Hoc Mobile Wireless Networks: Protocols and Systems. Prentice Hall, Englewood Cliffs (2001)
4. Murthy, C.S.R., Manoj, B.S.: Ad Hoc Wireless Networks: Architectures and Protocols. Prentice-Hall, Englewood Cliffs (2004)
5. Lang, D.: Routing Protocols for Mobile Ad Hoc Networks: Classification, Evaluation and Challenges. VDM Verlag (2008)
6. Asakura, K., Oishi, J., Watanabe, T.: An Ad-hoc Unicursal Protocol for Human Communication in Disaster Situations. In: the 2nd International Symposium on Intelligent Interactive Multimedia Systems and Services (KES-IIMSS 2009), pp. 511–521 (2009)
7. Mase, K.: Communications Supported by Ad Hoc Networks in Disasters. Journal of the Institute of Electronics, Information, and Communication Engineers 89(9), 796–800 (2006) (in Japanese)
8. Singh, S., Woo, M., Raghavendra, C.S.: Power-aware Routing in Mobile Ad Hoc Networks. In: The 4th Annual ACM/IEEE Int'l Conf. on Mobile Computing and Networking (MOBICOM), pp. 181–190 (1998)

9. Li, D., Jia, X., Liu, H.: Energy Efficient Broadcast Routing in Static Ad-hoc Wireless Networks. IEEE Trans. on Mobile Computing 3(2), 144–151 (2004)
10. Cormen, T.H., Leiserson, C.E., Rivest, R.L., Stein, C.: Introduction to Algorithms. The MIT Press, Cambridge (2009)
11. U.S. Geological Survey: Hazard Mapping Images and Data, `http://earthquake.usgs.gov/hazards/products/`
12. The Geographical Survey Institute: Hazard Map Portal Site (in Japanese), `http://www.gsi.go.jp/geowww/disapotal/index.html`
13. Nagoya City: Earthquake Maps for Your Town (in Japanese), `http://www.city.nagoya.jp/kurashi/category/20-2-5-6-0-0-0-0-0-0.html`
14. Issariyakul, T., Hossain, E.: Introduction to Network Simulator NS2. Springer, Heidelberg (2008)
15. Zeng, X., Bagrodia, R., Gerla, M.: GloMoSim: a Library for Parallel Simulation of Large-scale Wireless Networks. In: The 12th Workshop on Parallel and Distributed Simulations (PADS 1998), pp. 154–161 (1998)
16. Barr, R., Hass, Z.J., Renesse, R.: JiST: An Efficient Approach to Simulation Using Virtual Machines. Software – Practice and Experience 35(6), 539–576 (2005)
17. Barr, R., Hass, Z.J., Renesse, R.: Scalable Wireless Ad Hoc Network Simulation. In: Handbook on Theoretical and Algorithmic Aspects of Sensor, Ad hoc Wireless, and Peer-to-Peer Networks, pp. 297–311. CRC Press, Boca Raton (2005)
18. The Cabinet Office, government of Japan: A Manual for Assessment of Earthquake Damage (in Japanese), `http://www.bousai.go.jp/manual/`

Author Index